Lecture Notes in Physics

The Lecture Notes in Physics

The series Lecture Notes in Physics (LNP), founded in 1969, reports new developments in physics research and teaching – quickly and informally, but with a high quality and the explicit aim to summarize and communicate current knowledge in an accessible way. Books published in this series are conceived as bridging material between advanced graduate textbooks and the forefront of research to serve the following purposes:

• to be a compact and modern up-to-date source of reference on a well-defined topic;

• to serve as an accessible introduction to the field to postgraduate students and nonspecialist researchers from related areas;

• to be a source of advanced teaching material for specialized seminars, courses and schools.

Both monographs and multi-author volumes will be considered for publication. Edited volumes should, however, consist of a very limited number of contributions only. Proceedings will not be considered for LNP.

Volumes published in LNP are disseminated both in print and in electronic formats, the electronic archive is available at springerlink.com. The series content is indexed, abstracted and referenced by many abstracting and information services, bibliographic networks, subscription agencies, library networks, and consortia.

Proposals should be sent to a member of the Editorial Board, or directly to the managing editor at Springer:

Dr. Christian Caron
Springer Heidelberg
Physics Editorial Department I
Tiergartenstrasse 17
69121 Heidelberg/Germany
christian.caron@springer-sbm.com

Christian Klein   Olaf Richter

Ernst Equation and Riemann Surfaces

Analytical and Numerical Methods

 Springer

Authors

Christian Klein   Olaf Richter†
Max-Planck-Institut
für Mathematik
in den Naturwissenschaften
Inselstr. 22-26
04103 Leipzig, Germany
Email: klein@mis.mpg.de

C. Klein, O. Richter, *Ernst Equation and Riemann Surfaces*,
Lect. Notes Phys. 685 (Springer, Berlin Heidelberg 2005), DOI 10.1007/11540953

ISSN 0075-8450

ISBN 978-3-642-06677-1   e-ISBN 978-3-540-31513-1

Springer is a part of Springer Science+Business Media
springeronline.com
© Springer-Verlag Berlin Heidelberg 2010
Printed in The Netherlands

Preface

Exact solutions to Einstein's equations have been useful for the understanding of general relativity in many respects. They have led to physical concepts as black holes and event horizons and helped to visualize interesting features of the theory. In addition they have been used to test the quality of various approximation methods and numerical codes. The most powerful solution-generating methods are due to the theory of Integrable Systems. Since stars and galaxies in thermodynamical equilibrium lead to stationary axisymmetric spacetimes, it is of special physical interest that the Einstein equations in this case are equivalent to the completely integrable Ernst equation. In this context the most prominent solution is the Kerr metric describing a rotating black hole. Rich classes of solutions to integrable equations can be constructed with methods from the theory of Riemann surfaces which were originally introduced to generate periodic solutions to integrable wave equations such as the Korteweg–de Vries equation. The corresponding solutions to the Ernst equation, which contain the Kerr solution as a limit, are not periodic and are related to deformations of the underlying Riemann surface. In this volume we study these solutions to the Ernst equation in detail and discuss physical and mathematical aspects of this class both analytically and numerically.

Olaf Richter and myself had been working in this field for almost a decade, starting with our common stay in Jena. He contacted me two years ago that he intended to write a comprehensive review of the topic on the occasion of his habilitation which should serve as the basis for a monograph. Since the time appeared to have come for a summary of the current state of the art of the field, I agreed to join the project. But in November 2003, I got the shocking and completely unexpected news of Olaf Richter's passing away. This left me with the sad duty to finish our common project alone. The present volume is based on Olaf Richter's habilitation thesis. Roughly half of the material is taken after reworking from this thesis, and the rest is from later publications and other sources.

This work would not have been possible without the support of the Max Planck Institute for Mathematics in the Sciences, and there especially of Prof. E. Zeidler. It was supported in part by the Deutsche Forschungsgemeinschaft, the Marie Curie program of the European Union, the Schloessmann foundation, and the Max Planck Society. For helpful discussions and hints

I thank J. Bičák, A. Bobenko, S. Bonazzola, P. Breitenlohner, B. Carter, H. Farkas, P. Forgács, J. Frauendiener, E. Gourgoulhon, R. Kerner, D. Korotkin, D. Maison, V. Matveev, M. Niedermaier, J. Novák, H. Pfister, H. Samtleben, V. Shramchenko, and E. Zeidler. My special thanks go to my coauthors J. Frauendiener, D. Korotkin, O. Richter, and V. Shramchenko for their collaboration and their patience. I am grateful to P. Forgács, J. Frauendiener, T. Grava, M. King, D. Korotkin, D. Maison, H. Pfister, and E. Zeidler for critically reading part or all of the manuscript and for providing useful comments. I appreciate very much the effort of R. Beig for carefully reading the proofs and suggesting many improvements, and H. Schlegel for technical support.

Leipzig   *Christian Klein*
November 2004

Contents

1 Introduction

1.1 General Remarks on Integrability

In classical mechanics a well defined concept of integrability exists which is related to the Hamiltonian description of mechanics. If the phase space of a mechanical system is $2n$-dimensional, n integrals of motion in involution are sufficient for a complete description of the dynamics of the system. In this case the initial conditions specify the integrals of motion and thus the complete time evolution of the system. The task is to find such a system of integrals of motion. An important example for an integrable system in classical mechanics is the motion of a spinning top in the gravitational field of the Earth, the rotation of a rigid body about a fixed point (see e.g. [1], Chap. 6 and [2]). For a top where the orthogonal frame is attached to the body and where the axes coincide with the axes of inertia, the Hamilton function has the form

$$H = \frac{1}{2}(I_1 M_1^2 + I_2 M_2^2 + I_3 M_3^2) + \Gamma_1 p_1 + \Gamma_2 p_2 + \Gamma_3 p_3 ; \qquad (1.1)$$

here M is the angular momentum of the body, p is the gravitational field, the constant vector Γ gives the center of mass where the origin is the fixed point, and I_1^{-1}, I_2^{-1} and I_3^{-1} are the main moments of inertia of the body. The Hamilton equations $F_{,t} = \{H, F\}$ lead to the Kirchhoff equations

$$\boldsymbol{p}_{,t} = \boldsymbol{p} \times \boldsymbol{\omega} , \quad \boldsymbol{M}_{,t} = \boldsymbol{M} \times \boldsymbol{\omega} + \boldsymbol{p} \times \boldsymbol{u} , \quad \omega^i = \frac{\partial H}{\partial M_i} , \quad u^i = \frac{\partial H}{\partial p_i} , \quad (1.2)$$

where $\boldsymbol{p}_{,t} = \frac{\partial \boldsymbol{p}}{\partial t}$, and where $\times$ denotes the vector product in $\mathbb{R}^3$. One has the Poisson brackets

$$\{M_i, M_j\} = \varepsilon_{ijk} M_k , \quad \{M_i, p_j\} = \varepsilon_{ijk} p_k , \quad \{p_i, p_j\} = 0 , \quad i, j, k = 1, 2, 3 , \qquad (1.3)$$

where ε_{ijk} is the totally antisymmetric tensor with $\varepsilon_{123} = 1$. The Casimir functions $\sum_{i=1}^{3} p_i^2$ and $\sum_{i=1}^{3} p_i M_i$ are trivial integrals of motion for the Poisson brackets (1.3). Thus the (reduced) phase space in this example is four-dimensional and one needs one further conserved quantity in addition to the energy. For the Euler and Lagrange tops it is known for a long time that they possess such an additional integral of motion. It was somewhat surprising when in the late 19th century S. Kovalevskaya [3] found another integrable spinning top which is now named after her. It is interesting to note that this work is also one of the first applications in mathematical physics of the multi-dimensional theta functions being discussed in this volume. Some years later Goryachev and Chaplygin [4] discovered another integrable spinning top which has the interesting property to be integrable only for one particular orbit.

Whereas in classical mechanics a well defined mathematical theory of integrability has been developed, see [5–8], for partial differential equations the situation is more delicate. It is, therefore, not surprising that several approaches to this concept exist. Below we will mention some of these concepts though none of them yields for itself a completely satisfactory theory of integrability in infinite dimensions.

The modern theory of integrable systems began with the seminal work of Gardner, Greene, Kruskal and Miura [9] who introduced a method of integration for the Korteweg–de Vries (KdV) equation [10], see also [11, 12] and Sect. 1.2 below,

$$u_{,xxx} + 6uu_{,x} = 4u_{,t} \ ,$$

a non-linear partial differential equation for the scalar function u. The KdV equation plays an important role in the asymptotic description of one-dimensional wave phenomena in physical models. For instance it can be obtained via asymptotic expansions around simple wave motion of the one-dimensional Euler equations for shallow water. The key observation in [9] was that the spectrum of the Schrödinger operator can be stationary for a time dependent potential u if this potential solves the KdV equation. Thus it was possible to use powerful techniques developed for the Schrödinger equation to construct solutions to the KdV equation. The so-called inverse scattering method for instance allows for the reconstruction of the potential in the Schrödinger equation from the scattering data. Because of the close relation between the KdV and the Schrödinger equation, the inverse scattering method can be used to generate solutions to KdV for given solutions to the Schrödinger equation.

Later it was realized that this method is not limited to the KdV equation. It can be used to construct solutions to equations which can be treated as the integrability condition for an overdetermined linear differential system, the so-called Lax representation [13] which happened to be related to the Schrödinger equation in the case of KdV. For completely integrable equations this linear system contains an additional parameter, the so-called spectral parameter which reflects an underlying infinite dimensional symmetry group of the differential equation. Methods from the theory of complex functions can be used to construct classes of solutions for the linear system with prescribed analyticity properties with respect to the spectral parameter which lead to solutions to the non-linear partial differential equation to be studied. For a review of this subject see the monographs [12] and [14] to [17]. An example are solitons and the almost periodic solutions to equations like KdV which were constructed in terms of theta functions, certain transcendental functions associated to Riemann surfaces which will be considered in detail in this volume.

The existence of an associated linear differential system alone is not sufficient to ensure the integrability of an equation. There are equations associated with such a system for which the inverse scattering method is not applicable. The vacuum Einstein equations illustrate this fact: they can be written as the consistency condition of the field equations of a massless spin–3/2 particle (see [18]), but are not integrable in any accepted sense since it is known that there are solutions with chaotic behavior. The important point is that an integrable equation must have a Lax representation with a spectral parameter which is not the case for the spin–3/2 equations above. An open problem is how to find this representation for a given equation. There is an approach due to Estabrook and Wahlquist [19] for constructing a Lax pair, but this approach can rarely be used to decide whether there exists a Lax representation for a given equation.

Another criterion for integrability which again does not work in the general case is the so-called Painlevé test [20] to [24] and [12]. This method is based on the following observation for ordinary differential equations: It is well known that the singularities of linear differential equations are independent of the integration constants. However non-linear equations may have solutions with movable singularities, i. e. singularities depending on the constants of integration. If this is not the case for singularities other than poles, the equation is said to satisfy the Painlevé property. A partial differential equation with symmetry group G satisfies the Painlevé test if the reduced system of ordinary differential equations for the G-invariant solutions has the Painlevé property. A typical example for a one-dimensional wave equation is the ordinary differential equation obtained for travelling waves which depend only on the variable $\xi = x - ct$, $c = const$. The Ernst equation which we will discuss in the following can be reduced to the Painlevé VI equation, see [25,26]. For a Painlevé analysis of the vacuum Einstein equations we refer the reader to [27]. We remark that integrability and Painlevé property are not equivalent, as a simple example discussed in [28] shows. Nevertheless, a differential equation having this property has good chances to be considered as integrable. It is interesting to note that S. Kovalevskaya found the 'Kovalevski top' by requiring that its equations of motion should have the Painlevé property.

Another approach to integrability is based on a remarkable relationship between self-duality and integrability, see [28]. It is well known that the finite-action solutions of the Yang–Mills equations on $\mathbb{E}^4$ with gauge group G can be obtained by stereographic projection of G-bundles over the one-point compactification S^4 of $\mathbb{E}^4$ [29], the instantons. If the gauge group is SU(2), SU(3) or U(2) then any weakly stable instanton is known to be either self-dual or anti-self-dual. Thus, in order to find finite-action solutions to the Yang–Mills equations on $\mathbb{E}^4$ with gauge group SU(2), one has to construct connections in SU(2)-bundles over S^4 with (anti-)self-dual curvature. Fortunately, the Penrose–Ward transform (see [30]) relates such connections in a one-to-one fashion with holomorphic bundles over the three-dimensional complex projective space $\mathbb{CP}^3$. Since holomorphic bundles over $\mathbb{CP}^3$ can be constructed explicitly by several methods, e. g., the Atiyah–Drinfeld–Hitchin–Manin construction [31], all solutions with finite action to the (anti-)self-dual Yang–Mills equations (ASDYM) are in principle known. The fact that the space of solutions to the ASDYM equations is well understood yields an interesting approach to integrable equations by considering equations which are obtained as symmetry reductions of these equations. It is an important fact that most of the known integrable equations in two and three dimensions turn out to be symmetry reductions of the ASDYM equations and allow for a reduction of the Penrose–Ward transform, see [32]. Thus, the ASDYM equations can be used to provide a unifying description of integrable systems. However, there is no agreement about its universality. For example, the Kadomtsev-Petviashvili equation needs to be "ruthlessly hacked and stretched to fit the Procrustean bed of self-duality" [33]. Similarly, the Landau–Lifshitz equation does not fit into this picture.

Thus there is so far no universal concept to the integrability of partial differential equations as in the case of classical mechanics. However there exist powerful solution techniques for the equations known to be integrable. For instance rich classes of solutions can be obtained in terms of theta functions on Riemann surfaces. It is the objective of this volume to study exact solutions to the Einstein equations with astrophysical relevance. Given our current knowledge, there is little hope to find a general solution to the Einstein equations as is the case for the Poisson equation in Newtonian gravity since the former do not seem to be integrable in general. Nonetheless classes of astrophysically interesting solutions could be constructed in special cases as the Schwarzschild and Kerr black holes. Though the solution generating methods are limited to situations with symmetries, they are nonetheless of great importance. Exact solutions help to visualize relativistic effects and have led to the development of new physical concepts as the formation of a black hole in the collapse of a relativistic star and the existence of event horizons. In addition they can be used as testbeds for numerical and perturbative methods to solve the Einstein equations.

It is known that isolated matter configurations in thermodynamical equilibrium can be approximately considered as perfect fluid bodies. Such configurations lead within the framework of general relativity to stationary axisymmetric spacetimes. It turns out that the vacuum Einstein equations in this case are equivalent to the Ernst equation [34] (for details see Sect. 1.3 and Chap. 2)

$$(\Re\mathcal{E})\Delta\mathcal{E} = (\nabla\mathcal{E})^2 \qquad (1.4)$$

for a complex potential $\mathcal{E}(\varrho,\zeta)$ (ϱ, ζ are cylindrical coordinates in $\mathbb{R}^3$). This implies that powerful methods from the theory of integrable systems, especially from the theory of Riemann surfaces, can be applied to construct rich classes of solutions to this astrophysically interesting case. It is the purpose of this volume to study in detail a large class of solutions which can be given in terms of theta functions.

A hyperbolic version of the Ernst equation describes the collision of gravitational waves, see [35,36]. The elliptic version is a reduction of the so-called Bogomolny equations for the SU(2) Yang–Mills system and could be used to construct multi-monopole solutions to these equations, see [37] to [41]. It is also equivalent to the Yang equation [42] which can be obtained as a symmetry reduction of the self-dual SU(2) Yang–Mills equations. Mathematically it corresponds to a completely integrable sigma model [43] to [48], a harmonic map from the four-dimensional spacetime into the hyperbolic plane, see e.g. [49]. It is also closely related to so-called Bianchi surfaces (see [50,51]), two-dimensional surfaces in three-dimensional euclidean space with prescribed curvature.

As already mentioned the Ernst equation is completely integrable. Interestingly it shows all the integrability properties listed above: it has a Lax representation, has the Painlevé property and can be obtained as a symmetry reduction of the ASDYM equations. The associated Lax pair is closely related to the Lax pair for Bianchi surfaces. The special feature of this linear differential system is the fact that the spectral parameter varies on a Riemann surface with moving branch points. This means that the branch points are parametrized by the physical coordinates. The spectral parameter is thus not independent of the physical coordinates as in the case of the KdV equation. This leads to important consequences if one wants to study solutions in terms of theta functions which were found for the Ernst equation by Korotkin [52]. In contrast to solutions to evolution equations of KdV-type which are comprehensively studied in [1], the theta-functional solutions to the Ernst equation are not periodic or quasi-periodic. The reason is that the underlying Riemann surface is not 'static' in the case of the Ernst equation, since the branch points are parametrized by the physical coordinates. Thus in contrast to the cases studied in [1], the dependence of the theta functions on the branch points is important.

It is the purpose of this volume to illustrate the consequences of these deformations of the underlying algebraic curves on the solutions and the different properties of the solutions compared to the theta-functional solutions of KdV. The idea is to present the necessary mathematical and numerical tools to study this question. From an analytical point of view this includes variational formulas for Riemann surfaces and identities for theta functions as well as the discussion of singular algebraic curves. From a numerical point of view, the 'dynamical' nature of the Riemann surfaces requires an efficient code since all characteristic quantities of the Riemann surface have to be determined at each point of the spacetime. We study a physically relevant subclass of Korotkin's solutions and discuss in detail the physical and mathematical aspects with the aim to provide the necessary tools for concrete applications.

1.2 The Korteweg–de Vries Equation

A starting point for the inverse scattering method for the KdV equation is the Lax representation. We consider the following system of equations [53]

$$-\psi_{,xx} = u\psi - k^2\psi\,,$$
$$4\psi_{,t} = 4\psi_{,xxx} + 6u\psi_{,x} + 3u_{,x}\psi\,,$$

where $k \in \mathbb{C}$ is the spectral parameter. Note that the first equation is just the Schrödinger equation for the motion of a particle in a time dependent potential $u(x,t)$. Obviously, these equations form an overdetermined linear differential system for the function ψ. With

$$\psi_1 \doteq \psi_{,x} + k\psi\,,\quad \Phi = \begin{pmatrix}\psi\\\psi_1\end{pmatrix}\,,$$

we may put the above system of equations into the Lax form,

$$\Phi_{,x} = U(k)\,\Phi\,,$$
$$\Phi_{,t} = V(k)\,\Phi\,,\tag{1.5}$$

where U and V depend on the spectral parameter k and are given by

$$U(k) = -k\begin{pmatrix}1 & 0\\0 & -1\end{pmatrix} + \begin{pmatrix}0 & 1\\-u & 0\end{pmatrix}\,,$$
$$V(k) = -4k^3\begin{pmatrix}1 & 0\\0 & -1\end{pmatrix} + 4k^2\begin{pmatrix}0 & 1\\-u & 0\end{pmatrix} - k\begin{pmatrix}2u & 0\\u_{,x} & -2u\end{pmatrix} + \begin{pmatrix}-u_{,x} & 2u\\2u^2 - u_{,xx} & u_{,x}\end{pmatrix}\,.$$

The integrability condition $\Phi_{,xt} = \Phi_{,tx}$ of system (1.5) is just the KdV equation. Because of this condition, which implies vanishing curvature of Φ, the above linear differential system is also called the zero-curvature representation of the KdV equation. This system is an example for an autonomous linear system where the spectral parameter is independent of the physical coordinates. As already mentioned the existence of the spectral parameter indicates the presence of an infinite dimensional symmetry group for this equation.

The above linear differential system was used in [9] to solve initial value problems for the KdV equation for fast decreasing initial data in the following way: for $t = 0$ the Schrödinger equation is solved with $u(x,0)$ as the potential. Since the spectrum of the above Schrödinger operator turns out to be independent of t, the time dependence of the eigenfunctions is such that it is straight forward to obtain $\psi(x,t)$ for a given $\psi(x,0)$. The corresponding solution to the KdV equation then follows by reconstructing the corresponding potential $u(x,t)$ from the solution of the Schrödinger equation, a problem known as the inverse scattering problem. For fast decreasing data this problem was solved earlier by Gel'fand, Levitan and Marchenko, see [54, 55] and also [53]. It turned out that many important equations of physics as the sine–Gordon equation, the non-linear Schrödinger equation and the Landau–Lifshitz equation can be solved with the inverse scattering method, see [1, 53].

The inverse scattering method cannot be directly applied to the case of periodic potentials. It is known that the Schrödinger operator with a periodic potential has a spectrum with gaps. If the number of these gaps is finite, there is a natural interpretation of the gaps as branch cuts of a closed algebraic curve. In the case of the KdV equation, this curve is *hyperelliptic*,

$$\mu^2 = \prod_{j=1}^{2g+1} (\lambda - \lambda_j) \,, \tag{1.6}$$

where g is the genus of the curve, i.e., μ is the square root of a polynomial in λ (for $g = 1$ the curve is called elliptic). Notice that this curve is 'static' in the sense that the branch points λ_j of the curve are independent of the physical coordinates. The corresponding solution to the KdV equation can be expressed in terms of the associated theta function via the Its–Matveev formula [56, 57],

$$u = 2\partial_{xx} \ln \Theta(\boldsymbol{U}x + \boldsymbol{W}t + \boldsymbol{D}) + 2c \,, \tag{1.7}$$

where $\boldsymbol{U}$, $\boldsymbol{W}$ and c are characteristic objects of the underlying Riemann surface, and where $\boldsymbol{D}$ is a g-dimensional vector. For details of the history of finite gap solutions of the KdV equation see [1, 58]. The finite gap solutions (1.7) are either periodic or almost periodic, for examples see Sect. 3.5.

Solutions in terms of theta functions on a 'dynamic' curve, i.e., a curve with branch points depending on the physical coordinates play a role if one considers the long time behavior of an arbitrary solution to the KdV equation. After a rescaling of the time t and the spatial coordinate x to t/ε, x/ε, the KdV equation takes the form

$$\varepsilon^2 u_{,xxx} + 6uu_{,x} = 4u_{,t} \,,$$

which becomes the Hopf equation in the limit $\varepsilon \to 0$. Initial data describing a wave packet will lead in finite time to a point of gradient catastrophe of the solution for the Cauchy problem of the latter equation. For small ε the solution of such a Cauchy problem of KdV, develops near the point of gradient catastrophe of the solution of the Hopf equation, a zone of fast oscillations of wavelength of order epsilon. This oscillatory behavior can be approximately described by the exact periodic solution of the KdV equation where the spectral parameters are not constants but evolve according to the Whitham equations [59, 60]. This picture has been proposed by A. G. Gurevich and L. P. Pitaevskii [61] and has been rigorously proven in the works of P. Lax, D. Levermore [62] and S. Venakides [63].

1.3 The Ernst Equation

As already noted, the Einstein equations do not appear to be generically integrable. To find exact solutions, spacetimes with symmetries have to be considered, see [35] for a review. As Geroch has shown [64], already a spacetime with one (non-null) Killing vector has an interesting $SL(2, \mathbb{R})$ symmetry on the space of solutions of the field equations, see Sect. 2.1. In the case of two Killing vectors there are two physically interesting cases: colliding plane gravitational waves (two spacelike Killing vectors) and rotating bodies (one spacelike and one timelike Killing vector). The field equations reduce in both cases to a single complex differential equation, the (elliptic (+) respectively hyperbolic (−)) Ernst equation [34]

$$(\Re\mathcal{E})\left\{\frac{\partial^2\mathcal{E}}{\partial\zeta^2} \pm \frac{1}{\varrho}\frac{\partial}{\partial\varrho}\left(\varrho\frac{\partial\mathcal{E}}{\partial\varrho}\right)\right\} = \left(\frac{\partial\mathcal{E}}{\partial\zeta}\right)^2 \pm \left(\frac{\partial\mathcal{E}}{\partial\varrho}\right)^2$$

for a complex function $\mathcal{E} = \mathcal{E}(\varrho, \zeta)$, the Ernst potential depending on Weyl's canonical cylindrical coordinates ϱ and ζ. The real and the imaginary part of the Ernst potential are denoted with f and b respectively. It is interesting to note that the development of our understanding of the two classes went almost parallel: if one restricts oneself to real solutions one has to solve a linear differential equation and in both cases this class is known for a long time. The solutions are either the Einstein–Rosen waves [65] which describe collinearly polarized gravitational waves or the Weyl solutions [66], i. e., static and axisymmetric fields like the Schwarzschild solution for a static black hole. Furthermore, in both cases there was an interesting complex solution known before a Lax pair for the Ernst equation was given: in 1963 Kerr [67] found the solution for a rotating black hole, and in 1971 the Penrose–Khan solution [68] was published. In fact it was Ernst's original motivation to provide a framework for a convenient representation of the Kerr solution. It turns out that the Ernst potential for the Kerr metric is just an algebraic function of the coordinates ϱ, ζ, see [35],

$$\mathcal{E} = \frac{e^{-i\varphi}r_+ + e^{i\varphi}r_- - 2m\cos\varphi}{e^{-i\varphi}r_+ + e^{i\varphi}r_- + 2m\cos\varphi}, \tag{1.8}$$

where $r_\pm = \sqrt{(\zeta \pm m\cos\varphi)^2 + \varrho^2}$. The mass of the black hole is m, the angular momentum $J = m^2\sin\varphi$, and the horizon is located on the axis between $-m\cos\varphi$ and $m\cos\varphi$. For $\varphi = 0$ the Kerr solution reduces to the static and spherically symmetric Schwarzschild solution. In this volume we will mainly discuss the elliptic version of the Ernst equation.

The Ernst equation is equivalent to an $SL(2,\mathbb{R})$ sigma model which can be seen by the following consideration: let $\mathcal{J}$ be the $SL(2,\mathbb{R})$ matrix

$$\mathcal{J} = \frac{1}{f}\begin{pmatrix} 1 & -b \\ -b & f^2 + b^2 \end{pmatrix}. \tag{1.9}$$

With this matrix the elliptic Ernst equation is equivalent to the non-autonomous sigma model equation (for details see Sect. 2.1)

$$(\varrho\mathcal{J},_\varrho\mathcal{J}^{-1}),_\varrho + (\varrho\mathcal{J},_\zeta\mathcal{J}^{-1}),_\zeta = 0. \tag{1.10}$$

This symmetry property was apparently first observed in [69]. Sigma models [43] are used in many branches of physics as solid state physics, supergravity and super-symmetry theories, see [44,45] and the monographs [46,47]. One useful aspect of these models is the fact that many two-dimensional classical sigma models are completely integrable, see [48]. Sigma models represent harmonic maps ([49] and Sect. 2.1). In particular the Ernst equation represents a harmonic map from the 4 dimensional Lorentz manifold $\mathcal{M}$ into the hyperbolic plane $\mathcal{H}$. This fact was used by Weinstein [70] to [73] to prove existence and uniqueness for multi-black hole solutions where the horizons are located on the axis being separated by conical singularities, so-called Weyl struts. Mars and Senovilla [74] used these techniques to prove uniqueness of solutions for perfect fluid configurations.

The complete integrability of the Ernst equation was shown by Maison [75, 76] and Belinski–Zakharov [77], for alternative linear systems which are gauge equivalent see for instance [78] to [80]. One possible form of the linear system is

$$\Phi,_\xi = \frac{\mathcal{J},_\xi \mathcal{J}^{-1}}{1-\gamma}\Phi, \quad \Phi,_{\bar\xi} = \frac{\mathcal{J},_{\bar\xi}\mathcal{J}^{-1}}{1+\gamma}\Phi, \tag{1.11}$$

where $\xi = \zeta - i\varrho$, where $\Phi \in SL(2,\mathbb{R})$, and where

$$\gamma = \frac{2}{\xi - \bar\xi}\left(K - \frac{\xi + \bar\xi}{2} + \sqrt{(K-\xi)(K-\bar\xi)}\right), \tag{1.12}$$

where K is the spectral parameter. An interesting feature of these linear systems is that they are not autonomous, i.e., that the spectral parameter K is a coordinate on a 'dynamical' Riemann surface where the branch points are parametrized by the physical coordinates. The above system is thus defined on a whole family of surfaces of genus zero given by $\mu_0^2 = (K-\xi)(K-\bar\xi)$.

A key point in the construction of linear systems for the Ernst equation was Geroch's observation that the space of solutions of the Ernst equation allows for an infinite-dimensional symmetry group [81]. He conjectured that all asymptotically flat, stationary, axisymmetric vacuum solutions can be generated from Minkowski space by an element of this symmetry group which is now named after him. This conjecture was proven in [82]. We remark that a similar statement holds for the hyperbolic Ernst equation, see [83]. The relation between the Geroch group and the above mentioned $SL(2,\mathbb{R})$ symmetry, observed for solutions with one Killing vector, was clarified in [78]. There it was found that this group is the central extension of a group of holomorphic functions with values in $SL(2,\mathbb{R})$. In [84] the infinitesimal form of the Geroch group was shown to be the affine Kac–Moody algebra $A_1{}^{(1)}$. For the group theoretical properties of the Ernst equation see also [85] to [88].

To understand the geometric origin of the linear systems it is helpful to consider the above mentioned close relationship between integrability and self-duality. It was observed in [37, 89] that the Yang equation, an equation which is mathematically equivalent to the Ernst equation, can be understood as a symmetry reduction of the ASDYM equations on flat Minkowski space, the symmetry being given by two Killing vectors which correspond to rotation and time translation. It turned out that the Penrose–Ward transform factors through this symmetry, i. e., the Yang equation respectively the Ernst equation may be treated by the same methods as the ASDYM equations. This gives the possibility to describe the solutions to the stationary, axisymmetric vacuum equations in terms of fibre bundles over the symmetry reduced twistor space which was done by Mason and Woodhouse, see [88]. The equivalence of the Yang equation and the Ernst equation can also be used to construct multi-monopole solutions to the static axisymmetric Yang–Mills–Higgs equations, see [37] to [41].

Another geometric origin of the linear systems of the Ernst equation we are going to address in this volume is the close relation to Bianchi surfaces [50]. Bianchi studied two-dimensional surfaces in $\mathbb{R}^3$ with prescribed Gaussian curvature. For the case of negative constant curvature such surfaces are described by the sine–Gordon equation for a scalar function u,

$$u_{,xt} = \sin u \ . \tag{1.13}$$

The Gauss–Weingarten equations provide in a natural way a linear system for the sine–Gordon equation, where, however, a spectral parameter has to be introduced. For other classes of Bianchi surfaces the Gaussian curvature is not constant; the Gauss–Weingarten equation for Bianchi surfaces are closely related to the Ernst system, see [51]: both equations can be obtained as a symmetry reduction of the same differential system, where the Bianchi equation corresponds to a $SU(2)$ reduction and the Ernst equation to $SU(1,1)$ reduction.

Due to the non-autonomous character of the linear system (1.11), solutions in terms of theta functions first given by Korotkin [52,90] have different properties than the periodic solutions of equations as KdV. The solutions to the elliptic Ernst equation are given on a family of hyperelliptic surface corresponding to the algebraic curve

$$\mu^2 = (K - \xi)(K - \bar{\xi}) \prod_{j=1}^{g} (K - E_j)(K - F_j), \qquad (1.14)$$

where the branch points E_i, F_i are independent of the physical coordinates and subject to the reality condition $E_i, F_i \in \mathbb{R}$ or $E_i = \bar{F}_i$. The solutions depend only on the Weyl coordinates via the branch points of the Riemann surface. Thus the solutions are neither periodic or quasi-periodic since the modular dependence and not the dependence on the argument of the theta functions plays the main role in this context. For a study of the elliptic ($g = 1$) case see [91] to [93], for a more recent exposition see [94, 95]. We note that a similar approach is possible for the self-dual $SU(2)$ invariant Einstein equations, see [96]. It is the purpose of this volume to investigate and illustrate this class of solutions in cases of physical relevance, and to establish the differences with respect to the well-known periodic finite gap solutions discussed in [1].

To this end we consider boundary value problems for the Ernst equation. It is generally believed that most of the stars and galaxies can be described in good approximation as fluid bodies in thermodynamical equilibrium. In the framework of general relativity, this implies for isolated bodies, i.e., for asymptotically flat settings (see e. g. [97,98]) that the corresponding spacetimes are stationary and axisymmetric. This stresses the importance of the study of stationary axisymmetric spacetimes. A relativistic treatment is necessary for rapidly rotating and massive compact objects like pulsars, neutron stars and black holes as the one in the center of the Milky Way [99].

Though the importance of global solutions describing stationary axisymmetric fluid bodies is generally accepted, the complicated structure of the Einstein equations with matter gives little hope that such solutions can be found in the near future. Heilig has established existence theorems for rotating stars in Newton theory in the vicinity of known stationary perfect fluid solutions in [100] and for relativistic rotating stars in the vicinity of Newtonian solutions in [101]. Explicit solutions in the perfect fluid region, which are discussed as candidates for an interior solution, could only be given for special and somewhat unphysical equations of state [102] to [104] (in [105] it was shown that the Wahlquist solution cannot be the interior solution for a slowly rotating star). As indicated above, in the exterior vacuum region powerful solution generating techniques are at hand which do not seem to be applicable for general ideal fluids. For surfacelike distributions of the matter as disks which are discussed as models in astrophysics for certain types of galaxies and the matter in accretion disks around black holes, see [106], the matter equations reduce to ordinary differential equations. These provide boundary data for the vacuum field equations which correspond to a boundary value problem as studied in [107, 108]. In the case of a disk it is thus possible to find global solutions of the Einstein equations by solving a boundary value problem. The matter enters in these models only in form of boundary conditions for the vacuum equations.

One approach to solve boundary value problems for integrable equations with the help of the associated linear system is to translate the physical boundary conditions into a so-called Riemann–Hilbert problem (see Sect. 3.2) as was done in [109] to [111] for a rigidly rotating disk of dust. Here the boundary conditions at the disk are related to a jump discontinuity of the matrix of the linear system in the plane of the spectral parameter. The corresponding matrix Riemann–Hilbert problem on the Riemann sphere $\mathbb{CP}^1$ is equivalent to a linear integral equation, see [53]. Neugebauer and Meinel were able to reduce the matrix problem for the rigidly rotating disk of dust to a scalar Riemann–Hilbert problem on a hyperelliptic Riemann surface which can be solved explicitly via quadratures. By making use of gauge transformations of the linear system we were able to show in [112] that this is possible in general if the boundary value problem leads to a Riemann–Hilbert problem with rational jump data. Up to now there is, however, no direct way to infer the jump data from the physical boundary value problem in general. The explicit form of the hyperelliptic solutions, however, offers a different approach to boundary value problems: one can try to identify the free parameters in the solutions, a real valued function and a set of complex parameters, the branch points of the hyperelliptic Riemann surface, from the problem one wants to solve.

To this end we study Korotkin's solutions [52] as the solution of a generalized scalar Riemann–Hilbert problem on a hyperelliptic Riemann surface as in [112]. We present a complete discussion of the singularity structure of the corresponding Ernst potentials. It is possible to identify a subclass of solutions that are everywhere regular except at some contour, which can possibly be related to the surface of an isolated body, where the Ernst potential is bounded. These solutions are asymptotically flat and equatorially symmetric, and thus show all the features one might expect from the exterior solution for an isolated relativistic ideal fluid. They can have a Minkowskian and an extreme relativistic limit in which the body is 'hidden' behind a horizon, and in which the exterior solution becomes the extreme Kerr solution.

To provide concrete examples we study as in [113] a boundary value problem for a surface layer which consists of two streams of matter rotating with the same angular velocity in opposite direction. From an astrophysical point of view a Newtonian treatment of galaxies is sufficient as long as no black holes are involved which are genuinely relativistic objects. In this sense the study of disks has to be seen as a preparatory step for the study of systems consisting of black holes plus disks. Here we are mainly interested in illustrating features of the hyperelliptic solution. Counter-rotating disks can be seen as models for galaxies with counter-rotating matter components which are probably the results of the collision of two counter-rotating galaxies. There is experimental evidence [114] for galaxies with a large amount of counter-rotation: S0 galaxies (which are the link between elliptical and spiral galaxies in the Hubble fork of galaxies) can have up to 50 % counter-rotation. The first galaxy of this type, NGC 4550 in the Virgo cluster, has been discovered in 1992 [116] and has 30–40 % counter-rotation. Infinite disks with counter-rotating matter were discussed as sources of the Kerr metric in [117]. Physical properties for an explicit solution [118] as mass and angular momentum and the energy density are discussed following [119]. We give plots for the metric functions in typical situations and discuss interesting limiting cases.

1.4 Outline of the Content of the Book

In Chap. 2 we study a dimensional reduction of the vacuum Einstein equations in the presence of a single Killing vector, an approach that can be seen as a special case of Kaluza–Klein reductions of higher dimensional gravity theories, see e.g. [64,120]. Ehlers had already noted in [121] that a quotient space metric where the action of the Killing vector is divided out leads to simplified field equations on the orbit space of the Killing vector. It was found that the spatial part of the Killing vector is dual to a scalar potential called the twist potential. It is convenient to combine the norm of the Killing vector with the twist to a complex scalar potential since the integrability condition for the twist can be combined with the Einstein equation for the norm of the Killing vector to a generalized Ernst equation [34]. The remaining Einstein equations have the form of three-dimensional Einstein equations with an $SL(2, \mathbb{R})$ non-linear sigma model [69] as a matter source. This corresponds to a harmonic map from the spacetime $\mathcal{M}$ into the hyperbolic plane $\mathcal{H}$. In the presence of a second Killing vector the symmetry group of the sigma model is considerably enlarged. We consider here only the stationary axisymmetric case in Weyl coordinates [35]. In this case the symmetry group of the sigma model becomes the infinite dimensional Geroch group [78,81]. The Einstein equations reduce to the complex Ernst equation which decouples from the remaining metric functions. The latter can be obtained via quadratures for a given Ernst potential. The complete integrability of the Ernst equation is shown following Maison [75] and Belinski–Zakharov [77] in the form of an overdetermined linear differential system for an $SL(2, \mathbb{R})$ valued function Φ. We discuss the relation to the Bianchi surfaces studied in [51]. The Ernst equation is equivalent to the Yang equation [42] which can be obtained as a symmetry reduction of the (anti-)self-dual Yang–Mills equations. This equivalence is used to establish a relation to twistor theory. We show how the multi-monopole solutions to the $SU(2)$ Yang–Mills equations arise, see [37].

In Chap. 3 we study the Riemann–Hilbert problem for the Ernst equation. The simplest example for a Riemann–Hilbert problem is to find a function in the complex plane which has a prescribed discontinuity at a closed smooth contour. For analytic jump data this problem can be always solved explicitly in terms of the Cauchy integral. In the case of the Ernst equation, the Riemann–Hilbert problem is formulated for the matrix Φ of the associated linear differential system. As in [112] we exploit a gauge freedom to transform the matrix Riemann–Hilbert problem to a scalar problem on some Riemann surface. If the surface is non-compact, existence of solutions to this problem can be shown [122, 123]. If the surface is compact, the solutions form a subclass of Korotkin's solutions [52] and can be given in terms of hyperelliptic theta functions. We establish the relation of these solutions to Krichever's approach [124] to algebro-geometric solutions of integrable equations via the monodromy matrix. An algebraic form of the solutions [125, 126] free of theta functions is considered together with the so-called Picard–Fuchs equations. An alternative derivation of these solutions [127] based on an identity for theta functions due to Fay [128] is presented which leads also to explicit formulas for the metric. We summarize basic features of theta-functional solutions to the KdV and a two-dimensional generalization, the Kadomtsev–Petviashvili equation.

In Chap. 4 the hyperelliptic solutions to the Ernst equation are discussed following [118, 129]. We study potential singularities of the solutions as coinciding branch points, the axis of symmetry and ring and disk singularities which was first done in [118, 129]. A subclass of solutions is identified which are only singular at the zeros of the theta function in the denominator, asymptotically flat and equatorially symmetric. We discuss interesting limiting cases as the Minkowskian and the ultrarelativistic limit. Reductions of the theta-functional solutions in terms of theta functions on surfaces of lower genus are presented on the axis of symmetry and in the equatorial plane. We study the 'solitonic' limit in which the Riemann surface degenerates which leads for instance to the Kerr black hole or the static limit, where the Ernst potential is real and belongs to the static Weyl class.

In Chap. 5 we discuss boundary value problems for the Ernst equation. If one is interested in global solutions, two-dimensional matter distributions as shells and disks have to be studied. We consider the case of dust disks as in [106, 113]. To solve boundary value problems in terms of hyperelliptic solutions we study differential relations for the Ernst potential at a boundary on a given surface which follow from the Picard–Fuchs equations. As an example of this approach we discuss the derivation of the solution for the counter-rotating disk [113, 130].

To numerically evaluate the transcendental theta functions efficient numerical algorithms are necessary. This is especially true in the case of the dynamical Riemann surfaces studied here since the calculation of the so-called periods, which is typically the most time consuming operation, has to be performed for each spacetime point. In addition the underlying algebraic curve the deformation of which is being studied becomes singular at various points as the axis of symmetry. Thus a very efficient code of high precision is needed. In Chap. 6 we present a code based on spectral methods [131] which is tested by local and global methods. The code is used to test a code [132] based on spectral methods to directly solve the Einstein equations numerically.

As an example for physical properties of hyperelliptic solutions, we discuss the counter-rotating disk solution in Chap. 7 as in [119]. We consider various interesting limiting cases as the Newtonian, the static, the ultra-relativistic and the limit of a one-component disk [125]. The mass and the angular momentum of the disk as well as its mass density are discussed. The metric functions as well as the ergoregions, where there can be no stationary observer with respect to infinity are presented.

In Chap. 8 we mention open problems in the context of algebro-geometric solutions to the Ernst equation. This includes solutions of higher genus than 2 and a relation between the quantities characterizing the Riemann surfaces, the branch points, and physical quantities as the mass, the angular momentum and the equation of state in the matter. Black holes can be included in the formalism by considering partially degenerate Riemann surfaces, see [133, 134]. Since the stationary axisymmetric Einstein–Maxwell equations in the electro-vacuum are also completely integrable (an infinite dimensional symmetry group was discovered in [135]), theta-functional solutions can be constructed along the same lines, see [52]. The underlying Riemann surfaces are, however, three-sheeted in this case. Hyperelliptic solutions can be obtained by exploiting the $SU(2,1)/S[U(1,1) \times U(1)]$ invariance of the Einstein–Maxwell equations, i.e., by using a so-called Harrison transformation [136] as in [137].

In Chap. A of the appendix, where we fix the notation, we have included a brief review of the mathematics of compact Riemann surfaces needed in this volume. For details the reader is referred to textbooks by Farkas and Kra [138], Mumford [139], Fay [128] and Belokolos et al. [1]. A link to twistor theory as in [88] is established in Chap. B of the appendix.

2 The Ernst Equation

Ernst's original motivation in finding the Ernst equation [34] was to provide a simple scheme to construct the Kerr metric as a solution to the stationary axisymmetric Einstein equations in vacuum. In fact the Ernst potential for the Kerr solution is just an algebraic function in suitable coordinates, see (1.8). In this chapter we study a dimensional reduction of the vacuum Einstein equations in the presence of two Killing vectors which will lead to the Ernst equation. Due to the integrability of this equation large classes of physically interesting solutions can be constructed which will be the main subject of this book. The integrability of the Ernst equation has also played a role in the construction of multi-monopole solutions to the static axisymmetric Yang–Mills–Higgs (YMH) equations. We note the close relation to the theory of Bianchi surfaces. Bianchi considered surfaces in three-dimensional Euclidean space characterized by the harmonicity of $1/\sqrt{-\mathcal{K}}$ where $\mathcal{K}$ is the Gaussian curvature. In the case of constant negative Gaussian curvature, this leads e.g. to the pseudo-sphere which is a solution to the integrable sine–Gordon equation. If one studies Bianchi surfaces with non-constant Gaussian curvature one is led to equations which are closely related to the Ernst equation.

In Sect. 2.1 we will study vacuum spacetimes with a single non-null Killing vector. Using a quotient space formalism going back to Ehlers [121] and Geroch [64], we divide out the Killing action and study the field equations on its orbit space. This approach can be seen as a special case of Kaluza–Klein reductions of higher dimensional gravity theories. The spatial part of the Killing vector can be dualized to a scalar, the twist potential. It turns out that the Einstein equations are equivalent in this approach to equations for three-dimensional gravitation with matter where the matter corresponds to a $SL(2,\mathbb{R})$ sigma model. The symmetry properties of the sigma model imply that one can generate from a given solution new solutions via the action of the group $SL(2,\mathbb{R})$, the Ehlers transformations. However, these transformations do not transform asymptotically flat solutions into asymptotically flat solutions in the strong sense: the transformed solutions will either have a negative mass or a so-called Newman–Unti–Tamburini (NUT) parameter. Solutions with such a parameter, which corresponds to a magnetic monopole, are not physically acceptable, see e.g. [140]. In the presence of a second commuting Killing vector (we will mainly study the stationary axisymmetric case here) discussed in Sect. 2.2, the symmetry group becomes the infinite dimensional Geroch group. The Einstein equations reduce in Weyl coordinates to the Ernst equation, the remaining metric functions follow in terms of quadratures from a given Ernst potential. The infinite dimensional symmetry group of the Ernst equation is reflected by the fact that it can be treated e.g. as the integrability condition for an over-determined linear differential system for some $SL(2, \mathbb{R})$-valued matrix Φ which contains an additional parameter, the spectral parameter.

In Sect. 2.3 we study Bianchi surfaces mainly following [51]. The corresponding Gauß–Weingarten equations are very similar to the linear system of the Ernst equation and allow for the introduction of a spectral parameter which leads to a zero-curvature representation. The Bianchi equation and the Ernst equation correspond to $SU(2)$ respectively $SU(1,1)$ reductions of the same differential system. We discuss simple examples. In Sect. 2.4 we establish the equivalence of the Ernst and the Yang equation [42] which can be obtained as a symmetry reduction of the self-dual Yang–Mills equations. This allows to establish a link to twistor theory as in [88] which will be addressed in more detail in Chap. B of the appendix. In Sect. 2.5 we consider the static axisymmetric YMH equations for a massless Higgs boson in Manton's ansatz [141]. Following [37,40], we show the equivalence of the Bogomolny equations in this case to the Ernst or the Yang equation and discuss the Bogomolny–Prasad–Sommerfield (BPS) monopole as an example. In [38] to [41] the dressing method [53,77,142] was used to construct multi-monopole solutions to the YMH equations, where the monopoles are located in the origin. Spatially separated monopoles were shown to be non-axisymmetric in [143].

2.1 Dimensional Reduction and Group Structure

For basic knowledge of the differential geometry needed to study general relativity we refer the reader to the standard literature as [140] and [144] to [153]. We consider a four-dimensional manifold $(\mathcal{M}, \mathbf{g})$ with Lorentzian metric $\mathbf{g}$ of signature $+2$. The existence of a Killing vector can be used to establish a simplified version of the field equations by dividing out the group action. These quotient space metrics were first used in [121], see also [64]; here we will follow the approach of [120]. We use adapted coordinates in which the Killing vector Ξ is given by $\Xi = \partial_t$. We adopt here a coordinate dependent approach to simplify the expressions, but the decomposition can be performed independently of coordinates as in [64,70]. We also assume here that t is a timelike coordinate, i.e., ∇t is timelike. The construction works, however, for non-null Killing vectors. The vector Ξ is subject to the Killing equation

$$\mathcal{L}_\Xi g_{AB} = \mathcal{D}_B \Xi_A + \mathcal{D}_A \Xi_B = 0 \,, \tag{2.1}$$

where $\mathcal{L}$ denotes the Lie-derivative, where capital indices take the values $0,1,2,3$, and where $\mathcal{D}_A$ denotes the covariant derivative with respect to the metric **g**. The norm of the Killing vector will be denoted by f. The decomposition we are using is not defined at fixed points of the group action, i.e., the zeros of f, and the resulting equations will be singular at the set of zeros of f.

In contrast to a standard $3 + 1$-decomposition, the metric is written in this approach in the form

$$ds^2 = g_{AB}\mathrm{d}x^A\mathrm{d}x^B = -f(\mathrm{d}t + k_a\mathrm{d}x^a)(\mathrm{d}t + k_b\mathrm{d}x^b) + \frac{1}{f}h_{ab}\mathrm{d}x^a\mathrm{d}x^b ; \quad (2.2)$$

latin indices always take the values $1, 2, 3$ corresponding to the spatial coordinates. To establish the Einstein equations which are in vacuum just equivalent to the vanishing of the Ricci tensor, we use the standard definition of the curvature tensor $\mathcal{R}_{ABCD}$ via the commutator of covariant derivatives of an arbitrary vector which yields for the Killing vector

$$[\mathcal{D}_A, \mathcal{D}_B]\Xi_C = \mathcal{R}_{ABCD}\Xi^D . \quad (2.3)$$

Interchanging the first two indices in (2.3) leads to a term involving the Riemann tensor, interchanging the last two indices leads to a sign due to the Killing equation (2.1). If one interchanges these index pairs alternately until one arrives at their original position, one gets

$$2\mathcal{D}_A\mathcal{D}_B\Xi_C = (\mathcal{R}_{CABD} - \mathcal{R}_{BCAD} + \mathcal{R}_{ABCD})\Xi^D . \quad (2.4)$$

The symmetries of the Riemann tensor then imply the relation

$$\mathcal{D}_A\mathcal{D}_B\Xi_C = \mathcal{R}_{DABC}\Xi^D . \quad (2.5)$$

The vacuum Einstein equations $\mathcal{R}^C_{ACB} = 0$ thus lead to

$$\mathcal{D}^A\mathcal{D}_A\Xi_B = 0 . \quad (2.6)$$

The spacetime $\mathcal{M}$ can be decomposed in the quotient space $S = \mathcal{M}/\Xi$ and the Killing vector Ξ. The four dimensional tensor $\tilde{h}_{AB}$ defined by

$$\tilde{h}_{AB} \doteq g_{AB} + \frac{1}{f}\Xi_A\Xi_B = g_{AB} - \frac{\Xi_A\Xi_B}{\Xi_C\Xi^C} \quad (2.7)$$

has the components $\tilde{h}_{00} = \tilde{h}_{0a} = 0$, and $h_{ab} = \tilde{h}_{ab}f$ coincides with the previously defined spatial metric. The tensor is thus a projector orthogonal to Ξ_A, $\tilde{h}_{AB}\Xi^A = 0$. We can use $\tilde{h}_{ab}$ as a metric on the quotient space S. A tensor on S has the properties

$$\Xi^A T^{C\ldots D}_{A\ldots B} = 0 , \quad \ldots , \quad \Xi_D T^{C\ldots D}_{A\ldots B} = 0 , \quad \mathcal{L}_\Xi T^{C\ldots D}_{A\ldots B} = 0 . \quad (2.8)$$

The covariant derivative $\tilde{D}_a$ on S is related to the covariant derivative $\mathcal{D}_A$ on $\mathcal{M}$ via

$$\tilde{D}_e T^{c\ldots d}_{a\ldots b} = \tilde{h}^E_e \tilde{h}^A_a \ldots \tilde{h}^B_b \tilde{h}^c_C \ldots \tilde{h}^D_d \mathcal{D}_E T^{C\ldots D}_{A\ldots B} \ . \tag{2.9}$$

Thus we obtain for $\mathcal{R}_{AB} \Xi^A \Xi^B$

$$\frac{1}{2} \mathcal{D}^A \mathcal{D}_A f = (\mathcal{D}_A \Xi_B)(\mathcal{D}^A \Xi^B) \ . \tag{2.10}$$

Since $\tilde{h}^A_a \Xi_A = 0$ we have $\tilde{h}^A_a \mathcal{D}_B \Xi_A = \tilde{h}^A_a \mathcal{D}_B(\Xi_A/f)$. Thus $\tilde{h}^A_a \tilde{h}^B_b \mathcal{D}_A \Xi_B = f \tilde{h}^A_a \tilde{h}^B_b k_{[a,b]}$. Equation (2.3) for $\mathcal{R}_{aB} \Xi^B$ thus implies the Maxwell equation

$$\frac{1}{2} D_a(f^2 k^{[a,b]}) = 0 \ , \tag{2.11}$$

where D_a denotes the covariant derivative with respect to h_{ab}, and where $[ab]$ denotes anti-symmetrization in the indices. Notice that all indices here are raised and lowered with h_{ab}. Equation (2.11) implies that the dual $k^{[a,b]} \varepsilon_{abc}$ has vanishing curl and that it is thus the gradient of a scalar potential b. If we define this twist potential b via

$$\frac{1}{\sqrt{h} f^2} \varepsilon^{abc} \partial_c b \doteq k^{[a,b]} \ , \tag{2.12}$$

where h is the determinant of h_{ab}, then equation (2.11) is identically satisfied. The potentials f and b can be combined to the complex Ernst potential $\mathcal{E} = f + ib$ [34]. The equations for f and the integrability condition for b can then be combined to the generalized complex Ernst equation

$$f D_a D^a \mathcal{E} = D_a \mathcal{E} D^a \mathcal{E} \ . \tag{2.13}$$

To determine the relation between the Riemann tensor on S and on $\mathcal{M}$, we calculate $\tilde{D}_a \tilde{D}_b V_c$ where V_c is an arbitrary vector field on S. Thus we get with (2.9)

$$\tilde{D}_a \tilde{D}_b V_c = h^A_a h^B_b h^C_c \mathcal{D}_A(\tilde{h}^L_B \tilde{h}^M_C \mathcal{D}_L V_M) \tag{2.14}$$

which leads with (2.7) to

$$\tilde{D}_a \tilde{D}_b V_c = h^A_a h^B_b h^C_c \left(\mathcal{D}_A \mathcal{D}_B V_C + (\mathcal{D}_A \Xi_B) \Xi^L \mathcal{D}_L V_C / f \right.$$
$$\left. + \Xi^L (\mathcal{D}_A \Xi_C) \mathcal{D}_B V_L / f \right) \ . \tag{2.15}$$

To eliminate the derivatives of V_a in the last two terms in (2.15), we use for the second

$$\pounds_\Xi V_C = \Xi^L \mathcal{D}_L V_C - V^L \mathcal{D}_C \Xi_L = 0 \ , \tag{2.16}$$

and the fact that $\Xi_L V^L = 0$ for the third. Consequently we can write (2.15) as

$$\tilde{D}_a \tilde{D}_b V_c = h^A_a h^B_b h^C_c \left(\mathcal{D}_A \mathcal{D}_B V_C - (\mathcal{D}_A \Xi_B)(\mathcal{D}_C \Xi_L) V^L / f \right.$$
$$\left. - (\mathcal{D}_A \Xi_C)(\mathcal{D}_B \Xi_L) V^L / f \right) \ . \tag{2.17}$$

Since the vector V is arbitrary, we obtain a relation between the Riemann tensor $\mathcal{R}_{ABCD}$ on $\mathcal{M}$ and the Riemann tensor R_{abcd} on S (the Killing equation was used to establish the symmetry of the second term),

$$R_{abcd} = h_{[a}^{A}h_{b]}^{B}h_{[c}^{C}h_{d]}^{D}\left(\mathcal{R}_{ABCD} - 2(\mathcal{D}_A\Xi_B)(\mathcal{D}_C\Xi_D)/f \\ -2(\mathcal{D}_A\Xi_C)(\mathcal{D}_B\Xi_D)/f\right) . \tag{2.18}$$

The equations for the metric h_{ab} can thus be written in the form

$$R_{ab} = \frac{1}{2f^2}\Re(\mathcal{E}_{,a}\bar{\mathcal{E}}_{,b}) , \tag{2.19}$$

where R_{ab} is the three-dimensional Ricci tensor corresponding to h_{ab}. It is obvious that zeros of the norm of the Killing vector are singular points of the equations (2.13) and (2.19). Equations (2.19) can be interpreted as equations for three-dimensional gravitation with some matter model which turns out to be the well-known $SL(2,\mathbb{R})/SO(1,1)$ sigma model of pure gravity. Sigma models are a map $\Phi : (\mathcal{M}, \mathbf{g}) \to (\mathcal{N}, \mathbf{G})$ from a manifold $\mathcal{M}$ with metric $\mathbf{g}$ into a target manifold $\mathcal{N}, \mathbf{G}$,

$$S[\Phi] = \frac{1}{2}\int_{\mathcal{M}} \Phi_{,A}^{\mu}\Phi_{,B}^{\nu}G_{\mu\nu}(\Phi(x))g^{AB}(x)\sqrt{-g}d^4x, \tag{2.20}$$

where $g = \det(\mathbf{g})$. Sigma models correspond to so-called harmonic maps as will be shown below. In the case of the Einstein equations in the presence of a Killing vector, the field equations follow from the three-dimensional Lagrangian

$$L^{(3)} = \sqrt{h}\left(\frac{1}{2}R_a^a - \frac{1}{4f^2}h^{ab}(f_{,a}f_{,b} + b_{,a}b_{,b})\right) . \tag{2.21}$$

The metric

$$ds^2 = \frac{df^2 + db^2}{f^2} \tag{2.22}$$

is the invariant metric of $SL(2,\mathbb{R})/SO(1,1) \sim SU(1,1)$ which can be seen from a suitable parametrization of the coset space in terms of the coordinates f and b. It corresponds to the Poincaré metric of the upper-half plane. The action of the group $SL(2,\mathbb{R})$ on the Ernst potential is given in terms of a Möbius transformation. It is straight forward to check that the whole group is generated by the shift $\mathcal{E} \to \mathcal{E} + ic$ with $c \in \mathbb{R}$ and the inversion $\mathcal{E} \to 1/\mathcal{E}$, the Ehlers transformation [121]. Note that this transformation does not transform asymptotically flat spacetimes into asymptotically flat spacetimes which are characterized by the condition

$$\mathcal{E} = 1 - \frac{2m}{r} + O(r^{-2}) ; \tag{2.23}$$

here r is some suitably defined radial coordinate with the property $1/r \to 0$ at spatial infinity, and the real constant m is the Arnowitt–Deser–Misner

(ADM) mass, see e.g. [145]. An Ehlers transformed Ernst potential has a negative or an imaginary mass, a NUT parameter. Such spacetimes do not seem to have astrophysical relevance, see for instance [140] for a discussion of Taub–NUT spacetimes.

Metric (2.22) is also the metric of the hyperbolic plane modeled by the half-plane $\mathbb{H} = \{(f,b) : f > 0\}$. This is a Riemann surface with constant curvature -1, see also the section on Bianchi surfaces below. The *energy density* of the manifold S is given by

$$\varepsilon = \frac{h^{ab}}{f^2}(f_{,a}f_{,b} + b_{,a}b_{,b}) ,\qquad (2.24)$$

see [49,70]. The integral of the energy density is the *energy* of S. The critical points of this integral are the harmonic maps into the hyperbolic plane. They satisfy the Euler–Lagrange equations

$$D^a D_a f = \frac{D^a f D_a f - D^a b D_a b}{f} ,$$

$$D^a D_a b = 2\frac{D^a f D_a b}{f} ,\qquad (2.25)$$

which are obviously equivalent to the Ernst equation in this case. This aspect is discussed in detail by Weinstein in [70] to [73], where it was used to prove existence and uniqueness theorems for multi-black-hole solutions in the stationary axisymmetric case. In [74] these techniques were used to prove uniqueness of solutions for perfect fluid configurations.

2.2 The Stationary Axisymmetric Case

In the presence of a second Killing vector η commuting with Ξ,

$$\mathcal{L}_\Xi g_{AB} = \mathcal{L}_\eta g_{AB} = 0 , \quad [\Xi,\eta] = 0 ,\qquad (2.26)$$

the metric can be further simplified to the Weyl–Lewis–Papapetrou form with diagonal h_{ab} in situations which are asymptotically flat or which have a regular axis, see [154,155] and also [35]. We will only consider here such cases with stationary and axial symmetry, i.e., with $\eta = \partial_\phi$. In Weyl's cylindrical coordinates (ϱ, ζ, ϕ), we have

$$h_{ab} = \mathrm{diag}(\mathrm{e}^{2k}, \mathrm{e}^{2k}, W^2) .\qquad (2.27)$$

In addition we can choose $k_a = (0,0,a)$. All metric functions depend only on ϱ and ζ. The metric functions f, a and W^2 are Killing invariants,

$$f = -\Xi_A \Xi^A ,$$
$$af = -\Xi_A \eta^A ,$$
$$W^2 = -\Xi_A \Xi^A \eta_B \eta^B + (\Xi_A \eta^A)^2 .\qquad (2.28)$$

Obviously, the chosen metric does not change under conformal transformations in the (ϱ, ζ)-plane and basis transformations in the Lie algebra E_2 generated by the Killing vectors Ξ and η, i.e., coordinate transformations of the form

$$\begin{pmatrix} t' \\ \phi' \end{pmatrix} = \begin{pmatrix} \alpha & \beta \\ \gamma & \delta \end{pmatrix} \begin{pmatrix} t \\ \phi \end{pmatrix} , \tag{2.29}$$

with

$$\alpha\delta - \beta\gamma \neq 0 . \tag{2.30}$$

The coordinates are usually fixed by imposing regularity conditions on the symmetry axis $\varrho = 0$ and at spatial infinity. For a detailed treatment of this issue see e.g. [108]. On the regular part of the axis the metric should be elementary flat [148], i.e., the metric should be Minkowskian with vanishing first derivatives. Thus we require

$$g_{t\phi}(0, \zeta) = g_{\phi\phi}(0, \zeta) = 0 , \tag{2.31}$$

which implies

$$a(0, \zeta) = W(0, \zeta) = 0 , \tag{2.32}$$

and

$$f_{,\varrho}(0, \zeta) = a_{,\varrho}(0, \zeta) = 0 . \tag{2.33}$$

Furthermore, the ratio between the circumference of a circle around the symmetry axis and its radius should be 2π in the limit $\varrho \to 0$. More precisely,

$$\lim_{\varrho \to 0} \frac{\int_0^{2\pi} \sqrt{g_{\phi\phi}}\,d\phi}{\int_0^\varrho \sqrt{g_{\varrho\varrho}}\,d\varrho'} = 2\pi \lim_{\varrho \to 0} \frac{\sqrt{W^2/f - a^2 f}}{\int_0^\varrho e^k/\sqrt{f}\,d\varrho'} = 2\pi \frac{W_{,\varrho}}{e^k} \overset{!}{=} 2\pi , \tag{2.34}$$

by de l'Hôpital's rule. Thus,

$$W_{,\varrho}(0, \zeta) = e^{k(0,\zeta)} . \tag{2.35}$$

To summarize, on the axis we should have

$$\begin{aligned} f_{,\varrho}(0, \zeta) &= 0 , & a(0, \zeta) = a_{,\varrho}(0, \zeta) &= 0 , \\ W(0, \zeta) &= 0 , & e^{k(0,\zeta)} &= W_{,\varrho}(0, \zeta) . \end{aligned} \tag{2.36}$$

Remark 2.1. A transformation of the form $\phi' = \phi + \omega t$ with constant ω describes a change to a coordinate system rotating with ω with respect to the original system. The corresponding transformation of the metric coefficients is

$$\begin{aligned} g'_{tt} &= g_{tt} + 2\omega g_{t\phi} + \omega^2 g_{\phi\phi} , \\ g'_{t\phi} &= g_{t\phi} + \omega g_{\phi\phi} . \end{aligned}$$

To require that g should be Minkowskian on the symmetry axis does not fix ω.

In the following we are particularly interested in the study of the gravitational field of isolated matter configurations in an otherwise empty universe. Thus, we impose the condition of asymptotic flatness, i. e., that the metric is asymptotically Minkowskian

$$\lim_{r\to\infty} U = \lim_{r\to\infty} k = \lim_{r\to\infty} a = 0 , \quad \lim_{r\to\infty} (W - \varrho) = 0 , \quad (2.37)$$

where $e^{2U} = f$, where $r^2 = \varrho^2 + \zeta^2$, and where $U\varrho$ and $a\varrho^3$ remain bounded. Observe that ω can now be fixed by requiring that a vanishes at spatial infinity.

It is usually assumed that stationary axisymmetric fluid configurations are invariant under reflections at the equatorial plane and that the corresponding gravitational fields reflect this symmetry. In the following, we will often restrict ourselves to equatorially symmetric spacetimes, where the metric functions fulfil $g_{AB}(\varrho, \zeta) = g_{AB}(\varrho, -\zeta)$, i. e. a, e^{2U} and k are even functions in ζ.

Equations (2.19) for the $\phi\phi$ component imply

$$W_{,\varrho\varrho} + W_{,\zeta\zeta} = 0 . \quad (2.38)$$

Thus W is a harmonic function of ϱ and ζ which has to vanish on the axis, see (2.32). We can thus use W as a coordinate,

$$W = \varrho , \quad (2.39)$$

if $|\nabla W|^2 \neq 0$. Thus we use the metric

$$ds^2 = -f(dt + ad\phi)^2 + \frac{1}{f}(e^{2k}(d\varrho^2 + d\zeta^2) + \varrho^2 d\phi^2) . \quad (2.40)$$

For the following it proves helpful to use complex coordinates $\xi, \bar{\xi}$ which are related to ϱ, ζ via

$$\xi = \zeta - i\varrho . \quad (2.41)$$

Then equations (2.19) imply for the metric function k

$$k_{,\xi} = \frac{\xi - \bar{\xi}}{4f^2} \mathcal{E}_{,\xi} \bar{\mathcal{E}}_{,\xi} . \quad (2.42)$$

The integrability condition for k in (2.42) is just the Ernst equation which reads in this case

$$\mathcal{E}_{,\xi\bar{\xi}} - \frac{1}{2(\bar{\xi} - \xi)}(\mathcal{E}_{,\bar{\xi}} - \mathcal{E}_{,\xi}) = \frac{2}{\mathcal{E} + \bar{\mathcal{E}}} \mathcal{E}_{,\xi} \mathcal{E}_{,\bar{\xi}} . \quad (2.43)$$

For ϕ-independent $\mathcal{E}$ this is equivalent to

$$f\Delta\mathcal{E} = (\nabla\mathcal{E})^2 , \quad (2.44)$$

where Δ and ∇ are the standard operators in cylindrical coordinates. The duality relation (2.12) then reads

$$a_{,\xi} = \frac{i\varrho b_{,\xi}}{f^2} \; . \tag{2.45}$$

Again (2.45) is integrable because of the Ernst equation. Thus for a given solution to the Ernst equation, the metric functions a and k can be given in terms of quadratures.

We note that the spacetime is static if the Ernst potential is real and the function a vanishes. In this case the Ernst equation reduces to the axisymmetric Laplace equation for U, the so-called Euler–Darboux equation,

$$U_{,\varrho\varrho} + \frac{1}{\varrho}U_{,\varrho} + U_{,\zeta\zeta} = 0 \; . \tag{2.46}$$

The corresponding solutions belong to the Weyl class.

The above relations imply that solutions to the stationary axisymmetric Einstein equations in vacuum can be obtained by solving the Ernst equation only. The complete metric then follows in terms of line integrals. This is obviously an important simplification. The main advantage of the formulation of the Einstein equations in the Ernst picture is, however, the fact that the Ernst equation is completely integrable, i.e., that the symmetry group of the equation goes over in the presence of a second Killing vector from $SL(2,\mathbb{R})$ to the infinite dimensional Geroch group [81]. The relation between the Geroch group and the above mentioned $SL(2,\mathbb{R})$ symmetry for solutions with one Killing vector was established in [78]: The Geroch group is the central extension of a group of holomorphic functions with values in $SL(2,\mathbb{R})$. In [84] the infinitesimal form of the Geroch group was shown to be the affine Kac–Moody algebra $A_1{}^{(1)}$. For the group theoretical properties of the Ernst equation, see also [86] to [88].

Here we are mainly concerned with the analytical consequences of this infinite dimensional symmetry group. As in the case of the KdV equation (see the introduction and references given therein), the Ernst equation can be treated as the integrability condition of an over-determined linear differential system. One possible form of the linear system is (1.11) which is adapted to the $SL(2,\mathbb{R})$ symmetry of the equation. Here we use a gauge equivalent form of the system, see [80], which was first given in [156],

$$\Phi_{,\xi}\Phi^{-1} = \begin{pmatrix} \mathcal{M} & 0 \\ 0 & \mathcal{N} \end{pmatrix} + \sqrt{\frac{K-\xi}{K-\bar{\xi}}}\begin{pmatrix} 0 & \mathcal{M} \\ \mathcal{N} & 0 \end{pmatrix} \doteq U \; ,$$

$$\Phi_{,\bar{\xi}}\Phi^{-1} = \begin{pmatrix} \bar{\mathcal{N}} & 0 \\ 0 & \bar{\mathcal{M}} \end{pmatrix} + \sqrt{\frac{K-\xi}{K-\bar{\xi}}}\begin{pmatrix} 0 & \bar{\mathcal{N}} \\ \bar{\mathcal{M}} & 0 \end{pmatrix} \doteq V \; , \tag{2.47}$$

where

$$\mathcal{M} = \frac{\bar{\mathcal{E}},_\xi}{\mathcal{E} + \bar{\mathcal{E}}} , \quad \mathcal{N} = \frac{\mathcal{E},_\xi}{\mathcal{E} + \bar{\mathcal{E}}} . \tag{2.48}$$

The above system is defined on the Riemann surface $\mathcal{L}_0$ of genus 0 defined by $\mu_0^2 = (K - \xi)(K - \bar{\xi})$. This means that the branch points of the Riemann surface are parametrized by the physical coordinates, the Riemann surface is 'dynamical' in this sense; this is an important difference to equations like KdV. The solution of (2.47) can be normalized by the condition

$$\Phi(K = \infty^+, \xi, \bar{\xi}) = \begin{pmatrix} \bar{\mathcal{E}} & 1 \\ \mathcal{E} & -1 \end{pmatrix} , \tag{2.49}$$

which implies that the Ernst potential can be directly read off as one component of the matrix Φ at infinity in the upper sheet of $\mathcal{L}_0$ which is denoted by ∞^+ (this is the reason why we choose this system instead of the one in (1.11)). Consequently if one succeeds to construct a matrix Φ in a way that it solves the system (2.47), the corresponding Ernst potential is easily accessible. The next chapter will be devoted to the construction of a large class of solutions with the help of so-called Riemann–Hilbert techniques.

2.3 Bianchi Surfaces

In [75] the associated linear system of the Ernst equation was found by group theoretical methods applied to sigma models. The corresponding linear system in [156] was found by analogy to the sine–Gordon equation (SG). In this section we will establish the close relation between the Ernst equation and so-called Bianchi surfaces. It is well known that many equations which have been successfully studied with the inverse scattering method in the last 30 years have been obtained in the framework of surface theory more than 100 years ago. The most famous example for this is the SG equation

$$\Phi,_{xt} = \sin \Phi \tag{2.50}$$

which is equivalent to the Gauss–Codazzi system for surfaces with constant negative Gaussian curvature $\mathcal{K}$. The case of positive Gaussian or mean curvature $\mathcal{H}$ leads to the hyperbolic version of (2.50), the sinh–Gordon equation

$$\Phi,_{z\bar{z}} + \sinh \Phi = 0 . \tag{2.51}$$

Complete integrability of these equations is established by introducing a spectral parameter into the corresponding Gauss–Weingarten system which leads to a zero-curvature representation of the equations, see [157] for a review.

Bianchi [50] studied congruences C in $\mathbb{R}^3$, i.e., a two-parameter family of straight lines. In general a congruence has two focal surfaces S_1 and S_2.

Definition 2.2. *A congruence C is called a Bianchi congruence if the Gaussian curvatures of S_1 and S_2 at the points on the same straight line of the congruence are negative and coincide. The surfaces S_1 and S_2 are called Bianchi surfaces.*

These congruences and surfaces were apparently first introduced by Bianchi in [159], see [160] for a more modern exposition.

We consider the embedding of an arbitrary surface S in $\mathbb{R}^3$ defined by the vector function $\boldsymbol{F}(x,y) : \mathbb{R}^2 \to \mathbb{R}^3$. Let $\boldsymbol{N}(x,y)$ be a unit normal vector to S. As usual one defines for S the first and second fundamental form as

$$I = \langle \mathrm{d}\boldsymbol{F}, \mathrm{d}\boldsymbol{F} \rangle , \quad II = -\langle \mathrm{d}\boldsymbol{F}, \mathrm{d}\boldsymbol{N} \rangle . \tag{2.52}$$

Then the Gaussian curvature is given by

$$\mathcal{K} = \frac{\det II}{\det I} . \tag{2.53}$$

This curvature satisfies in the case of a Bianchi surface

$$\left[\frac{1}{\sqrt{-\mathcal{K}}} \right]_{,xy} = 0 , \tag{2.54}$$

where x and y are asymptotic coordinates on S, i.e., coordinates in which the second fundamental form is off-diagonal, see the concrete parametrization below in (2.57). Equation (2.54) implies

$$\mathcal{K} = -\frac{1}{\mathcal{P}^2(x,y)} , \quad \mathcal{P}(x,y) = F(x) + F(y) , \tag{2.55}$$

where $F(x)$ and $G(y)$ are arbitrary real valued functions.

If S has negative Gaussian curvature (2.55) with $\mathcal{P}(x,y) > 0$, we can always choose local coordinates (x,y) along the asymptotic lines ($\boldsymbol{F}(x = const, y)$ and $\boldsymbol{F}(x, y = const)$) along which the curvature is equal to 0. In this case the vectors $\boldsymbol{F}_{,x}$, $\boldsymbol{F}_{,y}$ as well as $\boldsymbol{F}_{,xx}$ and $\boldsymbol{F}_{,yy}$ are orthogonal to the normal vector $\boldsymbol{N}$. Then the first and second fundamental form can be parametrized in the following way,

$$I = \mathcal{P}^2(A^2\mathrm{d}x^2 + 2AB\cos\phi\,\mathrm{d}x\mathrm{d}y + B^2\mathrm{d}y^2) , \tag{2.56}$$
$$II = 2\mathcal{P}AB\sin\phi\,\mathrm{d}x\mathrm{d}y , \tag{2.57}$$

where $\mathcal{P}A = |\boldsymbol{F}_{,x}|$, $\mathcal{P}B = |\boldsymbol{F}_{,y}|$, and where ϕ is the angle between $\boldsymbol{F}_{,x}$ and $\boldsymbol{F}_{,y}$, i.e., between the asymptotic lines.

In the following we adopt a quaternionic description of the surfaces, see [51] and [161] for details. This description is based on the standard identification of $\mathbb{R}^3$ with the algebra $su(2)$,

$$X = (x_1, x_2, x_3) \in \mathbb{R}^3 \leftrightarrow X = \sum_{k=1}^{3} \frac{1}{2\mathrm{i}} x_k \sigma_k , \tag{2.58}$$

where σ_k are the Pauli matrices. We consider the following basis in $\mathbb{R}^3$:

$$\frac{\mathcal{P}A}{2i}(\cos\phi_1\sigma_1 + \sin\phi_1\sigma_2)\,, \quad \frac{\mathcal{P}B}{2i}(\cos\phi_2\sigma_1 + \sin\phi_2\sigma_2)\,, \quad \frac{\sigma_3}{2i}\,. \qquad (2.59)$$

There is an $SU(2)$ transformation $\tilde{\Phi}$ which maps this basis to the basis $(\boldsymbol{F}_{,x}, \boldsymbol{F}_{,y}, \boldsymbol{N})$,

$$\boldsymbol{F}_{,x} = \tilde{\Phi}^{-1}\frac{\mathcal{P}A}{2i}(\cos\phi_1\sigma_1 + \sin\phi_1\sigma_2)\tilde{\Phi}\,,$$

$$\boldsymbol{F}_{,y} = \tilde{\Phi}^{-1}\frac{\mathcal{P}B}{2i}(\cos\phi_2\sigma_1 + \sin\phi_2\sigma_2)\tilde{\Phi}\,,$$

$$\boldsymbol{N} = \tilde{\Phi}^{-1}\frac{\sigma_3}{2i}\tilde{\Phi}\,. \qquad (2.60)$$

The matrix $\tilde{\Phi}$ satisfies the Gauss–Weingarten equations

$$\tilde{\Phi}_{,x} = U\tilde{\Phi}\,, \quad \tilde{\Phi}_{,y} = V\tilde{\Phi}\,, \qquad (2.61)$$

where

$$U = \frac{i}{2}\left[-A\cos\phi_1\sigma_1 - A\sin\phi_1\sigma_2 + \left(\phi_{2,x} + \frac{\mathcal{P}_{,y}A}{2\mathcal{P}B}\sin(\phi_1 + \phi_2)\right)\sigma_3\right]\,,$$

$$V = \frac{i}{2}\left[B\cos\phi_2\sigma_1 - B\sin\phi_2\sigma_2 + \left(-\phi_{1,y} - \frac{\mathcal{P}_{,x}B}{2\mathcal{P}A}\sin(\phi_1 + \phi_2)\right)\sigma_3\right]\,. \qquad (2.62)$$

The compatibility condition for the Gauss–Weingarten system (2.61) is the following Gauss–Codazzi system of nonlinear equations for the real valued functions $A(x,y) > 0$, $B(x,y) > 0$ and $\phi \doteq \phi_1 + \phi_2$,

$$\phi_{,xy} + \frac{1}{2}\left(\frac{\mathcal{P}_{,x}B}{\mathcal{P}A}\sin\phi\right)_{,x} + \frac{1}{2}\left(\frac{\mathcal{P}_{,y}A}{\mathcal{P}B}\sin\phi\right)_{,y} - AB\sin\phi = 0\,,$$

$$A_{,y} + \frac{\mathcal{P}_{,y}}{2\mathcal{P}}A - \frac{\mathcal{P}_{,x}}{2\mathcal{P}}B\cos\phi = 0\,,$$

$$B_{,x} + \frac{\mathcal{P}_{,x}}{2\mathcal{P}}B - \frac{\mathcal{P}_{,y}}{2\mathcal{P}}A\cos\phi = 0\,. \qquad (2.63)$$

It turns out that one can introduce a single complex valued function $\mathcal{E} \doteq \tilde{\Phi}_{11}/\tilde{\Phi}_{12}$ instead of the three real valued functions (A, B, ϕ). The Gauss–Codazzi system (2.63) is equivalent to a second order differential system for the function $\mathcal{E}$,

$$(\mathcal{E} + \bar{\mathcal{E}}^{-1})\left(\mathcal{E}_{,xy} + \frac{\mathcal{P}_{,x}}{2\mathcal{P}}\mathcal{E}_{,y} + \frac{\mathcal{P}_{,y}}{2\mathcal{P}}\mathcal{E}_{,x}\right) = 2\mathcal{E}_{,x}\mathcal{E}_{,y}\,. \qquad (2.64)$$

The functions A, B and ϕ are given in terms of $\mathcal{E}$,

$$A = \frac{2|\mathcal{E}_{,x}|}{1 + |\mathcal{E}|^2}\,, \quad B = \frac{2|\mathcal{E}_{,y}|}{1 + |\mathcal{E}|^2}\,, \quad \phi = \pi + \arg\frac{\mathcal{E}_{,x}}{\mathcal{E}_{,y}}\,. \qquad (2.65)$$

Similarly one can express the fundamental forms and the moving frame in terms of the function $\mathcal{E}$, see [51]. Note that equation (2.64) can be written in terms of the Gauss map N which reads in terms of the function $\mathcal{E}$

$$N = \frac{1}{2i} \frac{1}{1 + |\mathcal{E}|^2} \begin{pmatrix} |\mathcal{E}|^2 - 1 & 2\bar{\mathcal{E}} \\ 2\mathcal{E} & 1 - |\mathcal{E}|^2 \end{pmatrix} , \tag{2.66}$$

in the form of

$$(\mathcal{P} N_{,x} N^{-1})_{,y} + (\mathcal{P} N_{,y} N^{-1})_{,x} = 0 . \tag{2.67}$$

The Gauss–Weingarten system (2.61) has almost the form of a Lax pair, but it lacks a spectral parameter. In the case of Bianchi surfaces, this parameter can be inserted as done by Bianchi [50]: with $\mathcal{P}(x,y) = F(x) + G(y)$ one can show that the equation (2.64) is the compatibility condition for the linear system (Ψ is a 2×2 matrix)

$$\Psi_{,x} = \mathcal{U}\Psi , \quad \Psi_{,y} = \mathcal{V}\Psi , \tag{2.68}$$

where

$$\mathcal{U} = \begin{pmatrix} \mathcal{A} & 0 \\ 0 & \mathcal{B} \end{pmatrix} + \sqrt{\frac{K - G(y)}{K + F(x)}} \begin{pmatrix} 0 & \mathcal{A} \\ \mathcal{B} & 0 \end{pmatrix} ,$$

$$\mathcal{V} = \begin{pmatrix} \mathcal{C} & 0 \\ 0 & \mathcal{D} \end{pmatrix} + \sqrt{\frac{K + F(x)}{K - G(y)}} \begin{pmatrix} 0 & \mathcal{C} \\ \mathcal{D} & 0 \end{pmatrix} . \tag{2.69}$$

The form of the functions $\mathcal{A}$, $\mathcal{B}$, $\mathcal{C}$ and $\mathcal{D}$ can be obtained from the condition

$$\Psi(x, y, K = \infty^+) = \begin{pmatrix} \mathcal{E} & 1 \\ \bar{\mathcal{E}}^{-1} & -1 \end{pmatrix} . \tag{2.70}$$

Gauging Ψ to be a $SU(2)$ matrix Φ with the property $\Phi(K = \infty^+) = \tilde{\Phi}$, the embedding of the Bianchi surface is given by

$$F = -2\Phi^{-1}\Phi_{,(1/K)}\big|_{K=\infty^+} , \quad N = \Phi^{-1}\frac{\sigma_3}{2i}\Phi\big|_{K=\infty^+} . \tag{2.71}$$

It is convenient to introduce locally the following types of Bianchi surfaces:

1. B_0-surfaces: $F = const$ and $G = const$. In this case we can put $F = 0$ and $G = 1$ without loss of generality which implies $K = -1$. For $\mathcal{A} = \mathcal{B} = 1$ the Gauss–Codazzi system is equivalent to the SG equation.
2. B_1-surfaces: $F(x) = const$, $G_{,y}(y) \neq 0$. Without loss of generality we can put locally $F = 0$ and $G = y$ which implies $K = -1/y^2$.
3. B_2-surfaces: $F_{,x}(x) \neq 0$, $G_{,y}(y) \neq 0$. Without loss of generality we can put locally $F = x$ and $G = y$, which implies $K = -1/(x + y)^2$.

In the latter case equation (2.64) reads

$$\left(\mathcal{E} + \bar{\mathcal{E}}^{-1}\right)\left(\mathcal{E}_{,xy} + \frac{\mathcal{E}_{,x} + \mathcal{E}_{,y}}{2(x+y)}\right) = 2\mathcal{E}_{,x}\mathcal{E}_{,y} \ . \tag{2.72}$$

This equation is obviously closely related to the Ernst equation. Both equations can be obtained as reductions of the following system for two complex potentials $\mathcal{E}_1$ and $\mathcal{E}_2$,

$$(\mathcal{E}_1 + \mathcal{E}_2)\left(\mathcal{E}_{1,xy} + \frac{\mathcal{E}_{1,x} + \mathcal{E}_{1,y}}{2(x+y)}\right) = 2\mathcal{E}_{1,x}\mathcal{E}_{1,y} \ ,$$

$$(\mathcal{E}_1 + \mathcal{E}_2)\left(\mathcal{E}_{2,xy} + \frac{\mathcal{E}_{2,x} + \mathcal{E}_{2,y}}{2(x+y)}\right) = 2\mathcal{E}_{2,x}\mathcal{E}_{2,y} \ . \tag{2.73}$$

The Bianchi equation (2.72) corresponds to the 'unitary' reduction $\mathcal{E}_2 = (\bar{\mathcal{E}}_1)^{-1}$, whereas the Ernst equation corresponds to the 'real' reduction $\mathcal{E} = \mathcal{E}_1 = \bar{\mathcal{E}}_2$. The system (2.73) has a zero curvature representation of the form (2.68) with

$$\Psi(K = \infty) = \begin{pmatrix} \mathcal{E}_1 & 1 \\ \mathcal{E}_2 & -1 \end{pmatrix} \ . \tag{2.74}$$

The reduction $\mathcal{E}_1 = (\bar{\mathcal{E}}_2)^{-1}$ leads to the system (2.68), the reduction $\mathcal{E}_1 = \bar{\mathcal{E}}_2$ leads to the system found in [156]. The matrix Ψ can be gauged in a way that it is in $SU(2)$ in the first case and in $SU(1,1)$ in the second case.

For illustration we will discuss some simple cases of Bianchi surfaces which correspond to solitonic solutions to the Bianchi equation. All formulas below are taken from [51] where the derivations and some of the figures are given. In the B_0 case the one-soliton solution of the SG equation leads to

$$F = \left(y - x + \Re\beta_1 \frac{2q}{1+q^2}\right)\frac{\sigma_1}{2i} - \frac{2q}{1+q^2}\Im\beta_1 \cos(x+y)\frac{\sigma_2}{2i}$$

$$- \frac{2q}{1+q^2}\Im\beta_1 \sin(x+y)\frac{\sigma_3}{2i}, \tag{2.75}$$

where with $\delta, q \in \mathbb{R}$

$$\beta_1 = \frac{1 - i\delta\exp(y/q - qx)}{1 + i\delta\exp(y/q - qx)} \ . \tag{2.76}$$

This is a pseudo-sphere surface parametrized by the real values q and δ. The case $\delta = q = 1$ is shown in Fig. 2.1. The case $q = 4$ and $\delta = 1$ leads to the infinite helix which is shown in Fig. 2.2.

The embedding for the one-soliton solution for B_1 surfaces can be written in the form ($y > \gamma_1$ positive)

$$F = (-x\sqrt{y} - 2i\omega_1\Re\beta_1)\frac{\sigma_1}{2i} + \left(2i\omega_1\Im\beta_1 \cos\frac{x}{\sqrt{y}}\right)\frac{\sigma_2}{2i} + \left(2i\omega_1\Im\beta_1 \sin\frac{x}{\sqrt{y}}\right)\frac{\sigma_3}{2i}, \tag{2.77}$$

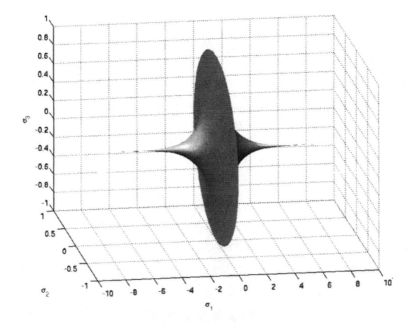

Fig. 2.1. Pseudo-sphere

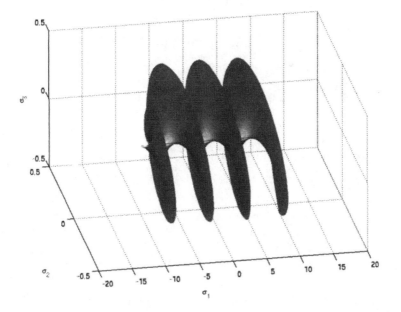

Fig. 2.2. Part of the infinite helix

where

$$\beta_1 = \frac{1 - i\delta \exp(i\mu_1 x/\sqrt{y})}{1 + i\delta \exp(i\mu_1 x/\sqrt{y})} \; , \quad \mu_1 = \sqrt{(\gamma_1 - y)/\gamma_1} \in i\mathbb{R} \; , \quad \omega_1 = \gamma_1 \mu_1 \in i\mathbb{R} \; .$$

(2.78)

The solution depends on the two real parameters δ and $\gamma_1 \in]0, y[$. A typical example of this surface can be seen in Fig. 2.3 for $\delta = \gamma_1 = 1$.

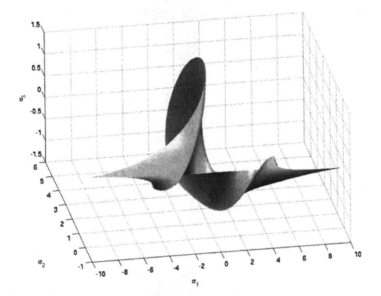

Fig. 2.3. One-soliton B_1-surface

There is no non-trivial one-soliton solution in the B_2 case. The embedding for the two-soliton solution has the form

$$F = \frac{4\Im(q_0 \bar{q}_1)}{1 + |q_1|^2} \frac{\sigma_1}{2i} + \frac{4\Im(q_0 + \sigma XY q_1)}{1 + |q_1|^2} \frac{\sigma_2}{2i} - \frac{4\Re(q_0 + \sigma XY q_1)}{1 + |q_1|^2} \frac{\sigma_3}{2i} \; , \quad (2.79)$$

where

$$q_0 = -i\sigma(\beta_1 X + \beta_2 Y) \; , \quad q_1 = -i(\beta_2 X + \beta_1 Y) \; . \quad (2.80)$$

The coordinates X and Y are related to the coordinates x, y via

$$X = \frac{1}{2i\sigma}(\omega_1 + \omega_2) \; , \quad Y = \frac{1}{2i\sigma}(\omega_1 - \omega_2) \; ,$$
$$\omega_1 = \sqrt{(\sigma + x)(\sigma - y)} \; , \quad \omega_2 = \bar{\omega}_1 \; . \quad (2.81)$$

These coordinates are more suitable than x, y to generate the whole surface since they are supposed to take positive and negative values of X and Y. If only one possible choice of sign in the ω_i is taken into account, only one of the four leafs of the surface will be obtained. The solution is parametrized by the real parameter σ and the real parameters α_1 and α_2 in

$$\beta_1 = \frac{1}{2}(e^{i\alpha_1} + e^{i\alpha_2}), \quad \beta_2 = \frac{1}{2}(e^{i\alpha_1} - e^{i\alpha_2}). \tag{2.82}$$

A typical example for these surfaces with $\sigma = 1$, $\alpha_1 = \pi/2$ and $\alpha_2 = 0$ is shown in Fig. 2.4, where $X, Y \in [-10, 10]$. The surface closes asymptotically for $X, Y \to \infty$, the two leafs disconnected by a small gap approach each other in this limit. The surface is however not smooth at this connection.

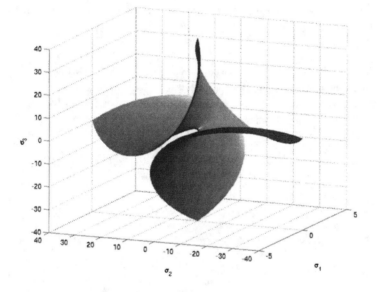

Fig. 2.4. Two-soliton B_2-surface

2.4 The Yang Equation

To understand the geometric origin of the zero-curvature representation of integrable equations, one approach consists in the study of symmetry reductions of the (anti)-self-dual Yang–Mills equations (ASDYM), see [28]. In the case studied here, the first observation is the equivalence between the Ernst equation and the Yang equation [42]. If we write the Weyl–Lewis–Papapetrou metric in the form

$$ds^2 = J_{\alpha\beta}\mathrm{d}x^\alpha \mathrm{d}x^\beta + e^{2(k-U)}(\mathrm{d}\varrho^2 + \mathrm{d}\zeta^2), \tag{2.83}$$

where $\alpha, \beta = t, \phi$ and

$$(J_{\alpha\beta}) = \begin{pmatrix} -f & -af \\ -af & \varrho^2/f - a^2 f \end{pmatrix} , \tag{2.84}$$

the matrix J is subject to the Yang equation

$$(\varrho J^{-1} J_{,\varrho})_{,\varrho} + \varrho (J^{-1} J_{,\varsigma})_{,\varsigma} = 0 . \tag{2.85}$$

Remark 2.3.
(i) With J being a solution to the Yang equation the matrix

$$J' \doteq AJB^{-1} , \tag{2.86}$$

with constant 2×2-matrices A and B also solves the Yang equation.
(ii) If J is a solution to the Yang equation then the matrices J^{-1}, and J^T are also solutions.

To establish the relation between the Yang equation and the Yang–Mills equations, we consider duality for the latter. Let M denote an affine four-dimensional space with t-times $+$ and s-times $-$ in the signature. Since $** = (-1)I$, the Hodge star operator on M has the eigenvalues ± 1 iff $s - t$ is even and $\pm i$ iff $s - t$ is odd.

Definition 2.4. *Let ω be a two-form on M. We call ω self-dual iff*

$$*\omega = \begin{cases} \omega & s - t - even \\ i\omega & s - t - odd \end{cases} . \tag{2.87}$$

Similarly, we call ω anti-self-dual iff

$$*\omega = \begin{cases} -\omega & s - t - even \\ -i\omega & s - t - odd \end{cases} . \tag{2.88}$$

For the two-form F of the field strength of the Maxwell field given by

$$(F_{\mu\nu}) = \begin{pmatrix} 0 & -E^1 & -E^2 & -E^3 \\ E^1 & 0 & B^3 & -B^2 \\ E^2 & -B^3 & 0 & B^1 \\ E^3 & B^2 & -B^1 & 0 \end{pmatrix} , \tag{2.89}$$

we find with

$$(*F)_{kl} = \sum_{i \leq j} F^{ij} \varepsilon_{ijkl} \tag{2.90}$$

for the components of the Hodge dual of the field strength

$$(*F_{ij}) = \begin{pmatrix} 0 & B^1 & B^2 & B^3 \\ -B^1 & 0 & E^3 & -E^2 \\ -B^2 & -E^3 & 0 & E^1 \\ -B^3 & E^2 & -E^1 & 0 \end{pmatrix} . \tag{2.91}$$

Thus, the only real solution to $*F = \pm iF$ is given by

$$E^1 = E^2 = E^3 = B^1 = B^2 = B^3 = 0 . \qquad (2.92)$$

The same holds for the ASDYM equations

$$*F = -F , \qquad (2.93)$$

where F is now the field strength of the non-Abelian gauge field.

On the other hand, for signatures of the metric of the form $(+ + ++)$, i.e. Euclidean space and $(+ + --)$, the so called ultra-hyperbolic spacetime, real solutions to the (anti-)self-dual Maxwell equations exist. In order to treat several signatures of the spacetime on the same footing, it is often convenient to consider the complexification $M^{\mathbb{C}}$ of M. Furthermore, let us define a local chart of $M^{\mathbb{C}}$ with double null coordinates $(w, \tilde{w}, z, \tilde{z})$ such that the line element $\mathrm{d}s^2$ on $M^{\mathbb{C}}$ has the following form

$$\mathrm{d}s^2 = 2\,(\mathrm{d}z\mathrm{d}\tilde{z} - \mathrm{d}w\mathrm{d}\tilde{w}) . \qquad (2.94)$$

Imposing certain reality conditions on the coordinates $(w, \tilde{w}, z, \tilde{z})$ yields real subspaces of $M^{\mathbb{C}}$, see [32].

(i) the four-dimensional Euclidean space $\mathbb{E}^4$: $\overline{w} = -\tilde{w}$, $\overline{z} = \tilde{z}$, i.e.,

$$\begin{pmatrix} \tilde{z} & w \\ \tilde{w} & z \end{pmatrix} = \frac{1}{\sqrt{2}} \begin{pmatrix} x^0 + ix^1 & -x^2 + ix^3 \\ x^2 + ix^3 & x^0 - ix^1 \end{pmatrix} ,$$

(ii) the four-dimensional Minkowski space M: $\overline{w} = \tilde{w}$, $\overline{z} = z$ and $\overline{\tilde{z}} = \tilde{z}$, i.e.,

$$\begin{pmatrix} \tilde{z} & w \\ \tilde{w} & z \end{pmatrix} = \frac{1}{\sqrt{2}} \begin{pmatrix} x^0 + x^1 & x^2 - ix^3 \\ x^2 + ix^3 & x^0 - x^1 \end{pmatrix} ,$$

(iii) the four-dimensional ultra-hyperbolic space U: $\overline{w} = \tilde{w}$, $\overline{z} = \tilde{z}$, i.e.,

$$\begin{pmatrix} \tilde{z} & w \\ \tilde{w} & z \end{pmatrix} = \frac{1}{\sqrt{2}} \begin{pmatrix} x^0 + ix^1 & x^2 - ix^3 \\ x^2 + ix^3 & x^0 - ix^1 \end{pmatrix} .$$

Here x^0, x^1, x^2, x^3 are real coordinates.

Remark 2.5.
In the case (iii) it is also possible to require $z, \tilde{z}, w$ and $\tilde{w}$ to be real in order to obtain a metric of the same signature.
The different real subspaces of M can be characterized as the set of fixed points of $M^{\mathbb{C}}$ with respect to the following anti-holomorphic involution

$$\sigma : M^{\mathbb{C}} \longrightarrow M^{\mathbb{C}} , \qquad (2.95)$$

defined by

$$\sigma\left(w, z, \tilde{w}, \tilde{z}\right) \doteq \left(-\bar{\tilde{w}}, \bar{\tilde{z}}, -\bar{w}, \bar{z}\right) ,$$
$$\sigma\left(w, z, \tilde{w}, \tilde{z}\right) \doteq \left(\bar{\tilde{w}}, \bar{\tilde{z}}, \bar{w}, \bar{z}\right) ,$$
$$\sigma\left(w, z, \tilde{w}, \tilde{z}\right) \doteq \left(-\bar{w}, \bar{z}, \bar{\tilde{w}}, \bar{\tilde{z}}\right) ,$$
$$\sigma\left(w, z, \tilde{w}, \tilde{z}\right) \doteq \left(\bar{\tilde{w}}, \bar{z}, \bar{w}, \bar{\tilde{z}}\right) . \tag{2.96}$$

Here the first definition just yields the Euclidean four-space $\mathbb{E}^4$, the following two the ultra-hyperbolic plane and the last one the Minkowski slice.

Let us turn, now, to the ASDYM equations (we consider here the anti-self-dual case, the treatment of the self-dual equations is analogous). We denote by D the covariant derivative

$$D = d + \Phi \tag{2.97}$$

of a connection on a complex vector bundle of rank n over a region U of spacetime. Let us denote by F the corresponding curvature two-form of the connection, i. e.,

$$F = F_{ij} dx^i \wedge dx^j ,$$

with

$$F_{ij} = \partial_i \Phi_j - \partial_j \Phi_i + [\Phi_i, \Phi_j] .$$

In double null coordinates $(z, \tilde{z}, w, \tilde{w})$ the condition of the anti-self-duality of F becomes

$$\partial_z \Phi_w - \partial_w \Phi_z + [\Phi_z, \Phi_w] = 0 , \tag{2.98}$$
$$\partial_{\tilde{z}} \Phi_{\tilde{w}} - \partial_{\tilde{w}} \Phi_{\tilde{z}} + [\Phi_{\tilde{z}}, \Phi_{\tilde{w}}] = 0 , \tag{2.99}$$
$$\partial_z \Phi_{\tilde{z}} - \partial_{\tilde{z}} \Phi_z - \partial_w \Phi_{\tilde{w}} + \partial_{\tilde{w}} \Phi_w + [\Phi_z, \Phi_{\tilde{z}}] - [\Phi_w, \Phi_{\tilde{w}}] = 0 . \tag{2.100}$$

With

$$D_w \doteq \partial_w + \Phi_w , \quad D_z \doteq \partial_z + \Phi_w , \quad D_{\tilde{w}} \doteq \partial_{\tilde{w}} + \Phi_{\tilde{w}} , \quad D_{\tilde{z}} \doteq \partial_{\tilde{z}} + \Phi_{\tilde{z}} ,$$

the above equations become

$$[D_z, D_w] = 0 ,$$
$$[D_{\tilde{z}}, D_{\tilde{w}}] = 0 ,$$
$$[D_z, D_{\tilde{z}}] - [D_w, D_{\tilde{w}}] = 0 ,$$

and defining

$$L \doteq D_w - \xi D_{\tilde{z}} ,$$
$$M \doteq D_z - \xi D_{\tilde{w}} ,$$

we find the following form of the ASDYM equations

$$[L, M] = 0 \,, \tag{2.101}$$

for all $\xi \in \mathbb{C}$.

Equations (2.98) and (2.99) can be integrated by the ansatz

$$\begin{aligned} \partial_w H = H \cdot \Phi_w \,, && \partial_z H = H \cdot \Phi_z \,, \\ \partial_{\tilde{w}} \hat{H} = \hat{H} \cdot \Phi_{\tilde{w}} \,, && \partial_{\tilde{z}} \hat{H} = \hat{H} \cdot \Phi_{\tilde{z}} \,, \end{aligned} \tag{2.102}$$

with

$$H \,, \quad \hat{H} : M^{\mathbb{C}} \longrightarrow Gl(n; \mathbb{C}) \,. \tag{2.103}$$

Under a gauge transformation

$$\Phi \longrightarrow \Phi' = g^{-1}\Phi g + g^{-1}\mathrm{d}g \,, \tag{2.104}$$

with $g \in Gl(n; \mathbb{C})$ the quantity $J \doteq H \cdot \hat{H}^{-1}$ remains unchanged, if we replace

$$H \longrightarrow H' = Hg \,, \quad \hat{H} \longrightarrow \hat{H}' = \hat{H}g \,. \tag{2.105}$$

Then (2.100) takes the form

$$\partial_w \left(J^{-1}\partial_{\tilde{w}} J \right) - \partial_z \left(J^{-1}\partial_{\tilde{z}} J \right) = 0 \,. \tag{2.106}$$

Let us now consider the Euclidean subspace of $M^{\mathbb{C}}$ with Weyl coordinates $(t, \phi, \varrho, \zeta)$ which are related to the double null coordinates by

$$\begin{aligned} z = \zeta - \mathrm{i}t \,, && \tilde{z} = \bar{z} \,, \\ w = -\varrho e^{-\mathrm{i}\phi} \,, && \tilde{w} = -\bar{w} \,, \end{aligned} \tag{2.107}$$

such that we have

$$\begin{aligned} \partial_z = \tfrac{1}{2}\left(\partial_\zeta + \mathrm{i}\partial_t\right) \,, && \partial_w = \tfrac{1}{2}\left(-e^{\mathrm{i}\phi}\partial_\varrho - \tfrac{\mathrm{i}e^{\mathrm{i}\phi}}{\varrho}\partial_\phi\right) \,, \\ \partial_{\tilde{z}} = \tfrac{1}{2}\left(\partial_\zeta - \mathrm{i}\partial_t\right) \,, && \partial_{\tilde{w}} = \tfrac{1}{2}\left(e^{-\mathrm{i}\phi}\partial_\varrho - \tfrac{\mathrm{i}e^{-\mathrm{i}\phi}}{\varrho}\partial_\phi\right) \,. \end{aligned} \tag{2.108}$$

Let the gauge potential Φ be given by

$$\Phi = \Phi_t \mathrm{d}t + \Phi_\phi \mathrm{d}\phi + \Phi_\varrho \mathrm{d}\varrho + \Phi_\zeta \mathrm{d}\zeta \,, \tag{2.109}$$

and assume that Φ is invariant under time translations and rotations

$$L_X \Phi = 0 = L_Y \Phi \,, \tag{2.110}$$

with $X = \partial_t$ and $Y = \partial_\phi$, i. e., the components of Φ do not depend on t and ϕ. Analogously, in double null coordinates this condition means that $\Phi_{,z}$, $\Phi_{,\tilde{z}}$, $\Phi_{,w}$ and $\Phi_{,\tilde{w}}$ do not depend on t and ϕ, but only on ϱ and ζ. Thus, for a gauge potential fulfilling (2.110) equation (2.106) takes the form

$$\partial_\varrho \left(J^{-1}\partial_\varrho J \right) + \frac{1}{\varrho}J^{-1}\partial_\varrho J + \partial_\zeta \left(J^{-1}\partial_\zeta J \right) = 0 \,, \tag{2.111}$$

what coincides with the Yang equation (2.85). Thus, the Yang equation is equivalent to the ASDYM equations on $\mathbb{E}^4$ for static and axisymmetric fields. However, a solution to (2.111) has to fulfil several additional conditions in order to yield a solution to the Yang equation:

(i) J has to be real and symmetric with $n = 2$ and
(ii) $\det J = -\varrho^2$,

what follows directly from the definition of J in equation (2.84). This equivalence can be used to construct solutions to the stationary axisymmetric Einstein equations with twistor methods which is discussed in Chap. B of the appendix.

Remark 2.6. Often equation (2.106) is called Yang equation.

An interesting consequence of the Yang equation being a symmetry reduction of the ASDYM equations (which was first observed by Witten [167] and Ward [168]) is the fact that the latter are known to be integrable. The Penrose–Ward transform (see e. g. [30]) establishes a natural one-to-one correspondence between anti-self-dual U(n)-gauge potentials over S^4 (up to gauge equivalence) and holomorphic vector bundles E of rank n over $\mathbb{CP}^3$ with a positive definite real form (up to isomorphism). An interesting feature of the Penrose–Ward transform is that it allows for symmetry reductions. It turns out that the Yang equation can be obtained as such a reduction. In fact, in [32] most of the known integrable non-linear equations have been solved by a symmetry reduction of the Penrose–Ward transform. For the stationary, axisymmetric Einstein equations this procedure has been worked out in [88], see also [163] and [164]. A detailed description of the procedure for the Kerr solution can be found in [165]. An important point is that in the solution process linear systems for the Yang equation are generated and get a geometric meaning, see Chap. B of the appendix for details.

Following [88], but see also [166], let us consider the holomorphicity condition for the sections $b = b(\varrho, \zeta, \lambda)$ of the holomorphic vector bundle $E \to R_V$,

$$\bar{\partial} b + \Psi b = 0 \ . \tag{2.112}$$

With $H = J$ and $\hat{H} = I$ this takes the form

$$\frac{1}{1+\lambda^2} J^{-1} \left(\frac{\partial J}{\partial \varrho} \bar{\partial}\varrho - \lambda \frac{\partial J}{\partial \zeta} \bar{\partial}\varrho + \frac{\partial J}{\partial \zeta} \bar{\partial}\zeta + \lambda \frac{\partial J}{\partial \varrho} \bar{\partial}\zeta \right) \cdot b + \frac{\partial b}{\partial \varrho} \bar{\partial}\varrho$$

$$+ \frac{\partial b}{\partial \zeta} \bar{\partial}\zeta + \frac{\partial b}{\partial \lambda} \bar{\partial}\lambda = 0 \ , \tag{2.113}$$

and using

$$\bar{\partial}\lambda = -\frac{\lambda\left[\left(\lambda^2 - 1\right) \bar{\partial}\varrho - 2\lambda\bar{\partial}\zeta\right]}{\varrho\left(\lambda^2 + 1\right)} \ , \tag{2.114}$$

we find the following system of equations

$$\frac{\partial b}{\partial \varrho} + \frac{\lambda - \lambda^3}{\varrho\left(\lambda^2 + 1\right)} + \frac{1}{1+\lambda^2} J^{-1} \left(\frac{\partial J}{\partial \varrho} - \lambda \frac{\partial J}{\partial \zeta} \right) \cdot b = 0 \ ,$$

$$\frac{\partial b}{\partial \zeta} + \frac{2\lambda^2}{\varrho\left(\lambda^2 + 1\right)} \frac{\partial b}{\partial \lambda} + \frac{1}{1+\lambda^2} J^{-1} \left(\frac{\partial J}{\partial \zeta} + \lambda \frac{\partial J}{\partial \varrho} \right) \cdot b = 0 \ . \tag{2.115}$$

By simple algebraic manipulations we obtain

$$\frac{\partial b}{\partial \varrho} + \frac{\lambda}{\varrho}\frac{\partial b}{\partial \lambda} + \lambda\frac{\partial b}{\partial \zeta} + J^{-1}\frac{\partial J}{\partial \varrho} \cdot b = 0 \,,$$

$$\frac{\partial b}{\partial \zeta} + \frac{\lambda^2}{\varrho}\frac{\partial b}{\partial \lambda} - \lambda\frac{\partial b}{\partial \varrho} + J^{-1}\frac{\partial J}{\partial \zeta} \cdot b = 0 \,. \qquad (2.116)$$

With

$$w = \frac{\varrho - \varrho\lambda^2 + 2\zeta\lambda}{2\lambda} \,, \qquad (2.117)$$

we find

$$\frac{\partial}{\partial \lambda} = \frac{-\varrho - \varrho\lambda^2}{2\lambda^2}\frac{\partial}{\partial w} \,. \qquad (2.118)$$

Restricting (2.116) to a leaf of the foliation, i. e. putting $w = $ const., we obtain with $s = s(\varrho, \zeta)$ the reduced system

$$(\partial_\varrho + \lambda\partial_\zeta)\, s + J^{-1}\partial_\varrho J \cdot s = 0 \,,$$

$$(\partial_\zeta - \lambda\partial_\varrho)\, s + J^{-1}\partial_\zeta J \cdot s = 0 \,. \qquad (2.119)$$

We have

Lemma 2.7. *The integrability condition for the linear system of equations (2.119) is the Yang equation.*

Proof. Rewriting (2.119) one obtains

$$\left(1 + \lambda^2\right)\partial_\varrho s = \left(\lambda J^{-1}\partial_\zeta J - J^{-1}\partial_\varrho J\right) s \,,$$

$$\left(1 + \lambda^2\right)\partial_\zeta s = -\left(\lambda J^{-1}\partial_\varrho J + J^{-1}\partial_\zeta J\right) s \,. \qquad (2.120)$$

For this system we write down the integrability condition $\partial_\zeta\partial_\varrho s = \partial_\varrho\partial_\zeta s$. Using

$$\partial_\varrho\left(\frac{1}{1 + \lambda^2}\right) = -\frac{2\lambda^2\left(1 - \lambda^2\right)}{\left(1 + \lambda^2\right)^2\left(2\varrho\lambda^2 + \varrho\left(1 - \lambda^2\right)\right)} \,, \qquad (2.121)$$

an analogous expression for $\partial_\zeta\left(1 + \lambda^2\right)^{-1}$ and ordering of terms with respect to powers in λ we find that the coefficients of the even powers of λ vanish iff

$$J^{-1}\partial_\varrho J + \varrho\left(\partial_\zeta\left(J^{-1}\partial_\zeta J\right) + \partial_\varrho\left(J^{-1}\partial_\varrho J\right)\right) = 0 \,, \qquad (2.122)$$

i. e., if the Yang equation is fulfilled. For the coefficients of the odd powers in λ we find that they vanish identically because of $\partial_\zeta\partial_\varrho J = \partial_\varrho\partial_\zeta J$.

$\square$

2.5 Multi-Monopoles
of the Yang–Mills–Higgs Equations

It was shown by 't Hooft [169] and Polyakov [170] that spontaneously broken gauge theories with a simple gauge group possess classical solutions that can be identified with magnetic monopoles. Such solutions can be interpreted as soliton-like particles. In the limit of vanishing Higgs potential when the Higgs field becomes massless, the field equations simplify considerably; the static minimal energy configurations are given as solutions to a set of first order equations, the Bogomolny equations. The analytic solution for the simply charged spherically symmetric 't Hooft–Polyakov monopole was found by Prasad and Sommerfield [171] and Bogomolny [172]. Since the monopoles in this case do not interact, multi-monopole solutions for the static axisymmetric case exist. The first exact solution for a doubly charged monopole were given by Ward [89] with twistor methods and by Forgács et al. [37] to [41]. Here we will follow the latter approach which used the equivalence of the Bogomolny equations in Manton's ansatz with the Ernst equation.

We consider an $SU(2)$ gauge theory with an isotriplet Higgs field in the limit of vanishing Higgs potential with Lagrangian density

$$L = -\frac{1}{4}F_{\mu\nu}^a F^{a\mu\nu} - \frac{1}{2}(D_\mu\Phi)^a(D^\mu\Phi)^a , \qquad (2.123)$$

where

$$F_{\mu\nu}^a = A_{\nu,\mu}^a - A_{\mu,\nu}^a - \varepsilon^{abc}A_\mu^b A_\nu^c , \quad (D_\mu\Phi)^a = \Phi_\mu^a - \varepsilon^{abc}A_\mu^b \Phi^c ; \qquad (2.124)$$

in this section indices $a, b, c = 1, 2, 3$ are isospin indices and greek indices spacetime indices. The Hamiltonian density for static configurations with vanishing electric fields ($A_0^a = 0$) is then

$$\mathcal{H} = \frac{1}{4}F_{ij}^a F^{aij} + \frac{1}{2}(D_i\Phi)^a(D^i\Phi)^a , \quad (a, i, j = 1, 2, 3) . \qquad (2.125)$$

The field equations in this case are solved by solutions of the Bogomolny equations [172]

$$F_{ij}^a = -\varepsilon_{ijk}(D_k\Phi)^a . \qquad (2.126)$$

Using the Bogomolny equations one can write the energy in the form

$$E = \int \mathcal{H}d^3x = \int (D_k\Phi)^a(D^k\Phi)^a d^3x = \frac{1}{2}\int \Delta|\Phi|^2 d^3x , \qquad (2.127)$$

where the last step makes use of the equations of motion of the Higgs field $D_k D^k\Phi = 0$. Asymptotically the Higgs field behaves as

$$|\Phi| = 1 - \frac{n}{r} + O(1/r^2) , \tag{2.128}$$

where n is the topological charge.

To describe multi-monopoles Manton [141] used an axially and mirror-symmetric ansatz which takes in cylindrical coordinates the form

$$A_0^a = 0 , \quad \Phi^a = (0, \phi_1, \phi_2) , \quad A_\phi^a = -(0, \eta_1, \eta_2) ,$$
$$A_z^a = -(W_1, 0, 0) , \quad A_\varrho^a = -(W_2, 0, 0) . \tag{2.129}$$

With this ansatz the Bogomolny equations read

$$\phi_{1,\varrho} - W_2\phi_2 = -\frac{1}{\varrho}(\eta_{1,z} - W_1\eta_2) ,$$

$$\phi_{2,\varrho} - W_2\phi_1 = -\frac{1}{\varrho}(\eta_{2,z} - W_1\eta_1) ,$$

$$W_{1,\varrho} - W_{2,z} = \frac{1}{\varrho}(\phi_1\eta_2 - \phi_2\eta_1) ,$$

$$\phi_{1,z} - W_1\phi_2 = \frac{1}{\varrho}(\eta_{1,\varrho} - W_2\eta_2) ,$$

$$\phi_{2,z} + W_1\phi_1 = \frac{1}{\varrho}(\eta_{2,\varrho} - W_2\eta_1) . \tag{2.130}$$

The system possesses a residual gauge invariance

$$W_i' = W_i + \Lambda_{,i} , \quad \begin{pmatrix} \phi_i' \\ \eta_i' \end{pmatrix} = \begin{pmatrix} \phi_i \\ \eta_i \end{pmatrix} \cos\Lambda + \varepsilon_{ij} \begin{pmatrix} \phi_j \\ \eta_j \end{pmatrix} \sin\Lambda . \tag{2.131}$$

It is possible to find a function Λ such that

$$W_1' = -\phi_1' , \quad W_2' = \frac{1}{\varrho}\eta_1' . \tag{2.132}$$

The remaining equations (2.131) imply

$$\phi_2 = -(\ln f)_{,z} , \quad \frac{1}{\varrho}\eta_2 = (\ln f)_{,\varrho} , \quad W_1 = -\frac{b_{,z}}{f} , \quad W_2 = -\frac{b_{,\varrho}}{f} , \tag{2.133}$$

where $\mathcal{E} = f + ib$ solves the Ernst equation. For details see [40].

There is a close relation between the static axisymmetric YMH equations for a massless Higgs fields and the stationary axisymmetric self-dual $SU(2)$ Yang–Mills equations discussed in the previous section. The self-duality equations

$$F_{\mu\nu}^a = *F_{\mu\nu}^a \tag{2.134}$$

reduce for fields independent of the Euclidean time to

$$F_{ij}^a = -\varepsilon_{ijk}(D_k A_0)^a . \tag{2.135}$$

The latter equation is equivalent to the Bogomolny equations (2.130) if A_0^a is interpreted as the Higgs field. This equivalence can be used to solve the YMH equations in the considered case in terms of the Yang or the Ernst equation as in the previous section.

Consequently the solution techniques available for the Ernst equation can be used also for the static axisymmetric multi-monopoles. The difference is, however, the asymptotic behavior. Whereas asymptotically flat configurations are typically studied in a relativistic setting, i.e., an asymptotically constant Ernst potential, the asymptotic condition (2.128) for the Higgs field implies an exponentially growing Ernst potential on the axis of symmetry. The simplest solution with this property is the BPS monopole [171, 172] for which we have $\mathcal{E} = (\varrho + iP)/F$ with

$$F = \frac{r}{\sinh r} + r \cosh r \coth r - z \sinh z , \quad P = z \cosh z - r \sinh z \coth z .$$
$$(2.136)$$

This implies for the Higgs field

$$\phi^2 = \phi_1^2 + \phi_2^2 = \left(\coth \mu - \frac{1}{\mu} \right)^2 , \quad \mu^2 = (w - z)^2 + \varrho^2 , \qquad (2.137)$$

where w is a real constant. For multi-monopole solutions to the Ernst equation see [40, 41].

3 Riemann–Hilbert Problem
and Fay's Identity

In Chap. 2 we have shown that the Ernst equation can be treated as the integrability condition of an overdetermined linear differential system for some matrix-valued function Φ. The important point is that this matrix depends on a spectral parameter. The existence of such a linear system can be used to generate large classes of solutions to the corresponding integrable equation. The idea is to construct a matrix Φ with certain analyticity properties with respect to the spectral parameter in a way that it solves the linear differential system for a corresponding Ernst potential. This means that the theory of complex functions can be used to obtain rich classes of matrices Φ and thus of Ernst potentials.

The methods we are using here are so-called Riemann–Hilbert techniques, i.e. we look for matrices Φ with prescribed singularities with respect to the spectral parameter, in particular jump discontinuities, poles and essential singularities. In general a matrix Riemann–Hilbert problem is equivalent to a linear integral equation for which it is difficult to construct explicit solutions. In the case of the Ernst equation this can be achieved for rational jump data, the corresponding solutions being Korotkin's [52] hyperelliptic solutions in terms of multidimensional theta functions. The characteristic feature of these solutions is that the underlying Riemann surface is 'dynamic' in the sense that the branch points depend on the physical coordinates. This is in contrast to theta-functional solutions to integrable evolution equations like KdV and KP where the Riemann surface is independent of the physical coordinates and where the latter enter only the argument of theta functions. The corresponding solutions are periodic or quasi-periodic. We briefly discuss hyperelliptic solutions to the KdV and the KP equation to illustrate the difference to the corresponding solutions to the Ernst equation which will be studied in detail in later chapters.

To construct theta-functional solutions to the KdV and KP equation, we use an approach due to Fay [128] (see also Mumford [139]) based on Fay's trisecant identity for theta functions. This identity can be seen as a generalization of a well-known identity for the so-called cross ratio function in plane projective geometry. Let $P_i \in \mathbb{CP}^1$, $i = 1, 2, 3, 4$, be four arbitrary disjoint points in $\mathbb{CP}^1$. Then the cross ratio function λ_{1234} is defined as

$$\lambda_{1234} = \frac{(P_1 - P_2)(P_3 - P_4)}{(P_1 - P_4)(P_3 - P_2)} \ .$$

(3.1)

It is obviously subject to the identity

$$\lambda_{1234} + \lambda_{1324} = 1 \ .$$

(3.2)

Interestingly this identity can be generalized to Riemann surfaces where theta functions are involved, see [173] to [176]. The important point is that this identity due to Fay [128] holds for four points in arbitrary position. In the limit of coinciding points it leads to identities for derivatives of theta functions. These identities can be used to prove that the given theta functional expressions solve the corresponding integrable equation. For details on the mathematics of Riemann surfaces and the notation used in this book, the reader is referred to appendix A and the literature given therein.

In Sect. 3.1 we start from a linear differential system for the Ernst equation and discuss the analytical properties the matrix Φ must have to be a solution of this system for a given Ernst potential. In Sect. 3.2 we formulate the Riemann–Hilbert problem for the matrix Φ in the plane of the spectral parameter. For illustration we discuss the Riemann–Hilbert problem in the complex plane which can be solved in terms of the Cauchy integral. We use a gauge freedom for the matrix Φ in the linear system to transform the matrix Riemann–Hilbert problem for the Ernst equation with analytic jump data to a scalar problem on a Riemann surface, see [112]. Existence of solutions is established for the case of non-compact surfaces [123]. For rational jump data the Riemann surface will be compact, and as shown in Sect. 3.3, the corresponding solutions to the Ernst equation can be given in terms of hyperelliptic theta functions. In Sect. 3.4 we discuss an algebraic form of these solutions as in [125, 126].

In Sect. 3.5 we use a degenerated form of Fay's identity to construct solutions in terms of theta functions to the KdV and the KP equation. The corresponding solutions are periodic or quasi-periodic. We show typical solutions on hyperelliptic Riemann surfaces for these equations and their solitonic limit where certain periods on the Riemann surface diverge (see [177, 178]). In this limit one obtains solitonic wave packets. To use Fay's identity for the Ernst equation we have to take into account the fact that the branch points of the Riemann surface depend on the physical coordinates. This is done in the form of Rauch's variational formulas for Riemann surfaces [179]. In Sect. 3.6 we show how to use these formulas to obtain theta-functional solutions to the Ernst equation. This approach also leads to explicit formulas for the complete metric in terms of theta functions, see [127].

3.1 Linear System of the Ernst Equation

In Chap. 2 we have shown that the Ernst equation can be treated as the integrability condition of an overdetermined linear differential system which contains a spectral parameter. Now we want to exploit this feature for the construction of explicit solutions to the Ernst equation. There are several forms of this linear system discussed in the literature ([75, 77, 156]). They are related through gauge transformations (see [78–80]) and the choice of a specific form of the linear system is equivalent to a gauge fixing. We will use the form of [156] given in (2.47). The system is defined on the Riemann surface $\mathcal{L}_0$ given by

$$\mu_0^2(K) = (K - \xi)(K - \bar{\xi}) .\tag{3.3}$$

On $\mathcal{L}_0$ we have an involutive map σ

$$\mathcal{L}_0 \ni P = (K, \pm\mu_0(K)) \to \sigma(P) \equiv P^\sigma = (K, \mp\mu_0(K)) \in \mathcal{L}_0 ,\tag{3.4}$$

and an anti-holomorphic involution τ

$$\mathcal{L}_0 \ni P = (K, \pm\mu_0(K)) \to \tau(P) \equiv \bar{P} = (\bar{K}, \pm\mu_0(\bar{K})) \in \mathcal{L}_0 .\tag{3.5}$$

A point P on $\mathcal{L}_0$ with projection K into the complex plane will be denoted by $P = (K, \pm\mu_0(K)) = K^\pm$ where the sign of the root is fixed in the vicinity of infinity, $\lim_{K\to\infty} \mu_0(K)/K = 1$ on the plus-sheet.

As we will show below, linear systems as in (2.47) are useful for the construction of solutions to the Ernst equation. To this end one investigates the singularity structure of the matrices $\Phi_{,\xi}\Phi^{-1}$ and $\Phi_{,\bar{\xi}}\Phi^{-1}$ with respect to the spectral parameter and infers a set of conditions for the matrix Φ (it has to be at least twice differentiable with respect to ξ and $\bar{\xi}$) that satisfies the linear system (2.47). This is done (see e. g. [52]) in

Theorem 3.1. *Let Φ ($P \in \mathcal{L}_0$) be a 2×2-matrix valued function on $\mathcal{L}_0$ with the following properties:*

I. *$\Phi(P)$ ($P \in \mathcal{L}_0$) is holomorphic and invertible at the branch points ξ and $\bar{\xi}$ such that the logarithmic derivative $\Phi_{,\xi}\Phi^{-1}$ diverges as $(K - \xi)^{\frac{1}{2}}$ at ξ and $\Phi_{,\bar{\xi}}\Phi^{-1}$ as $(K - \bar{\xi})^{\frac{1}{2}}$ at $\bar{\xi}$.*

II. *All singularities of Φ on $\mathcal{L}_0$ (poles, essential singularities, zeros of the determinant of Φ, branch cuts and branch points) are regular which means that the logarithmic derivatives $\Phi_{,\xi}\Phi^{-1}$ and $\Phi_{,\bar{\xi}}\Phi^{-1}$ are holomorphic in the neighborhood of the singular points (this implies they have to be independent of ξ and $\bar{\xi}$). In particular $\Phi(P)$ should have*

    a) *regular singularities at the points $A_i \in \mathcal{L}_0$ ($i = 1, \ldots, n$) which do not depend on ξ, $\bar{\xi}$,*

    b) *regular essential singularities at the points S_i ($i = 1, \ldots, m$) which do not depend on ξ and $\bar{\xi}$,*

    c) *jump discontinuities at a set of (orientable, piecewise smooth) contours $\Gamma_i \subset \mathcal{L}_0$ ($i = 1, \ldots, l$) independent of ξ and $\bar{\xi}$, which are related on both sides of the contours by*

$$\Phi_-(P) = \Phi_+(P)\mathcal{G}_i(P)|_{P\in\Gamma_i} ,\tag{3.6}$$

where $\mathcal{G}_i(P)$ are matrices independent of ξ and $\bar{\xi}$ with Hölder-continuous components and non-vanishing determinant.

III. Φ satisfies the reduction condition

$$\Phi(P^\sigma) = \sigma_3 \Phi(P)\gamma , \qquad (3.7)$$

where σ_3 is the third Pauli matrix, and where γ is an invertible matrix independent of ξ and $\bar{\xi}$.

IV. The normalization and reality condition

$$\Phi(P = \infty^+) = \begin{pmatrix} \bar{\mathcal{E}} & 1 \\ \mathcal{E} & -1 \end{pmatrix} \qquad (3.8)$$

is fulfilled.

Then the function $\mathcal{E}$ in (3.8) is a solution to the Ernst equation.

A proof of this Theorem may be obtained by comparing the above matrix Φ with the linear system (2.47).

Proof. Because of I, Φ and Φ^{-1} can be expanded in a series in $t = \sqrt{K - \xi}$ and $t' = \sqrt{K - \bar{\xi}}$ in a neighborhood of $P = \xi$ and $P = \bar{\xi} \neq \xi$ respectively at all points ξ, $\bar{\xi}$ which do not belong to the singularities given in II. This implies that

$$\Phi_{,\xi}\Phi^{-1} = \frac{\alpha_0}{t} + \alpha_1 + \alpha_2 t + \cdots ,$$

where the α_i are independent of t. We recognize that, because of I and II, $\Phi_{,\xi}\Phi^{-1} - \alpha_0/t$ is a holomorphic function. The normalization condition IV implies that this quantity is bounded at infinity. According to Liouville's theorem, it is a constant. Since Φ, Φ^{-1} and $\Phi_{,\xi}$ are single-valued functions on $\mathcal{L}_0$, they must be functions of K and μ_0. Therefore, we have

$$\Phi_{,\xi}\Phi^{-1} = \beta_0 \sqrt{\frac{K - \bar{\xi}}{K - \xi}} + \beta_1 .$$

The matrix β_0 has to be independent of K and μ_0 since $\Phi_{,\xi}\Phi^{-1}$ must have the same number of zeros and poles on $\mathcal{L}_0$. The structure of the matrices β_0 and β_1 follows from III. From the normalization condition IV, it follows that $\Phi_{,\xi}\Phi^{-1}$ has the structure of (2.47). The corresponding equation for $\Phi_{,\bar{\xi}}\Phi^{-1}$ can be obtained in the same way.

$\square$

For a given Ernst potential $\mathcal{E}$, the matrix Φ in the above theorem is not uniquely determined. This reflects the fact that the gauge is not completely fixed in the linear system (2.47). If we choose without loss of generality $\gamma = \sigma_1$ (the first Pauli matrix), the remaining gauge freedom can be seen from

Corollary 3.2. *Let $\Phi(P)$ be a matrix subject to the conditions of Theorem 3.1, and $C(K)$ be a 2×2-matrix that only depends on $K \in \mathbb{C}$ with the properties*

$$C(K) = \alpha_1(K)I + \alpha_2(K)\sigma_1 ,$$
$$\alpha_1(\infty) = 1 , \quad \alpha_2(\infty) = 0 . \tag{3.9}$$

Then the matrix $\Phi'(P) = \Phi(P)C(K)$ also satisfies the conditions of Theorem 3.1 and $\Phi'(\infty^+) = \Phi(\infty^+)$.

It is this freedom to which we refer in the following when we speak of the gauge freedom of the linear system.

In other words: matrices Φ which are related via the multiplication from the right by a matrix C of the above form lead to the same Ernst potential though their singularity structure may be vastly different (the functions α_i need not be holomorphic).

It is interesting to note that the metric function a can be obtained from a given matrix Φ fulfilling the conditions of the above Theorem without solving equation (2.45), see [52]. We get

Proposition 3.3. *Let δ be a local parameter in the vicinity of ∞^-. Then*

$$Z \doteq (a - a_0)e^{2U} = \mathrm{i}(\Phi_{11} - \Phi_{12})_{,\delta} , \tag{3.10}$$

where a_0 is a constant that is fixed by the condition that $a = 0$ on the regular part of the axis and at spatial infinity, and where $\Phi_{,\delta}$ denotes the linear term in the expansion of Φ in δ divided by δ.

The proof follows from the linear system (2.47).

Proof. It is straightforward to check the relation

$$(\Phi^{-1}\Phi_{,\delta})_{,\xi} = \Phi^{-1}(\Phi_{,\xi}\Phi^{-1})_{,\delta}\Phi . \tag{3.11}$$

With (2.47), we get

$$\left(\Phi^{-1}\Phi_{,\delta}\right)_{21,\xi} = \frac{\mathrm{i}\varrho(\bar{\mathcal{E}} - \mathcal{E})_{,\xi}}{(\mathcal{E} + \bar{\mathcal{E}})^2} , \tag{3.12}$$

from which (3.10) follows together with (2.45).

$\square$

Notice that a_0 is not gauge independent (in the sense of the above corollary) whereas a is.

Theorem 3.1 can be used to construct solutions to the Ernst equation by determining the structure and the singularities of Φ in accordance with conditions I–IV. From the Theorem it can be seen that the only possible singularities of the Ernst potential can occur where these conditions are not fulfilled, i. e. where Φ cannot be normalized or where ξ coincides with one of the singularities of Φ, e. g. a contour Γ. The latter is particularly interesting if one wants to solve boundary value problems for the Ernst equation, since the boundary can be chosen as a singularity of the Ernst potential (the potential needs to be only continuous there in the case of a Dirichlet problem and to be differentiable in the case of a von Neumann problem). In this case the contour Γ has to be chosen in a way that $\xi \in \Gamma$ just corresponds to the contour in the $(\xi, \bar{\xi})$-plane where the boundary values are prescribed. For example a disk of radius $\varrho_0 = 1$ in the plane $\zeta = 0$ is given by the contour

$$\Gamma_\xi = \{0 \le \varrho \le \varrho_0 , \quad \zeta = 0\} . \tag{3.13}$$

The corresponding contour Γ on $\mathcal{L}_0$ is the covering of the imaginary axis in the upper sheet between $-\mathrm{i}$ and i. The Ernst potential will not be continuous at this contour, but its boundary values will be bounded. Notice however that the Ernst potential will not be singular in any case if ξ coincides with a singularity of Φ since the latter may be e. g. pure gauge. Theorem 3.1 merely ensures that the solution will be regular at all other points.

Therefore, we will concentrate in the present work on a Riemann–Hilbert problem which can be formulated in the following way:

Let Γ be a set of (orientable piecewise smooth) contours $\Gamma_k \subset \mathcal{L}_0$ ($k = 1, \dots, l$) such that with $P \in \Gamma$ also $\bar{P} \in \Gamma$ and $P^\sigma \in \Gamma$. Let $\mathcal{G}_k(P)$ be matrices on Γ_k with Hölder–continuous components and non-vanishing determinant subject to the reality condition $\mathcal{G}_{ii}(\bar{P}) = \bar{\mathcal{G}}_{ii}(P)$ for the diagonal elements, and $\mathcal{G}_{ij}(\bar{P}) = -\bar{\mathcal{G}}_{ij}(P)$ for the off-diagonal elements, and $\mathcal{G}(P^\sigma) = \sigma_1 \mathcal{G}(P)\sigma_1$. Both Γ and $\mathcal{G}$ have to be independent of ξ, $\bar{\xi}$. The matrix Φ has to be everywhere regular except at the contour Γ where the boundary values on both sides of the contours (denoted by $\Phi_\pm$) are related via

$$\Phi_-(P) = \Phi_+(P)\mathcal{G}_i(P)|_{P \in \Gamma_i} . \tag{3.14}$$

It may be easily checked that a matrix Φ constructed in this way fulfils the conditions of Theorem 3.1.

3.2 Solutions to the Ernst Equation via Riemann–Hilbert Problems

In this section we review basic facts of the Riemann–Hilbert problem in the complex plane and on the Riemann sphere. Then we solve the Riemann–Hilbert problem for the Ernst equation by using a gauge transformation. In the case of rational jump data, the Ernst potential will be given in terms of hyperelliptic theta functions.

3.2.1 Riemann–Hilbert Problems on the Complex Plane and the Riemann Sphere

To solve Riemann–Hilbert problems, one basically uses the same methods as in the simplest case, the problem in the complex plane for a scalar function

ψ, see e. g. [180, 181]. If Γ_K is a simply connected closed smooth contour and G a nonzero analytic function on Γ_K in $\mathbb{C}$, the function ψ that is holomorphic except at the contour Γ_K where

$$\psi_- = \psi_+ G \qquad (3.15)$$

is obviously given by the Cauchy integral,

$$\psi(K) = F(K) \exp\left(\frac{1}{2\pi i} \int_{\Gamma_K} \frac{\ln G \mathrm{d}X}{X - K}\right), \qquad (3.16)$$

where F is an arbitrary holomorphic function, and where the principal value of the logarithm has to be taken. The well-known analytic properties of the Cauchy integral ensure that condition (3.15) is satisfied. Formula (3.16) shows that the solution to a Riemann–Hilbert problem of the above form is only determined up to a holomorphic function. Since F is holomorphic, the solution will be uniquely determined (due to Liouville's theorem) by a normalization condition $\psi(\infty) = \psi_0$ for $\infty \notin \Gamma_K$. Uniqueness is lost if one allows for additional poles since F in (3.16) has then to be replaced by a meromorphic function. We note that the above conditions on the contour may be relaxed: G may be a non-zero Hölder continuous function, and Γ may consist of a set of piecewise smooth orientable contours which are not closed. The Plemelj formula, see e. g. [182], assures that formula (3.16) still gives the solution to (3.15). A normalization condition will however only establish uniqueness of the solution if $G = 1$ at the endpoints of Γ.

Riemann–Hilbert problems on the sphere $\mathcal{L}_0$ which occur in the case of the Ernst equation can be treated in much the same way as the problems in the complex plane. The basic building block for the solutions is the differential of the third kind $\mathrm{d}\omega_{K^+K^-}(X)$ that corresponds to the differential $\mathrm{d}X/(X - K)$ in the complex plane, i. e. a differential that can be locally written as $F(X, K)\mathrm{d}X$ where $F(X, K)$ is holomorphic except for $X = K^\pm$ where the residue is ± 1. On the Riemann sphere $\mathcal{L}_0$ this differential is given by

$$\mathrm{d}\omega_{K^+K^-}(X) = -\frac{\mu_0(K) + \mu_0(X)}{2\mu_0(X)} \mathrm{d}X . \qquad (3.17)$$

If we make the ansatz

$$\Phi - \Phi_0 = \frac{1}{2\pi i} \int_\Gamma \chi(X) \mathrm{d}\omega_{K^+K^-}(X) , \qquad (3.18)$$

where the 2×2-matrix χ is defined on Γ, and where Φ_0 is holomorphic, we get for (3.6) at the contour Γ with the Plemelj formula

$$\Phi^\pm = \pm \frac{1}{2}\chi + \frac{1}{2\pi i} \int_\Gamma \chi(X) \mathrm{d}\omega_{K^+K^-}(X) .$$

Thus the Riemann–Hilbert problem (3.6) is equivalent to the integral equations

$$\frac{1}{2}\chi(\mathcal{G}+1) + \frac{1}{2\pi i}\int_\Gamma \chi(X)d\omega_{K+K^-}(X)(\mathcal{G}-1) = 0. \tag{3.19}$$

For simplicity, we will only consider in the following the case where the projection of the contour Γ into the complex plane has a simply connected component Γ_K.

3.2.2 Gauge Transformations of the Riemann–Hilbert Problem

Apparently, the Riemann–Hilbert problem can be used to generate solutions to the Ernst equation that contain four real-valued free functions, the components of the matrix $\mathcal{G}$. The remarks on the gauge freedom of the matrix Φ indicate however that two of them are related to gauge transformations. It seems plausible that one can choose a gauge in which $\mathcal{G}$ has only two independent components. Neugebauer and Meinel [183] used the form

$$\mathcal{G}' = \begin{pmatrix} \alpha & 0 \\ \beta & 1 \end{pmatrix} \tag{3.20}$$

for the jump matrices on the contours in the upper sheet. To demonstrate that this choice is possible one has to show the existence of a gauge transformation of the form (3.9) which transforms a general Riemann–Hilbert problem for the Ernst equation to the above form. This leads to the Riemann–Hilbert problem for the matrix C in the gauge transformation (3.9) on the contour Γ of the form

$$\mathcal{G}C_- = C_+\mathcal{G}' \tag{3.21}$$

with $\mathcal{G}'$ as given in (3.20). It can be shown by a simple calculation that this problem has a, not necessarily unique, solution. This form of the gauge transformation does not change the singularity structure of Φ (i. e. Φ will only be singular at Γ). The reduction condition III of Theorem 3.1 leads to

$$\mathcal{G}_2 = \begin{pmatrix} 1 & \beta \\ 0 & \alpha \end{pmatrix}$$

on the contour Γ_2 in the $--$sheet. Because of the reality conditions for $\mathcal{G}$, this implies that solutions to the Ernst equation following from (3.6) contain two real valued functions which correspond to α and β. The reduction and reality properties of Φ (see Theorem 3.1) make it possible to consider only one component of the matrix, e. g. Φ_{12} from which the Ernst potential follows as $\mathcal{E} = \Phi_{12}(\infty^+)$. With

$$\Phi_{12} = \psi_0 + \frac{1}{4\pi i}\int_{\Gamma_1} \frac{\mu_0(K) + \mu_0(X)}{2\mu_0(X)(X-K)} \mathcal{Z}(X,\xi,\bar{\xi})dX \,,$$

we obtain the Ernst potential for given α and β where $\mathcal{Z}$ is the solution of the integral equation

$$-\frac{\alpha+1}{2}\mathcal{Z} = \frac{\alpha-1}{4\pi i}\int_{\Gamma_1} \frac{\mu_0(K)+\mu_0(X)}{\mu_0(X)(X-K)}\mathcal{Z}(X,\xi,\bar{\xi})\mathrm{d}X$$

$$-\frac{\beta}{4\pi i}\int_{\Gamma_1} \frac{\mu_0(X)-\mu_0(K)}{\mu_0(X)(X-K)}\mathcal{Z}(X,\xi,\bar{\xi})\mathrm{d}X , \qquad (3.22)$$

and where ψ_0 follows from the normalization condition $\Phi_{12}(\infty^-)=1$.

Explicit solutions can in general only be obtained for diagonal $\mathcal{G}$, i.e. for $\beta=0$. In this case we get with the above formulas $\mathcal{E}=\bar{\mathcal{E}}=\mathrm{e}^{2U}$ with

$$U = -\frac{1}{4\pi i}\int_{\Gamma_1} \frac{\ln\alpha}{\sqrt{(K-\zeta)^2+\varrho^2}}\mathrm{d}K . \qquad (3.23)$$

Thus, all solutions are real in this case which implies that they belong to the static Weyl class. Since the Ernst equation reduces to the Euler–Darboux equation for U if $\mathcal{E}$ is real, the function U in (3.23) solves the axisymmetric Laplace equation. In fact one can show (see Theorem 5.1) that the contour integral there is equivalent to the Poisson integral with a distributional density. It can also be directly seen from the expression (3.23) that the dependence on the physical coordinates ϱ and ζ is exclusively via the branch points of the family of surfaces $\mathcal{L}_0$.

However, if one drops the condition that the gauge transformed matrix Φ' should have the same singularity structure as the original matrix in (3.6), it is possible to reduce the above matrix Riemann–Hilbert problem further. This leads us to

Theorem 3.4. *Let the conditions for (3.6) hold and let the projection of the contour Γ into the complex plane consist of one simple smooth arc. Let, furthermore, the components of the matrix $\mathcal{G}$ be quotients of holomorphic functions. Then the Riemann–Hilbert problem (3.6) is gauge equivalent to a problem with the diagonal matrix $\mathcal{G}' = \mathrm{diag}(G,1)$ (on the contour in the upper sheet) on a two sheeted covering of the Riemann surface $\mathcal{L}_0$ where G is a quotient of holomorphic functions on Γ.*

Proof. The proof uses again the explicit construction of the gauge transformation which takes in our case the form

$$(\mathcal{G}_{11}+\mathcal{G}_{12}+\mathcal{G}_{21}+\mathcal{G}_{22})(\alpha_1+\alpha_2)^- = (G+1)(\alpha_1+\alpha_2)^+ ,$$
$$(\mathcal{G}_{11}-\mathcal{G}_{12}-\mathcal{G}_{21}+\mathcal{G}_{22})(\alpha_1-\alpha_2)^- = (G+1)(\alpha_1-\alpha_2)^+ ,$$
$$(\mathcal{G}_{11}-\mathcal{G}_{12}+\mathcal{G}_{21}-\mathcal{G}_{22})(\alpha_1-\alpha_2)^- = (G-1)(\alpha_1+\alpha_2)^+ ,$$
$$(\mathcal{G}_{11}+\mathcal{G}_{12}-\mathcal{G}_{21}-\mathcal{G}_{22})(\alpha_1+\alpha_2)^- = (G-1)(\alpha_1-\alpha_2)^+ . \qquad (3.24)$$

As already mentioned, this system will in general not have a solution if the α_i have to be holomorphic except at Γ. Therefore, we make the ansatz

$$\frac{\alpha_1 + \alpha_2}{\alpha_1 - \alpha_2} = \lambda \exp\left(\frac{1}{2\pi i} \int_\Gamma \frac{dK'}{K - K'} \ln \frac{\mathcal{G}_{11} + \mathcal{G}_{12} + \mathcal{G}_{21} + \mathcal{G}_{22}}{\mathcal{G}_{11} - \mathcal{G}_{12} - \mathcal{G}_{21} + \mathcal{G}_{22}}\right), \qquad (3.25)$$

where λ is a possibly multi-valued function of $K \in \bar{C}$ alone. With this ansatz we can solve the above system and determine G and λ,

$$\lambda = \sqrt{\frac{\mathcal{G}_{11} - \mathcal{G}_{12} + \mathcal{G}_{21} - \mathcal{G}_{22}}{\mathcal{G}_{11} + \mathcal{G}_{12} - \mathcal{G}_{21} - \mathcal{G}_{22}}}, \qquad (3.26)$$

$$\frac{G + 1}{G - 1} = \sqrt{\frac{(\mathcal{G}_{11} + \mathcal{G}_{22})^2 - (\mathcal{G}_{12} + \mathcal{G}_{21})^2}{(\mathcal{G}_{11} - \mathcal{G}_{22})^2 - (\mathcal{G}_{12} - \mathcal{G}_{21})^2}}. \qquad (3.27)$$

In an abuse of notation, we have here denoted the analytic continuation of the $\mathcal{G}_{ij}$ (which is obvious since the functions are assumed to be quotients of holomorphic functions) with the same symbol as the functions that were originally only defined on Γ (this is still the case for the function G). Writing λ in the form $\lambda^2 = F/H$ where F and H are holomorphic functions (which is possible by assumption), one can recognize that the whole system has to be considered on the Riemann surface $\hat{\mathcal{L}}$ given by

$$\hat{\mu}^2(K) = F(K)H(K). \qquad (3.28)$$

This is a two-sheeted covering of the two-sheeted surface $\mathcal{L}_0$ on which the spectral parameter varies, and thus a four-sheeted covering of the complex plane. It is on this surface that the gauge transformed matrix Φ' and the function λ are single-valued.

□

The content of the Theorem may be put into this form: it is always possible to transform the Riemann–Hilbert problem with 'holomorphic' jump data to diagonal form. The price one has to pay for this is the occurrence of a four-sheeted Riemann surface since the gauge transformation would be multi-valued otherwise. The condition that the projection of Γ into the complex plane consists of only one arc can be replaced by the condition that the analytic continuations of the $\mathcal{G}_{ij}$ coincide on all contours Γ_k.

Let us, now, turn to the solution of the Riemann–Hilbert problem on the covering surface $\hat{\mathcal{L}}$. Of course, the structure of $\hat{\mathcal{L}}$ depends crucially on the jump data of the original Riemann–Hilbert problem. We will first consider the case of $\hat{\mathcal{L}}$ being non-compact and then turn to the compact case.

3.2.3 The Non-compact Case

It is well known, see [181], that there is a close relation between Riemann–Hilbert problems on Riemann surfaces and holomorphic vector bundles over them. The idea is to use vector bundles to relate the Riemann–Hilbert problem on the Riemann surface to the corresponding problem in the complex place. To this end we want to make use of the basic result of fibre bundle theory that a vector bundle over a manifold $\hat{\mathcal{L}}$ is completely determined by a triple $\left(\hat{\mathcal{L}}, \{U_i\}, \{t_{ij}\}\right)$. Here the U_i are an open cover and the t_{ij} are the transition functions, defined on intersections $U_i \cap U_j$. Let $\hat{\mathcal{L}}$ be equipped with a covering $\mathcal{N} = \{U_i, i \in I\}$, where I denotes some set of indices and let us suppose that there exists a number N, the covering constant, such that any point $P \in \hat{\mathcal{L}}$ belongs to no more than N domains of the covering, see [181]. We assume that the contour Γ is compact and closed, dividing $\hat{\mathcal{L}}$ into, in general, non-compact domains. To simplify the discussion we are looking as in [184] for solutions $\phi(P)$ with finite Dirichlet integral, i. e.

$$\int\limits_{\hat{\mathcal{L}} \backslash \Delta(\Gamma)} \overline{\mathrm{d}\phi} \wedge \mathrm{d}\phi < \infty \,, \tag{3.29}$$

where $\Delta(\Gamma)$ denotes some neighborhood of Γ.

Let $\{\phi_i(P)\}$ be solutions to (3.15) in the domains U_i, different from zero. In the domains U_i the original scalar Riemann–Hilbert problem is reduced to a problem on the complex plane $\mathbb{C}$, which can be solved as in Sect. 3.2.1. In other words, the functions $\phi_i(P)$ are non vanishing on U_i and fulfil (3.15) on the intersection $\Gamma \cap U_i$. If $\Gamma \cap U_i = \emptyset$ then $\phi_i(P)$ is a holomorphic function in U_i. Let us now define some functions $t_{ij}^{\pm}(P)$ in $U_i \cap U_j$ by

$$t_{ij}^{\pm}(P) = \frac{\phi_{j\pm}(P)}{\phi_{i\pm}(P)} \,, \tag{3.30}$$

for $P \in U_i \cap U_j$. We have

$$t_{ij}^{-}(P) = \frac{\phi_{j-}(P)}{\phi_{i-}(P)} = \frac{\phi_{j+}(P)G(P)}{\phi_{i+}(P)G(P)} = \frac{\phi_{j+}(P)}{\phi_{i+}(P)} = t_{ij}^{+}(P) \,, \tag{3.31}$$

i. e. the functions $t_{ij}(P) \doteq t_{ij}^{+}(P) = t_{ij}^{-}(P)$ do not jump at the contour Γ. Since we have on the intersection $U_i \cap U_j \cap U_k$

$$t_{ij}^{-}(P) t_{jk}^{-}(P) t_{ki}^{-}(P) =$$
$$\frac{\phi_{j-}(P)}{\phi_{i-}(P)} \frac{\phi_{k-}(P)}{\phi_{j-}(P)} \frac{\phi_{i-}(P)}{\phi_{k-}(P)} = \frac{\phi_{j+}(P)}{\phi_{i+}(P)} \frac{\phi_{k+}(P)}{\phi_{j+}(P)} \frac{\phi_{i+}(P)}{\phi_{k+}(P)} = 1 \,, \tag{3.32}$$

these functions fulfil the consistency conditions of a vector bundle. Thus, we may associate to the Riemann–Hilbert problem (3.15) a vector bundle. It is remarkable that this bundle is, due to a theorem by Grauert [185], a trivial one. From this Theorem it follows that for non-compact $\hat{\mathcal{L}}$ the line bundle B_G associated to the Riemann–Hilbert problem (3.15) has the form

$$B_G \simeq \hat{\mathcal{L}} \times \mathbb{C} \,. \tag{3.33}$$

Since the $\phi_i(P)$ are non-vanishing the functions $t_{ij}(P)$ take values in $\mathbb{C}^*$. Therefore, we may define a complex bundle with structure group $\mathbb{C}^*$ and standard fibre $\mathbb{C}$, i.e. a line bundle B_G over the compact Riemann surface $\hat{\mathcal{L}}$ by the 5-tupel $(\hat{\mathcal{L}}, \{U_i\}, \{t_{ij}\}, \mathbb{C}, \mathbb{C}^*)$.

Thus we have shown that for $\hat{\mathcal{L}}$ being non-compact there is a simple geometric characterization of the Riemann–Hilbert problem in terms of fibre bundles over $\hat{\mathcal{L}}$. Due to its local properties the bundle approach allows to reduce the scalar Riemann–Hilbert problems on $\hat{\mathcal{L}}$ to problems on the complex plane $\mathbb{C}$ which can be explicitly solved as shown above.

3.2.4 The Compact Case

Let us now turn to the case that $\hat{\mathcal{L}}$ is compact of genus g. Then $\hat{\mathcal{L}}$ is given by an equation of the form

$$\hat{\mu}^2 = \prod_{i=1}^{g} (K - E_i)(K - F_i) \,, \tag{3.34}$$

where E_i and F_i are obviously independent of the physical coordinates. This equation represents a two-sheeted covering of the Riemann sphere and thus a four-sheeted covering of the complex plane. A point $\hat{P} \in \hat{\mathcal{L}}$ can be given in the form $\hat{P} = (K, \mu_0(K), \hat{\mu}(K))$. The Hurwitz diagram of $\hat{\mathcal{L}}$ is shown in Fig. 3.1. As in the non-compact case we may associate to a Riemann–Hilbert problem

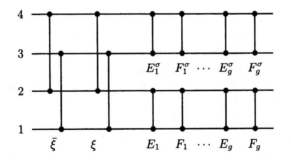

Fig. 3.1. The Hurwitz diagram of the covering surface shows the Riemann surface as seen from the side

(3.15) a line bundle over this surface in the case of compact $\hat{\mathcal{L}}$. Contrary to the non-compact case there is no analogue of Grauert's theorem. However, it has been shown by Zverovich [180] that on compact Riemann surfaces the solutions to scalar Riemann–Hilbert problems can be written elegantly in terms of theta functions on this surface.

3.3 Hyperelliptic Solutions of the Ernst Equation

In this section, we want to give an explicit construction of the matrix Φ in accordance with Theorem 3.1 for the case of $\hat{\mathcal{L}}$ being a compact Riemann surface. This limitation will allow us to give an explicit solution of the problem in terms of theta functions. Recall that this is possible, as the discussion in 3.2.2 has shown, if the original matrix Riemann–Hilbert problem has rational jump data.

Notice that there is an automorphism σ of $\hat{\mathcal{L}}$ inherited from $\mathcal{L}_0$ which ensures $E_i^\sigma = E_i$ and $F_i^\sigma = F_i$. The orbit space $\mathcal{L} = \hat{\mathcal{L}}/\sigma$ is then, see [1], again a Riemann surface, namely a hyperelliptic surface given by

$$\mu^2 = (K - \xi)(K - \bar{\xi}) \prod_{i=1}^{g} (K - E_i)(K - F_i) \,, \qquad (3.35)$$

where the fixed points of σ lead to additional branch points of $\hat{\mathcal{L}}/\sigma$. We impose the reality condition $E_i, F_i \in \mathbb{R}$ or $E_i = \bar{F}_i$ on the branch points in order to satisfy the reality condition of Theorem 3.1.

For the moment, we fix the physical coordinates ξ and $\bar{\xi}$ in a way that $\varrho \neq 0$ and that ξ and $\bar{\xi}$ do not coincide with the singular points of Φ in order to ensure that the first condition of Theorem 3.1 is valid. In the next chapters we will study the dependence of the found solution on ξ and $\bar{\xi}$. In order to give the solution to this special case of the generalized Riemann–Hilbert problem, we use the theory of theta functions associated to a Riemann surface (see [1, 186]) and the solution of the Riemann–Hilbert problem on a Riemann surface, as given in [180]. As we will need only hyperelliptic Riemann surfaces of the form (3.35), we restrict ourselves to this case. For a brief review of the basic notions of differential geometry on Riemann surfaces we refer the reader to Appendix A and references cited therein.

The following theorem gives the solution to the generalized Riemann–Hilbert problem on the hyperelliptic Riemann surface $\mathcal{L}$.

Theorem 3.5. *Let ξ be a fixed complex constant ($\varrho \neq 0$) not coinciding with the singularities of ψ or the branch points E_i or F_i. Let $\Omega(P_H)$ be a linear combination of normalized Abelian integrals of the second kind (with singularities p independent of ξ and $\bar{\xi}$) and third kind (with in addition singularities at all real branch points with residues $\pm\frac{1}{2}$), satisfying $\bar{\Omega}(P_H) = \Omega(\bar{P}_H)$. Let Γ be a piecewise smooth contour on $\mathcal{L}$ decomposing into a finite set of connected components $\{\Gamma_j\}$ ($j = 1, \ldots, N$), each of which is homeomorphic to the interval $(0, 1)$. Let*

$$G(t) = \sum_{j=1}^{N} \alpha(t, \Gamma_j) G_j(t) \,, \quad \ln G(t) = \sum_{j=1}^{N} \alpha(t, \Gamma_j) \ln G_j(t) \,, \quad t \in \Gamma \setminus \mathfrak{L} \,.$$

$$(3.36)$$

where the $G_j(t)$ are Hölder continuous functions, where $\mathfrak{L}$ denotes the divisor of end points of the Γ_i, and where

$$\alpha(t, \Gamma_j) = \begin{cases} 1 \; if \; t \in \Gamma_j \\ 0 \; otherwise \end{cases} , \qquad (3.37)$$

$(j = 1, \ldots, N)$. Then the function

$$\psi(P_H) = \psi_0 \frac{\Theta(\omega(P_H) - \omega(\mathfrak{D}) + u + b - K)}{\Theta(\omega(P_H) - \omega(\mathfrak{D}) - K)}$$

$$\times \; \exp\left\{ \Omega(P_H) + \frac{1}{2\pi i} \int_\Gamma \ln G(\tau) d\omega_{P_H P_0}(\tau) \right\} , \qquad (3.38)$$

has a jump at Γ of the form

$$\psi^+(t) = G(t)\psi^-(t) , \quad t \in \Gamma ; \qquad (3.39)$$

here

$$\mathfrak{D} = P_1 + \cdots + P_g$$

is a fixed non-special divisor on $\mathcal{L}$ which is subject to the reality condition: either $P_i \in \mathbb{R}$ or with $P_i \in \mathfrak{D}$ we have $\bar{P}_i \in \mathfrak{D}$ or P_i is a branch point E_i or F_i. b is the vector of b-periods of Ω with components

$$b_i = \oint_{b_i} d\Omega , \qquad (3.40)$$

$i = 1, \ldots, g$, the u_i are given by

$$u_i = \int_\Gamma \ln G d\omega_i , \qquad (3.41)$$

and ψ_0 is a normalization constant. The paths of integration have to be the same for all integrals.

Proof. We want to prove that $\psi(P_H)$ is a single valued function on $\mathcal{L}$. If we choose a different path of integration for the integrals in the exponent and the Abel map $\omega(P_H)$ and denote the corresponding integrals by a prime, the primed and unprimed integrals are connected via

$$\Omega'(P_H) = \Omega(P_H) + \oint_S d\Omega , \qquad (3.42)$$

(similarly for the other integrals) where S is a closed contour on $\mathcal{L}$ which may be decomposed in the homology basis $(a_1, \ldots, a_g, b_1, \ldots, b_g)$ as follows

$$S = \sum_{i=1}^{g} m_i\, a_i + \sum_{i=1}^{g} n_i\, b_i \,, \tag{3.43}$$

with $m_i, n_i \in \mathbb{Z}$. Then we have, e.g. for Ω and ω

$$\Omega(P_H) \to \Omega(P_H) + \sum_{i=1}^{g} n_i b_i = \Omega(P_H) + \langle N, b \rangle \,,$$

$$\omega(P_H) \to \omega(P_H) + 2\pi i M + BN \,, \tag{3.44}$$

where $M = (m_1, \ldots, m_g)$, $N = (n_1, \ldots, n_g) \in \mathbb{Z}^g$. Under this transformation, the original quotient of theta functions in (3.38) will be multiplied by

$$\exp\left(-\langle N, b \rangle\right)\,, \tag{3.45}$$

but this term is just compensated by the contour integral over S in the exponent. The same argument holds for the line integral over the contour Γ since the u_i are its b-periods. This shows that $\psi(P_H)$ is a single valued function on $\mathcal{L}$.

From the properties of the theta function, we also find that $\psi(P)$ has g simple poles at the points $P_1, \ldots, P_g$ and g simple zeros. Additional poles, zeros and essential singularities can be obtained by a suitable choice of Abelian integrals of the second kind (essential singularities) and third kind (zeros and poles). We remark that the assumption $\bar{\Omega}(P) = \Omega(\bar{P})$ had to be introduced in order to satisfy the reality condition of Theorem 3.1.

□

Remark 3.6. Without loss of generality we can choose $\mathfrak{D}$ to consist only of branch points since $\mathfrak{D}$ gives the poles of ψ due to the zeros of the theta function in the denominator. This can always be compensated by a suitable choice of the zeros and poles of ψ which arise from the integrals of the third kind in Ω. All $P_i \in \mathfrak{D}$ shall have multiplicity 1 and be chosen in a way that $\Theta_{pq}(x)$ with $[p, q] = \omega(\mathfrak{D}) + \mathcal{K}$, where $\mathcal{K}$ is the Riemann vector, has the same reality properties as the Riemann theta function $\Theta(x)$.

Our next aim is to define a matrix-valued function $\Phi(P)$ on $\mathcal{L}_0$, satisfying the conditions of Theorem 3.1, with the help of the above solution to the scalar Riemann–Hilbert problem on the hyperelliptic surface $\mathcal{L}$. To this end we define a further function on $\mathcal{L}$ by

$$\chi(P_H) = \chi_0 \frac{\Theta(\omega(P_H) + u + b - \omega(\bar{\xi}) - \omega(\mathfrak{D}) - \mathcal{K})}{\Theta(\omega(P_H) - \omega(\mathfrak{D}) - \mathcal{K})}$$

$$\times \exp\left(\Omega(P_H) + \frac{1}{2\pi i} \int_{\Gamma} \ln G d\omega_{P_H P_0}\right), \tag{3.46}$$

where χ_0 is again a normalization constant. We recall that $\omega(\bar{\xi}) = -i\pi(1, \ldots, 1)$ in the used cut system of Fig. A.4. It can be easily seen that the analytic behavior of $\chi(P_H)$ is identical to that of $\psi(P_H)$, except that it changes sign at every a-cut. χ is thus not a single valued function on $\mathcal{L}$. However, it is single valued on $\hat{\mathcal{L}}$ which can be viewed as two copies of $\mathcal{L}$ cut along $[\xi, \bar{\xi}]$ and glued together along this cut. We define a vector $\mathbf{X}$ on $\hat{\mathcal{L}}$ by fixing the sign in front of χ in the vicinity of the points $\xi^\pm = (K_0, 0, \pm\hat{\mu}(\xi)) \in \hat{\mathcal{L}}$,

$$\mathbf{X}(\hat{P}) = \begin{pmatrix} \psi(\hat{P}) \\ \pm\chi(\hat{P}) \end{pmatrix}, \quad \hat{P} \sim \xi^\pm. \tag{3.47}$$

By means of this vector, we can construct a matrix Φ on $\mathcal{L}_0$ by

$$\Phi(P) = (\mathbf{X}(K, \mu_0(K), +\hat{\mu}(K)), \mathbf{X}(K, \mu_0(K), -\hat{\mu}(K))), \tag{3.48}$$

where the signs are again fixed in the vicinity of $\xi^\pm$.

It may be readily checked that this ansatz is in accordance with the reduction condition (3.7) (this is in fact the reason why one has to define the function χ in the way (3.46)). The behavior at the singularities is as required in condition II: For the contour Γ and the singularities of the Abelian integrals Ω, this is obvious. At the branch points E_i and F_i, one gets the following behavior: at points P_i of the divisor $\mathfrak{D}$, the components of Φ have a simple pole, and the determinant diverges as $(K - P_i)^{-\frac{1}{2}}$, if this branch point is not a singularity of an integral of the third kind in Ω or lies on the contour Γ. If the latter condition holds at the remaining branch points, the components are regular there but the determinant vanishes as $(K - P_i)^{\frac{1}{2}}$. If the branch points coincide with one of the singularities of the integrals in the exponent in (3.38), this merely changes the singular behavior of Φ and its determinant there. Condition II of theorem 3.1 is however obviously satisfied.

Since Φ in (3.48) is only a function of P, it will not be regular at the cuts $[E_i, F_i]$. At the a-cuts encircling non-real branch points, we get $\Phi^-|_{a_i} = \Phi^+|_{a_i}\sigma_1$, whereas we have $\Phi^-|_{a_i} = -\Phi^+|_{a_i}\sigma_2$ at the a-cuts encircling real branch points. The logarithmic derivatives of Φ with respect to ξ and $\bar{\xi}$ are however holomorphic at all these points. One can recognize that the behavior at the non-real branch points is related to a gauge transformation of the form (3.9). This means that one can find a gauge transformed matrix Φ' that is completely regular at these points if the integrals in the exponent are regular there. With

$$\alpha_1 = \frac{1}{2}(1 + \lambda), \quad \alpha_2 = \frac{1}{2}(1 - \lambda), \tag{3.49}$$

and

$$\lambda = \prod_{i=1}^{g} \sqrt{\frac{K - P_i}{K - \bar{P}_i}},$$

where

$$\mathfrak{D} = \sum_{i=1}^{g} P_i,$$

this may be checked by direct calculation. The real branch points, however, cannot be related to gauge transformations.

Normalizing ψ and χ (if possible) in a way that $\psi(\infty_H^-) = 1$ and $\chi(\infty_H^-) = -1$, one can see that Φ is then in accordance with all conditions of Theorem 3.1 since the reality condition follows from the reality properties of the theta functions and the Riemann–Hilbert problem. It follows from the modular properties of the theta function that Φ is at least differentiable with respect to ξ and $\bar{\xi}$ at points where ξ does not coincide with the singularities of the integrals in the exponent or the remaining branch points of $\mathcal{L}$. Let the paths between $[\xi, \infty^-]$ and $[\xi, \infty^+]$ be the same in all integrals and let them have the same projection into the complex plane (i. e. one is the involuted of the other). Then the results may be summarized in

Theorem 3.7. *Let* $\Theta_{pq}(\omega(\infty^-) + u) \neq 0$. *Then the function*

$$\mathcal{E}(\xi, \bar{\xi}) = \frac{\Theta_{pq}(\omega(\infty^+) + u + b)}{\Theta_{pq}(\omega(\infty^-) + u + b)} \exp\left\{ \Omega|_{\infty}^{\infty^+} + \frac{1}{2\pi i} \int_{\Gamma} \ln G(\tau) d\omega_{\infty^+ \infty^-}(\tau) \right\}$$

(3.50)

is a solution to the Ernst equation.

The metric function af follows with (3.10) as a corollary,

Corollary 3.8. *The metric function* af *is given by*

$$Z = (a - a_0)f = i \left(\ln \frac{\Theta_{pq}(\omega(\infty^-) + u + b + \omega(\bar{\xi}))}{\Theta_{pq}(\omega(\infty^-) + u + b)} \right)_{,\delta} .$$

(3.51)

Remark 3.9. We remark that Φ in (3.48) consists of eigenvectors of the monodromy matrix (see [1, 124])

$$L\boldsymbol{X}(K, \mu_0(K), \pm\hat{\mu}(K)) = \hat{\mu}\boldsymbol{X}(K, \mu_0(K), \pm\hat{\mu}(K)) ,$$

which can be introduced as follows. For a given linear system (2.47), we define the monodromy matrix L as a solution to the system

$$L_{,\xi} = [U, L] , \quad L_{,\bar{\xi}} = [V, L] .$$

(3.52)

For a known solution Φ of (2.47), L can be directly constructed in the form

$$L(K) = -\hat{\mu}(K)\Phi\mathcal{C}\Phi^{-1} ,$$

(3.53)

where $\mathcal{C}$ is an arbitrary constant matrix with $\det \mathcal{C} = -1$ and where $\hat{\mu}$ does not depend on the physical coordinates. Since Φ is analytic in K, there is a solution to (3.52) with the same properties. It follows from (3.52) that the coefficients of the characteristic polynomial

$$Q(\mu, K) = \det (L(K) - \hat{\mu}I)$$

are independent of the physical coordinates. Without loss of generality we may assume $\operatorname{Tr}L(K) = 0$. Then L has the structure

$$L = \begin{pmatrix} A(K) & B(K) \\ C(K) & -A(K) \end{pmatrix} .$$

(3.54)

The equation $Q(\hat{\mu}, K) = 0$, i.e.

$$\hat{\mu}^2 = A^2 + BC ,$$

(3.55)

is then the equation of an algebraic curve which in general will have infinite genus.

3.4 Finite Gap Solutions and Picard–Fuchs Equations

Korotkin [52] originally gave his solutions in the slightly different form of the so called Baker–Akhiezer function (see [1]) of the Ernst system. The Baker–Akhiezer function has essential singularities and poles, and gives periodic or quasiperiodic solutions to integrable nonlinear evolution equations as KdV and KP. There the essential singularity is uniquely determined by the structure of the differential equation. In contrast to these equations, the solutions (3.50) are in general neither periodic nor quasiperiodic, and the essential singularity can be nearly arbitrarily chosen. The form of the solution to the Riemann–Hilbert problem shows that one might even think of 'putting the singularities densely on a line and integrate over the integrals with some measure': an Abelian integral Ω_p of the second kind with a pole of first order at p can be used as an analogue to the Cauchy kernel. A contour integral over this kernel with some measure, $\int_\Gamma \ln G \Omega_p dp$, is thus just another way to write down the solution to a Riemann–Hilbert problem on a Riemann surface. The above solutions to the Ernst equation constructed via Riemann–Hilbert techniques are thus a subclass of Korotkin's solutions.

It is possible to give an algebraic representation of the solutions (3.50) which was first used in [125], see also [126]. We define the divisor $\mathfrak{K} = \sum_{i=1}^{g} K_i$ as the solution of the Jacobi inversion problem $(i = 1, \ldots, g)$

$$\int_{\mathfrak{D}}^{\mathfrak{K}} \frac{\tau^{i-1} d\tau}{\mu(\tau)} = \frac{1}{2\pi i} \int_\Gamma \ln G \frac{\tau^{i-1} d\tau}{\mu(\tau)} \doteq \tilde{u}_i ,$$

(3.56)

where the divisor $\mathfrak{D} = \sum_{i=1}^{g} E_i$. With the help of these divisors and formula (A.62) for integrals of the third kind in [187], we can write (3.50) in the form

$$\ln \mathcal{E} = \int_{\mathfrak{D}}^{\mathfrak{K}} \frac{\tau^g d\tau}{\mu(\tau)} - \frac{1}{2\pi i} \int_\Gamma \ln G \frac{\tau^g d\tau}{\mu(\tau)} ,$$

(3.57)

Since the $\tilde{u}_i$ in (3.56) are subject to the recursive relation

$$\tilde{u}_{n+1,\xi} = \tilde{u}_{n,\xi} + \frac{1}{2}\tilde{u}_n \,, \tag{3.58}$$

we get

$$\sum_{n=1}^{g} \frac{(K_n - \xi)K_n^j}{\mu(K_n)} K_{n,\xi} = 0 \,, \quad j = 0, ..., g-2 \tag{3.59}$$

and

$$(\ln \mathcal{E})_{,\xi} = \sum_{n=1}^{g} \frac{(K_n - \xi)K_n^{g-1}}{\mu(K_n)} K_{n,\xi} \,. \tag{3.60}$$

Solving for the $K_{n,\xi}$, $n = 1, \ldots, g$, we obtain

$$K_{n,\xi} = (\ln \mathcal{E})_{,\xi} \frac{\mu(K_n)}{K_n - \xi} \frac{1}{\prod_{m=1,m\neq n}^{g}(K_n - K_m)} \,. \tag{3.61}$$

Additional information follows from the reality of the $\tilde{u}_i$ which implies

$$\omega(\mathfrak{K}) - \omega(\mathfrak{D}) = \omega(\bar{\mathfrak{K}}) - \omega(\bar{\mathfrak{D}}) \,. \tag{3.62}$$

Using Abel's theorem (see Sect. A.3 of the appendix and references therein) on the condition (3.62), we obtain the relation for an arbitrary $K \in \mathbb{C}$

$$(1-x^2)\prod_{i=1}^{g}(K - K_i)(K - \bar{K}_i) = \prod_{i=1}^{g}(K - E_i)(K - \bar{E}_i) - (K - \xi)(K - \bar{\xi})Q_2^2(K) \,, \tag{3.63}$$

where with purely imaginary x_i, x

$$Q_2(K) = x_0 + x_1 K + ... + x_{g-2}K^{g-2} + xK^{g-1} \,. \tag{3.64}$$

Since (3.63) has to hold for all $K \in \mathbb{C}$, it is equivalent to $2g$ real algebraic equations for the K_i if the x_i are given. Using (A.47) and (3.57) we find

$$\frac{\mathcal{E}}{\bar{\mathcal{E}}} = \frac{1+x}{1-x} \tag{3.65}$$

which implies $x = ibe^{-2U}$.

The underlying reason for the above differential relations is that the Ernst potential $\mathcal{E}$ is studied on a family of Riemann surfaces parametrized by the moving branch points ξ and $\bar{\xi}$. The periods on this surface (i. e. integrals along closed curves) are subject to differential identities, the so called Picard–Fuchs equations. It is a general feature of the periods of rational functions [188–190] that they satisfy a differential system of finite order with Fuchsian singularities. An elegant way to find the Picard–Fuchs system explicitly is via the notion of the Manin connection in the bundle $H^1_{\mathrm{DR}}(\Sigma_g) \to \Sigma_g$ on a Riemann surface Σ_g of genus g, see [191]. The investigation turns out to be particularly simple if one uses the following standard form of the (hyperelliptic) Riemann surface Σ_g (all hyperelliptic surfaces of genus g are conformally equivalent to this standard form, i.e. they can be transformed to this form by a Möbius transformation)

$$y^2 = (x - z) \prod_{i=1}^{2g} (x - E_i) \doteq (x - z) P(x) = (x - z) \sum_{j=0}^{2g} a_j x^j , \qquad (3.66)$$

where the E_i do not depend on z. Using $j_0 = dx/y$, $j_1 = x j_0, \dots, j_{2g-1} = x^{2g-1} j_0$ as the basis for the de Rham cohomology $H_{\mathrm{DR}}^1(\Sigma_g)$, we obtain for the matrix M_n^m $(m, n = 0, \dots, 2g - 1)$ of the Manin connection, defined by

$$\frac{\partial j_n}{\partial z} = M_n^m j_m ,$$

the following expression

$$M_n^m = \begin{cases} \dfrac{z^n}{2P(z)} \left((m+1) a_{m+1} + z^{-m-1} \displaystyle\sum_{j=0}^{m} a_j z^j \right) & \text{for } 0 \le m < n \\[4ex] \dfrac{z^n}{2P(z)} \left((m+1) a_{m+1} - \displaystyle\sum_{j=0}^{2g-1-m} a_{m+1+j} z^j \right) & \text{for } n \le m \le 2g - 1 \end{cases} .$$

$$(3.67)$$

One finds that the periods satisfy a similar recursive condition as (3.58). An analogous consideration can be performed for the $\bar{\xi}$-dependence of the periods. One finds that the integrability condition of the Picard–Fuchs systems is just the axisymmetric Laplace equation.

On the other hand, with the help of some boundary conditions (for instance at $|\xi| \to \infty$), the $\tilde{u}_n$ can be uniquely determined from the system (3.58). Thus the class of solutions discussed by Meinel and Neugebauer may be phrased in the following form: if an arbitrary solution of the Laplace equation is given, one can calculate the functions $\tilde{u}_n$ with (3.58) and the boundary condition, and ends up with a solution to the Ernst equation of the form (3.50) which is a subclass of Korotkin's solutions, see also [126].

3.5 Theta-functional Solutions to the KdV and KP Equation

The Kadomtsev–Petviashvili (KP) equation [192] for the real valued potential u depending on the three real coordinates (x, y, t) can be written in the form

$$3u_{,yy} + (6uu_{,x} + u_{,xxx} - 4u_{,t})_{,x} = 0 . \qquad (3.68)$$

The completely integrable equation has a physical interpretation as describing the propagation of weakly two-dimensional waves of small amplitude in shallow water as well as similar physical processes, see [1]. It can be seen as a two-dimensional generalization of the KdV equation

$$6uu_{,x} + u_{,xxx} - 4u_{,t} = 0 , \qquad (3.69)$$

where u depends only on x and t. Algebro-geometric solutions to the KP equation can be given on an arbitrary Riemann surface, see [1, 193]. The corresponding solutions can be seen as a typical example for the periodic or quasi-periodic solutions to nonlinear evolution equation. To compare with the corresponding solutions of the Ernst equation, we will briefly discuss solutions on hyperelliptic Riemann surfaces of the form

$$\mu^2 = \prod_{i=1}^{g+1}(K - E_i)(K - F_i) , \quad E_i, F_i \in \mathbb{R} , \quad i = 1, \ldots, g+1 , \qquad (3.70)$$

here, for details the reader is referred to [1].

Perhaps the most elegant way to establish explicit expressions for theta-functional solutions to various integrable equations is an identity for theta functions due to Fay [128]. This identity can be seen as a generalization of the identity for the cross ratio function (3.2) on $\mathbb{CP}^1$ to a Riemann surface, an interpretation being discussed in detail in [173] to [176].

Use the vectors U, V and W as defined in (A.69) as the expansion of the Abel map $\int_P^{P'} d\omega$ for $P' \sim P$ as a series in the difference τ of the local parameters. We put

$$u = 2(D_P^2 \ln \Theta(z) - 2e_1) , \qquad (3.71)$$

where e_1 is the function defined in (A.79), and get for relation (A.68) after differentiation with D_P^2

$$D_P(D_P^3 u + 6u D_P u - 2D_P'' u) + 3D_P' D_P' u = 0 . \qquad (3.72)$$

Because of (A.69) and (A.70) this equation is for $z = Ux + Vy + Wt + D$ equivalent to the KP equation, where D is some arbitrary real vector.

If we take $P \to \infty$ and if the hyperelliptic surface $\mathcal{L}$ is branched at infinity, the vector V vanishes. This implies that there is no y-dependence in the formula (3.71), and the KP equation reduces to the KdV equation (3.69).

3.5.1 Hyperelliptic and Solitonic Solutions

We use the cut-system of Fig. 3.2. An interesting limiting case of the theta-functional solutions on a genus g surface is the so-called solitonic limit, see [1, 53, 139]. In this case $E_i \to F_i$ for $i = 1, \ldots, g$ which leads to the g-soliton solution. Since g of the cuts collapse to double points, the diagonal elements of the Riemann matrix diverge in the used cut-system as $\mathbf{B}_{ii} \sim 2\ln(F_i - E_i)$ whereas all other a- and b-periods remain finite. The theta series thus breaks down to a sum containing only elementary functions. To obtain the standard

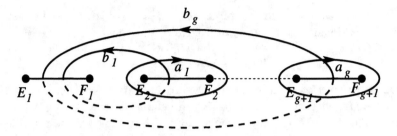

Fig. 3.2. Canonical cycles

form of the g-soliton, we choose the vector $\boldsymbol{D}$ in (3.71) to correspond to the half-integer characteristic

$$\frac{1}{2}\begin{bmatrix} 1 \dots 1 \\ 0 \dots 0 \end{bmatrix} . \tag{3.73}$$

In this section we present typical plots of hyperelliptic solutions to the KdV and the KP equation as examples. Such plots can already be found in [139, 177] and [194]. The figures below are generated with the code presented in [119, 131, 178] which is discussed in more detail in Chap. 6. We will show general situations as well as almost degenerate surfaces which are identical to their corresponding solitonic solution up to numerical accuracy. In all plots, the vector $\boldsymbol{D}$ in (3.71) is chosen to correspond to the characteristic (3.73).

To begin we want to discuss plots of genus 2 solutions to the KdV equation. We consider a hyperelliptic surface of the form (3.70) with branch points $[-2, -2 + \varepsilon, -1, -1 + \varepsilon, 3]$ (the surface is branched at infinity which can be formally achieved by omitting the term with F_{g+2}). In Fig. 3.3 we show the case for $\varepsilon = 10^{-14}$ which is identical to the 2-soliton solution within machine precision being of the order of 10^{-14}. The 2-soliton has the form

$$u - 2c = \frac{U_1^2 - U_2^2}{2} \frac{U_1^2 \cosh^2 \frac{z_2}{2} + U_2^2 \sinh^2 \frac{z_1}{2}}{(U_1 \cosh \frac{z_1}{2} \cosh \frac{z_2}{2} - U_2 \sinh \frac{z_1}{2} \sinh \frac{z_2}{2})^2} , \tag{3.74}$$

where

$$z_i = U_i x + W_i t , \quad U_1 = 2\sqrt{E_3 - E_1} , \quad U_2 = 2\sqrt{E_3 - E_2} , \tag{3.75}$$

and

$$W_1 = -\sqrt{E_3 - E_1}(E_3 + 2E_1) , \quad W_2 = -\sqrt{E_3 - E_2}(E_3 + 2E_2) , \quad c = -\frac{E_3}{2} . \tag{3.76}$$

The one dimensional waves in shallow water are depicted in dependence on x and t. It can be seen that a soliton coming from the right and traveling in negative x-direction has a collision with a soliton traveling in positive x-direction. After the collision the typical phase shift by π of u can be observed, otherwise the solitons are unaffected.

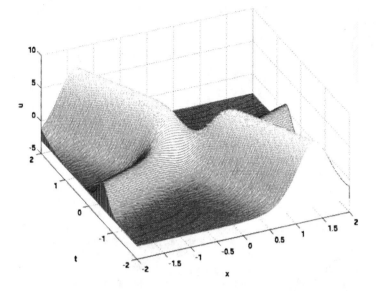

Fig. 3.3. 'Almost' solitonic genus 2 solution to the KdV equation

In Fig. 3.4 we show the case $\varepsilon = 1$. The almost periodic nature of the solution is clearly recognizable. The solution can be interpreted as an infinite train of solitons.

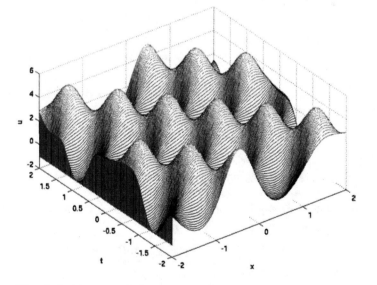

Fig. 3.4. Almost periodic genus 2 solution to the KdV equation

To obtain solutions on a surface of genus 6 we considered the surface with the branch points $[-6, -6+\varepsilon, -4, -4+\varepsilon, -2, -2+\varepsilon, 0, \varepsilon, 2, 2+\varepsilon, 4, 4+\varepsilon, 6]$. The situation for $\varepsilon = 1$ is shown in Fig. 3.5. In Fig. 3.6 we show a solution of genus 6 in an almost solitonic situation ($\varepsilon = 10^{-14}$). One can see the collision of 6 solitons at the center of the plot.

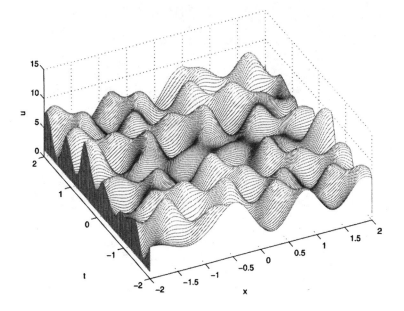

Fig. 3.5. Almost periodic genus 6 solution to the KdV equation

A genus 4 solution of the KP equation is shown for fixed t on the hyperelliptic surface with branch points $[-5, -4, -3, -3+\varepsilon, -1, -1+\varepsilon, 1, 1+\varepsilon, 3, 3+\varepsilon]$. In Fig. 3.7 we show the almost periodic situation for $\varepsilon = 1$. The almost solitonic case with $\varepsilon = 0.001$ is shown in Fig. 3.8.

3.6 Ernst Equation, Fay Identities and Variational Formulas on Hyperelliptic Surfaces

This section will be devoted to the proof of the following theorem using Fay's identities and Rauch's variational formulas:

Theorem 3.10. *Let the branch points E_m, F_m of the curve $\mathcal{L}$ (3.35) be $(\xi, \bar{\xi})$-independent. Then the function*

$$\mathcal{E} = \frac{\Theta_{pq}(\int_\xi^{\infty^-})}{\Theta_{pq}(\int_\xi^{\infty^-})}, \qquad (3.77)$$

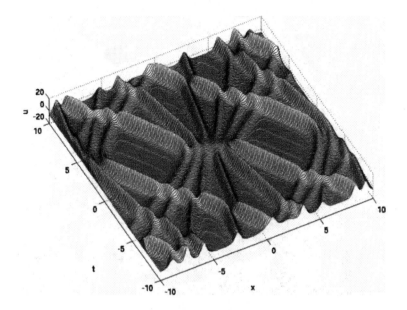

Fig. 3.6. Almost solitonic genus 6 solution to the KdV equation

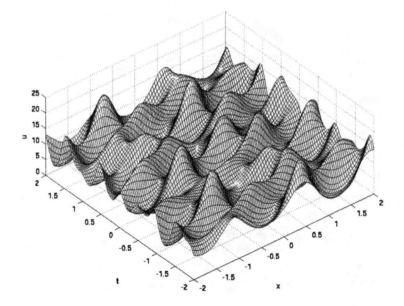

Fig. 3.7. Almost periodic genus 4 solution to the KP equation

where the theta function corresponds to the matrix of b-periods of the curve
$\mathcal{L}$, *and where an arbitrary* $(\xi, \bar{\xi})$-*independent non-singular characteristic* $[\boldsymbol{p}, \boldsymbol{q}]$
obeys the reality conditions

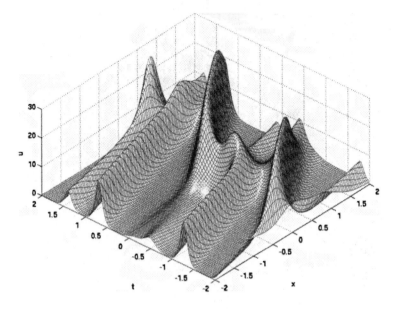

Fig. 3.8. Almost solitonic genus 4 solution to the KP equation

$$\mathbf{B}p + q \in \mathbb{R}^g ,$$

satisfies the Ernst equation (2.43) in the region of the ξ-plane, where $\Theta_{pq}(0) \neq 0$, and, in addition,

$$\Theta_{pq}(\int_\xi^{\infty^-}) \neq 0 .$$

$\int_\xi^a$ *denotes the line integral of the vector $d\boldsymbol{\omega}$ from ξ to a. The integration paths in the numerator and denominator are supposed to have the same projection onto $\mathbb{CP}^1$; therefore, $\int_\xi^{\infty^+} = -\int_\xi^{\infty^-}$.*

The relation of this form of the theta-functional solutions to the one given in (3.50) will be established below in Sect. 3.6.4.

The proof will consist of a series of auxiliary statements: we shall compute the derivatives of the Ernst potential with respect to $(\xi, \bar{\xi})$ and the action of the cylindrical Laplace operator

$$\Delta \equiv \partial^2_{\varrho\varrho} + \frac{1}{\varrho}\partial_\varrho + \partial^2_{\zeta\zeta} \equiv 4 \left(\partial^2_{\xi\bar{\xi}} - \frac{1}{2(\bar{\xi} - \xi)} \left(\partial_{\bar{\xi}} - \partial_\xi \right) \right) \qquad (3.78)$$

on the Ernst potential. We note that the real part of the Ernst potential can be written in a compact form:

Proposition 3.11. *The real part of the Ernst potential is given by the following expression:*

$$\mathcal{E} + \bar{\mathcal{E}} = 2Q \frac{\Theta_{pq}(0)\Theta_{pq}(\int_{\bar{\xi}}^{\xi})}{\Theta_{pq}(\int_{\xi}^{\infty^-})\Theta_{pq}(\int_{\bar{\xi}}^{\infty^-})} , \tag{3.79}$$

where the function

$$Q(\xi, \bar{\xi}) = \frac{1}{2} \frac{E(\xi, \bar{\xi})E(\infty^-, \infty^+)}{E(\xi, \infty^-)E(\bar{\xi}, \infty^+)} \tag{3.80}$$

does not depend on p, q. Taking into account that $\mathcal{E} \equiv 1$ if $p = q = 0$, we get an alternative form of the function Q in terms of theta functions with zero characteristics $(\Theta \equiv \Theta_{00})$:

$$Q = \frac{\Theta(\int_{\xi}^{\infty^-})\Theta(\int_{\bar{\xi}}^{\infty^-})}{\Theta(0)\Theta(\int_{\xi}^{\bar{\xi}})} . \tag{3.81}$$

This gives an elegant form of the metric function f in terms of theta functions.

Proof: The proof is an immediate corollary of Fay's identity (A.68) applied to the points $(\infty^+, \infty^-, \xi, \bar{\xi})$ if we note the following

Lemma 3.12. *The following relation holds:*

$$\frac{E(\infty^+, \bar{\xi})E(\infty^-, \xi)}{E(\infty^-, \bar{\xi})E(\infty^+, \xi)} = -1 . \tag{3.82}$$

Proof. To prove (3.82) we use formula ([128], p.21) which is valid for arbitrary four points P_i, $i = 1, 2, 3, 4$ on $\mathcal{L}$:

$$\ln \frac{E(P_2, P_4)E(P_1, P_3)}{E(P_1, P_4)E(P_2, P_3)} = \int_{P_3}^{P_4} d\omega_{P_2 P_1} . \tag{3.83}$$

Assuming $P_1 = \bar{\xi}$, $P_2 = \xi$, $P_3 = \infty^-$, $P_4 = \infty^+$, we get the integral $\int_{\infty^-}^{\infty^+} d\omega_{\xi, \bar{\xi}}$ along the path encircling the branch point ξ. On the hyperelliptic curve (3.35) with our choice of canonical cycles (Fig.1) the Abelian integral $\int d\omega_{\xi, \bar{\xi}}$ can be computed explicitly to give $\frac{1}{2} \ln \frac{\lambda - \xi}{\lambda - \bar{\xi}} + C$, where C is an arbitrary constant (indeed, this expression has the required structure of singularities at ξ and $\bar{\xi}$, and does not suffer any modification with respect to tracing along a-cycles shown in Fig. A.4; we remind that the local parameters around ξ and $\bar{\xi}$ are $\sqrt{\lambda - \xi}$ and $\sqrt{\lambda - \bar{\xi}}$, respectively). Therefore,

$$\int_{\infty^-}^{\infty^+} d\omega_{\xi, \bar{\xi}} = \frac{1}{2} \ln \frac{\lambda - \xi}{\lambda - \bar{\xi}} \Big|_{\infty^+}^{\infty^-} = \frac{1}{2} 2\pi i = \pi i ,$$

which gives (3.82).

□

To prove the above Theorem it is necessary to take the dependence of the branch points on the physical coordinates into account. Rauch's variational formulas [179] describe the dependence of the basic normalized holomorphic differentials $d\omega_\alpha$ and the matrix of b-periods $\mathbf{B}_{\alpha\beta}$ on the positions of the branch points, see section A.6 of the appendix. The formulas (A.80), (A.81), together with the heat equation for theta functions (A.56), imply the following dependence of hyperelliptic theta functions on the branch points:

Lemma 3.13. *The derivative of the hyperelliptic theta function $\Theta_{pq}(z)$ with a $\{\lambda_m\}$-dependent argument z with respect to a branch point λ_m is given by*

$$\partial_{\lambda_m}\Theta_{pq}(z) = \frac{1}{4}D_{\lambda_m}D_{\lambda_m}\Theta_{pq}(z) + \sum_\alpha \partial_{z_\alpha}\{\Theta_{pq}(z)\}\frac{dz_\alpha}{d\lambda_m}. \tag{3.84}$$

3.6.1 First Derivatives of the Ernst Potential

We will first give convenient relations for the first derivatives of the Ernst potential which where obtained in [129] with the use of the zero-curvature representation of the Ernst equation.

Proposition 3.14. *The first derivatives of the Ernst potential (3.77) are given by the following expressions:*

$$\mathcal{E}_\xi = \frac{c_2(\infty^-, \xi, \infty^+)}{2}\frac{\Theta_{pq}(0)}{\Theta_{pq}^2(\int_\xi^{\infty^-})}D_\xi\Theta_{pq}(0), \tag{3.85}$$

$$\mathcal{E}_{\bar\xi} = \frac{c_2(\infty^-, \bar\xi, \infty^+)}{2}\frac{\Theta_{pq}(\int_{\bar\xi}^\xi)}{\Theta_{pq}^2(\int_\xi^{\infty^-})}D_{\bar\xi}\Theta_{pq}\left(\int_\xi^{\bar\xi}\right), \tag{3.86}$$

where c_2 is the function (A.74) from the degenerated Fay identity (A.72).

Proof. Let us first note the following corollary of Rauch's variational formulas:

$$\frac{d}{d\xi}\int_\xi^{\infty^+}d\omega_\alpha(\tau) \equiv -\frac{d}{d\xi}\int_\xi^{\infty^-}d\omega_\alpha(\tau) = -\frac{1}{4}c_1(\infty^-, \xi, \infty^+)\frac{d\omega_\alpha(\xi)}{d\tau_\xi}. \tag{3.87}$$

where c_1 is as defined in (A.73) and where $\tau(P)$ denotes a local parameter in the vicinity of P. To prove (3.87) we notice that, according to (A.80), the derivative of a holomorphic differential with respect to a branch point is proportional to the bi-differential (the Bergmann kernel); consequently the integration of this differential gives a differential of the third kind, according to (A.73), (3.87).

The idea of the proof is to differentiate the Ernst potential with respect to ξ and to use (3.84) and (3.87) to relate these derivatives to directional derivatives of the theta functions. We get

$$(\ln \mathcal{E})_\xi = \frac{1}{4} \left\{ D_\xi D_\xi \ln \mathcal{E} + \left(D_\xi \ln \Theta_{pq} \left(\int_\xi^{\infty^-} \right) \right)^2 - \left(D_\xi \ln \Theta_{pq} \left(\int_\xi^{\infty^-} \right) \right)^2 \right.$$

$$\left. - c_1(\infty^-, \xi, \infty^+) D_\xi \ln \left(\Theta_{pq} \left(\int_\xi^{\infty^-} \right) \Theta_{pq} \left(\int_\xi^{\infty^-} \right) \right) \right\}. \qquad (3.88)$$

The resulting expression can be simplified with the help of Fay's identities. It follows from Fay's identity (A.72) with $z = \int_\xi^{\infty^-}$, $P_1 = \infty^-$, $P_2 = \xi$, $P_3 = \infty^+$ that [1]

$$D_\xi \ln \mathcal{E} = c_1(\infty^-, \xi, \infty^+) + c_2(\infty^-, \xi, \infty^+) \frac{\Theta_{pq}^2(0)}{\Theta_{pq}(\int_\xi^{\infty^-}) \Theta_{pq}(\int_\xi^{\infty^-})} ; \qquad (3.89)$$

applying the operator D_ξ once more to both sides of this identity, we get

$$D_\xi D_\xi \ln \mathcal{E} = c_2(\infty^-, \xi, \infty^+) D_\xi \left\{ \frac{\Theta_{pq}^2(0)}{\Theta_{pq}(\int_\xi^{\infty^-}) \Theta_{pq}(\int_\xi^{\infty^-})} \right\}.$$

Substituting this expression into (3.88), we arrive at the formula

$$(\ln \mathcal{E})_\xi = \frac{1}{4} c_2(\infty^-, \xi, \infty^+) D_\xi \left\{ \frac{\Theta_{pq}^2(0)}{\Theta_{pq}(\int_\xi^{\infty^-}) \Theta_{pq}(\int_\xi^{\infty^-})} \right\}$$

$$+ \frac{1}{4} D_\xi \ln \left\{ \Theta_{pq} \left(\int_\xi^{\infty^-} \right) \Theta_{pq} \left(\int_\xi^{\infty^-} \right) \right\} \{ D_\xi \ln \mathcal{E} - c_1(\infty^-, \xi, \infty^+) \}.$$

We use (3.89) again to simplify the last term which leads to (3.85). The expression (3.86) for $\mathcal{E}_{\bar\xi}$ can be proven analogously.

□

3.6.2 Action of the Laplace Operator on the Ernst Potential and Ernst Equation

The same techniques can be used to determine the second derivatives of the Ernst potential which enter the axisymmetric Laplace operator.

---

[1] It is worth noticing at this point that the action of the operator D_ξ on the Ernst potential has a priori nothing to do with the partial derivative of the Ernst potential with respect to ξ: according to the definition (A.71), $D_\xi \mathcal{E}$ is just a directional derivative of $\mathcal{E}$ with respect to q in the direction given by the values of the basic holomorphic differentials at the branch point ξ of the Riemann surface $\mathcal{L}$.

Theorem 3.15. *The action of the cylindrical Laplace operator (3.78) on the Ernst potential has the following form:*

$$\Delta \mathcal{E} = -2c_2(\infty^-,\xi,\infty^+)c_2(\xi,\bar{\xi},\infty^+)\frac{\Theta_{pq}(\int_{\xi}^{\infty^-})}{\Theta_{pq}^3(\int_{\xi}^{\infty^-})}D_{\bar{\xi}}\Theta_{pq}\left(\int_{\xi}^{\bar{\xi}}\right)D_{\xi}\Theta_{pq}(0) .$$

(3.90)

where the function c_2 is defined by (A.74).

To prove (3.90) we need to compute the derivatives with respect to $\bar{\xi}$ of all three multipliers in (3.85) with the help of the degenerated versions (A.72) and (A.75) of Fay's identities. These derivatives are given by the following proposition.

Proposition 3.16. *The following identities hold:*

$$4\left\{\ln\frac{\Theta_{pq}(0)}{\Theta_{pq}(\int_{\xi}^{\infty^-})}\right\}_{\bar{\xi}} = -c_2^2(\xi,\bar{\xi},\infty^-) + c_1^2(\xi,\bar{\xi},\infty^-)$$

(3.91)

$$-2c_2(\xi,\bar{\xi},\infty^-)D_{\bar{\xi}}\ln\Theta_{pq}\left(\int_{\xi}^{\bar{\xi}}\right)\frac{\Theta_{pq}(\int_{\xi}^{\bar{\xi}})\Theta_{pq}(\int_{\bar{\xi}}^{\infty^-})}{\Theta_{pq}(\int_{\xi}^{\infty^-})\Theta_{pq}(0)} ,$$

$$2\left(D_{\xi}\ln\Theta_{pq}(0)\right)_{\bar{\xi}} = d_2(\bar{\xi},\xi)\frac{\Theta_{pq}\left(\int_{\xi}^{\bar{\xi}}\right)}{\Theta_{pq}^2(0)}D_{\bar{\xi}}\Theta_{pq}\left(\int_{\xi}^{\bar{\xi}}\right) ,$$

(3.92)

$$\partial_{\bar{\xi}}\ln c_2(\infty^-,\xi,\infty^+) = -\frac{1}{2}\left(c_1^2(\xi,\bar{\xi},\infty^-) - c_2^2(\xi,\bar{\xi},\infty^-)\right) - \frac{1}{2(\bar{\xi}-\xi)} .$$

(3.93)

In the proof we need a corollary of formula (A.83):

Lemma 3.17. *The following relation holds:*

$$\frac{\pm 1}{\sqrt{\xi-\bar{\xi}}} = \frac{c_2(\infty^-,\xi,\infty^+)}{2Q} = c_2(\bar{\xi},\xi,\infty^+) .$$

(3.94)

The correct sign in (3.94) depends on the choice of all branches of the square roots in (3.94) and is unessential for our purposes.

Proof. To prove (3.94) it is sufficient to consider the ratio of two root functions (A.83): one with $\lambda_n = \xi$, $\lambda_m = \bar{\xi}$, and $a = \infty^+$ and another with $\lambda_n = \xi$, $\lambda_m = \bar{\xi}$ and $a \to \lambda_m$. Then the unknown function C in (A.83) drops out and we end up with (3.94).

□

The rest of the proof uses the same techniques as in the previous subsection for the first derivatives of the Ernst potential, for details see [127].

Proposition 3.16 leads to (3.90) if we take into account the

Lemma 3.18. *The following identity holds:*

$$\frac{1}{\xi - \bar{\xi}} \frac{c_2(\infty^-, \bar{\xi}, \infty^+)}{c_2(\infty^-, \xi, \infty^+)} + d_2(\bar{\xi}, \xi) = 0 . \tag{3.95}$$

Proof. We rewrite the left hand side in prime forms, using (3.94) for $(\xi - \bar{\xi})$. Then

$$\frac{1}{\xi - \bar{\xi}} \frac{c_2(\infty^-, \bar{\xi}, \infty^+)}{c_2(\infty^-, \xi, \infty^+)} = \frac{E(\infty^+, \bar{\xi})}{E^2(\bar{\xi}, \xi) E(\infty^+, \xi) d\tau_\xi} \frac{E(\infty^-, \xi)}{E(\infty^-, \bar{\xi}) d\tau_{\bar{\xi}}}$$

$$= -\frac{1}{E^2(\xi, \bar{\xi}) d\tau_\xi d\tau_{\bar{\xi}}} , \tag{3.96}$$

here we used that $\int_\xi^{\infty^+} = -\int_\xi^{\infty^-}$ and $\int_{\bar{\xi}}^{\infty^+} = -\int_{\bar{\xi}}^{\infty^-}$ and took into account that the prime form is proportional to a theta function with odd characteristic. The minus sign in (3.96) appears due to lemma 3.12.   □

To verify that (3.77) is a solution of the Ernst equation, one has to compare the action (3.90) of the Laplace operator on the Ernst potential with the expression

$$\frac{8\mathcal{E}_\xi \mathcal{E}_{\bar{\xi}}}{\mathcal{E} + \bar{\mathcal{E}}} = \frac{c_2(\infty^-, \xi, \infty^+) c_2(\infty^-, \bar{\xi}, \infty^+)}{Q} \frac{\Theta_{pq}(\int_{\bar{\xi}}^{\infty^-})}{\Theta_{pq}^3(\int_\xi^{\infty^-})} D_\xi \Theta_{pq}(0) D_{\bar{\xi}} \Theta_{pq} \left(\int_\xi^{\bar{\xi}} \right) \tag{3.97}$$

computed from (3.85), (3.86) and (3.79). The coincidence of these terms follows from the definitions of c_2 and Q.

3.6.3 Metric Functions for the Stationary Axisymmetric Vacuum

Explicit integration of equations (2.45) and (2.42) to obtain expressions for the metric functions a and k is rather non-trivial; for the algebro-geometric solutions (3.77) it was carried out explicitly, exploiting the zero-curvature representation, in the papers [52,95,129]. In the sequel we show how to achieve these results on the sole base of Fay's identities and Rauch's formulas.

It was shown in [52] with the help of the inverse scattering method that the function a, corresponding to the Ernst potential (A.82), is related to the logarithmic derivative of theta functions (3.98) which was alternatively expressed in [129] via theta functions themselves. One has the following

Proposition 3.19. *Let a_0 be a constant with respect to ξ and $\bar{\xi}$. Then the metric function ae^{2U} for the Ernst potential (A.82) is given by the expression:*

$$(a - a_0)e^{2U} = -\varrho \left(\frac{1}{Q} \frac{\Theta_{pq}(0)\Theta_{pq}(\int_\xi^{\infty^-} + \int_{\bar{\xi}}^{\infty^-})}{\Theta_{pq}(\int_\xi^{\infty^-})\Theta_{pq}(\int_{\bar{\xi}}^{\infty^-})} - 1 \right) . \tag{3.98}$$

Proof. We have to show that equation (2.45) is satisfied with the function a given by expression (3.98). With the function Z defined in (3.51), equation (2.45) is obviously equivalent to the equation

$$Z_{\bar{\xi}} = \frac{1}{\mathcal{E} + \bar{\mathcal{E}}} \left((Z + \varrho)\bar{\mathcal{E}}_{\bar{\xi}} + (Z - \varrho)\mathcal{E}_{\bar{\xi}} \right) . \tag{3.99}$$

This is done in complete analogy to the calculations of the first derivatives of the Ernst potential above, for details see [127].

Taking into account the relation (3.94), this implies

$$Z_{\bar{\xi}} = \frac{c_2(\xi, \bar{\xi}, \infty^-)\varrho}{2Q} \frac{\Theta_{pq}(0)\Theta_{pq}(\int_{\xi}^{\infty^-} + \int_{\xi}^{\infty^-})}{\Theta_{pq}^2(\int_{\bar{\xi}}^{\infty^-})} D_{\bar{\xi}} \ln \Theta_{pq}(0) \tag{3.100}$$

$$- \frac{c_2(\xi, \bar{\xi}, \infty^-)\varrho}{2Q} \frac{\Theta_{pq}(\int_{\xi}^{\bar{\xi}})\Theta_{pq}(\int_{\xi}^{\infty^-} + \int_{\xi}^{\infty^-})}{\Theta_{pq}^2(\int_{\bar{\xi}}^{\infty^-})} D_{\bar{\xi}} \ln \Theta_{pq} \left(\int_{\xi}^{\bar{\xi}} \right) .$$

Whereas the expression for $Z - \varrho$ follows directly from (3.98), we can write $Z + \varrho$, using Fay's identity (A.68), in the convenient form

$$Z + \varrho = \frac{\varrho}{Q} \frac{\Theta_{pq}(2\int_{\xi}^{\infty^-})\Theta_{pq}(\int_{\xi}^{\bar{\xi}})}{\Theta_{pq}(\int_{\xi}^{\infty^-})\Theta_{pq}(\int_{\bar{\xi}}^{\infty^-})} . \tag{3.101}$$

Relation (3.100) turns out to be equivalent to (3.99) if we use equalities (3.94), (3.79), (3.85) and (3.86).

$\square$

The metric function e^{2k} was calculated in [95] as the τ-function [195] of the Schlesinger system associated to the Ernst equation, see [203]. Here we shall prove the resulting formula using Fay's identities.

Theorem 3.20. *The metric function e^{2k} is given by*

$$e^{2k} = K \frac{\Theta_{pq}(0)\Theta_{pq}(\int_{\xi}^{\bar{\xi}})}{\Theta(0)\Theta(\int_{\xi}^{\bar{\xi}})} . \tag{3.102}$$

where K is a constant, and where as before $\int_{\xi}^{\bar{\xi}} \equiv -i\pi(1, \ldots, 1)$.

Proof. We have to show that (2.42) is satisfied with k given by (3.102). Taking into account the relations (3.85), (3.86), and (3.94), we obtain the following proposition we need to prove:

Proposition 3.21. *The following identity holds:*

$$\frac{1}{8} \left(D_{\xi} D_{\xi} \ln \frac{\Theta_{pq}(0)\Theta_{pq}(\int_{\xi}^{\bar{\xi}})}{\Theta(0)\Theta(\int_{\xi}^{\bar{\xi}})} + (D_{\xi} \ln \Theta_{pq}(0))^2 + \left(D_{\xi} \ln \Theta_{pq} \left(\int_{\xi}^{\bar{\xi}} \right) \right)^2 \right)$$

$$= \frac{1}{4} D_{\xi} \ln \Theta_{pq}(0) D_{\xi} \ln \Theta_{pq} \left(\int_{\xi}^{\bar{\xi}} \right) . \tag{3.103}$$

Proof: As the first step of the proof of identity (3.103) we observe that (3.103) can be rewritten in terms of the theta function without characteristics as follows:

$$D_\xi D_\xi \ln \frac{\Theta(V)\Theta(\int_\xi^{\bar\xi}+V)}{\Theta(0)\Theta(\int_\xi^{\bar\xi})} + (D_\xi \ln \Theta(V))^2 + \left(D_\xi \ln \Theta\left(\int_\xi^{\bar\xi}+V\right)\right)^2 \quad (3.104)$$

$$= 2D_\xi \ln \Theta(V) D_\xi \ln \Theta\left(\int_\xi^{\bar\xi}+V\right),$$

where $V \equiv Bp + q$ i.e. all exponential terms arising from relation (A.54) between the theta function with characteristics and the theta function with shifted argument drop out; therefore the statement (3.103) takes the form (3.104).

The idea of the proof of identity (3.104) is the following: we define a function F as the difference of the left-hand and the right-hand side of (3.104). We show that the derivatives of the function F with respect to any component p_α and any q_α of the vectors p and q vanish. Then the function F must be a constant with respect to p and q; thus it is sufficient to observe that this function vanishes at $p = q = 0$.

Function F depends only on the combination $V \equiv Bp + q$; therefore, all partial derivatives of F with respect to each p_α are linear combinations of the partial derivatives with respect to q_α; thus it is sufficient to prove that all partial derivatives of F with respect to q_α vanish.

In turn, to show that all partial derivatives of F with respect to q_α are equal to zero, it is sufficient to prove that $D_P F \equiv \sum_{\alpha=1}^g \frac{\partial F}{\partial q_\alpha} \frac{d\omega_\alpha(P)}{d\tau_P}$ vanishes for an arbitrary point $P \in \mathcal{L}$, taking into account the following

Lemma 3.22. *There exists a positive divisor $P_1 + \ldots + P_g$ of degree g on $\mathcal{L}$ such that the vectors $d\omega(P_1)/d\tau_{P_1}, \ldots, d\omega(P_g)/d\tau_{P_g}$ are linearly independent.*

Proof. Suppose the opposite, i.e. that $\det\{d\omega_\alpha(\beta)\}$ vanishes for any divisor $P_1 + \ldots + P_g$. Let us integrate this determinant along a basic cycle a_β with respect to the variable β for each β. On one hand, the result should equal 0 according to our assumption. On the other hand, we get the determinant of the unit matrix, which equals 1. This contradiction proves the lemma.

□

Thus for suitably chosen P, the vector $d\omega(P)/d\tau_P$ will take all values in $\mathbb{C}^g$. If one can show that $D_P F = 0$ for arbitrary P, this implies that F must be a constant.

Now let us calculate the D_P derivative of F (3.104) where $P \neq \xi$ is an otherwise arbitrary point on $\mathcal{L}$. With the help of Fay's identity (A.75) we can write down this derivative as follows:

$$D_P F = d_2(P,\xi) \frac{\Theta(\int_\xi^P +V)\Theta(\int_P^\xi +V)}{\Theta^2(V)} D_\xi \ln \frac{\Theta(\int_\xi^P +V)\Theta(\int_P^\xi +V)}{\Theta^2(\int_\xi^{\bar\xi} +V)} \quad (3.105)$$

$$+ d_2(P,\xi) \frac{\Theta(\int_{\bar\xi}^P +V)\Theta(\int_P^{\bar\xi} +V)}{\Theta^2(\int_\xi^{\bar\xi} +V)} D_\xi \ln \frac{\Theta(\int_{\bar\xi}^P +V)\Theta(\int_P^{\bar\xi} +V)}{\Theta^2(V)} \,.$$

The degenerated Fay identity (A.72) implies

$$D_\xi \ln \left\{ \Theta \left(\int_\xi^P +V \right) \Theta \left(\int_P^\xi +V \right) \right\} \quad (3.106)$$

$$= 2 D_\xi \ln \Theta \left(\int_\xi^{\bar\xi} +V \right) + c_2(\bar\xi,\xi,P) \frac{\Theta(V)}{\Theta(\int_\xi^{\bar\xi} +V)} \left(\frac{\Theta(\int_{\bar\xi}^P +V)}{\Theta(\int_\xi^P +V)} - \frac{\Theta(\int_P^{\bar\xi} +V)}{\Theta(\int_P^\xi +V)} \right) \,.$$

Substituting (3.106), together with the corresponding relation for $D_\xi \{\Theta(\int_{\bar\xi}^P +V)\Theta(\int_P^{\bar\xi} +V)\}$ into (3.105), we find that the D_P derivative of F is identically zero for all $P \neq \xi$. Consequently, the difference F between the r.h.s. and l.h.s of (3.104) must be a constant with respect to the characteristics $[p, q]$. Considering the case $[p, q] = [0, 0]$ we see that both sides of (2.42) are zero in this case. $\square$

3.6.4 Relation to the Previous Form of the Solutions

The relation to the previous form of the Ernst potential can be established as follows: Let $g = \tilde g + n$ and $p_{\tilde g+j} = h_j \in \mathbb{R}$, $q_{\tilde g+j} = 0$ for $j = 1,\ldots,n$. Consider the limit of collapsing branch cuts for $j > \tilde g$, i.e. $E_{\tilde g+j} \to F_{\tilde g+j}$. In this limit all quantities entering (3.77) can be expressed in terms of quantities (denoted with a tilde) on the surface $\tilde{\mathcal L}$ of genus $\tilde g$ given by $\tilde\mu^2 = (K-\xi)(K-\bar\xi) \prod_{i=1}^{\tilde g}(K - E_i)(K - F_i)$, the surface $\mathcal L$ with the collapsing cuts removed. The holomorphic differentials have the limit, see [128],

$$d\omega_i \to d\tilde\omega_i \,, \quad i = 1,\ldots,\tilde g\,, \quad d\omega_i \to d\tilde\omega_{E_i^- E_i^+} \,, \quad i = \tilde g+1,\ldots,n \,. \quad (3.107)$$

In other words the holomorphic differentials of $\mathcal L$ become holomorphic differentials on $\tilde{\mathcal L}$ and differentials of the third kind with poles at the collapsed branch cuts. Since the b-periods of differentials of the third kind can be expressed in terms of the Abel map of the poles, see e.g. (A.27), one can use formula (A.54) to get for (3.77)

$$\mathcal{E} = \frac{\tilde\Theta_{\tilde p \tilde q}(\tilde\omega(\infty^+) + \sum_{j=1}^n h_j(\omega(E_j^-) - \omega(E_j^+)))}{\tilde\Theta_{\tilde p \tilde q}(\tilde\omega(\infty^-) + \sum_{j=1}^n h_j(\omega(E_j^-) - \omega(E_j^+)))}$$

$$\times \exp\left(\sum_{j=1}^n h_j \int_{\infty^-}^{\infty^+} d\omega_{E_j^- E_j^+} \right) \,. \quad (3.108)$$

By taking the limit $\sum_{j=1}^{n} \to \int_{\Gamma}$ from a sum to a line integral over the E_j, we get after a partial integration (we assume $\ln G$ vanishes at the limits of integration) and the identification $h(K) = \partial_K(\ln G)$ formula (3.50) (without the integrals of the second kind which can be included in the form of distributional h) for (3.108) where we have used (A.22). For details of the above construction see [95].

4 Analyticity Properties and Limiting Cases

In Chap. 3 we have used the linear system for the Ernst equation to construct solutions via Riemann–Hilbert techniques. This was done on the Riemann surface of the spectral parameter, the physical coordinates were fixed in a way that they did not coincide with the singularities of the matrix of the linear system. In the present chapter we want to investigate the behavior of the found solutions in dependence of the physical coordinates, especially at the potential singularities which were so far excluded. This analysis allows us to identify a whole subclass of solutions which will be only singular at some contour which could be identified with the surface of some star or galaxy. Since solutions of astrophysical interest typically have an equatorial symmetry, a reflection symmetry at the equatorial plane, we will identify theta-functional solutions with this property. For the found subclass we investigate interesting limiting cases as the limit of large distance from the material source. This allows to identify asymptotically flat solutions which can describe isolated matter sources. We also study the static limit and the 'solitonic' limit, in which the Riemann surface degenerates. In this vicinity the solutions can be given in terms of elementary functions, they belong either to the static Weyl class or the multi-black hole solutions.

In Sect. 4.1 we study all possible singularities of the solutions in terms of hyperelliptic theta functions. Using techniques by Fay [128] and Yamada [196] for degenerate Riemann surfaces, we show that the solutions can be analytic in the case of coinciding branch points. In Sect. 4.2 we identify equatorially symmetric solutions and give reduction formulas in the equatorial plane and on the axis. In Sect. 4.3 we study certain limiting cases as the asymptotic limit and the solitonic limit. As an example for the latter we obtain the Kerr solution and discuss briefly some of its features.

4.1 The Singular Structure of the Ernst Potential

The construction of the solutions in the previous sections with the help of Theorem 3.1 also indicates where the resulting Ernst potential (3.50) may be singular: only at points ξ where the conditions of Theorem 3.1 do not hold. Notice that these conditions are sufficient for the regularity of $\mathcal{E}$ at all other points. It may turn out though that the Ernst potential is perfectly regular at points which had been so far excluded, e. g. in the case of singularities that are pure gauge. We will therefore discuss all possible singular points of the solutions (3.50).

It is helpful that this discussion can be performed on the Riemann surface $\mathcal{L}$, where the powerful hyperelliptic calculus can be used. The introduction of the four sheeted surface $\hat{\mathcal{L}}$ was necessary for the construction of the solutions. In addition it provides an understanding of the mathematical properties of corresponding solutions to the Ernst equation.

The possible singularities of $\mathcal{E}$ can be directly inferred from the potential in the form (3.50). The Ernst potential will be singular at the zeros of the denominator. It is possibly not regular at the points where ξ is identical to the singularities of Ω or when ξ is on Γ. Critical points of a different kind are the branch points E_i and F_i. If ξ coincides with one of these points, the Riemann surface $\mathcal{L}$ degenerates. The same happens at the axis where the branch points ξ and $\bar{\xi}$ coincide. This is a reminiscent of the singular behavior of the three-dimensional Laplace operator on the axis in the axisymmetric case. The main aim of the analysis below is to single out a class of solutions that may be interesting in the context of boundary value problems for the Ernst equation that describe e. g. the exterior of a rotating body. Thus we will not study the nature of the singularities (e. g. curvature singularities) but single out a large class of solutions where the Ernst potential is only discontinuous at a (closed) contour that could be identified with the surface of a body.

4.1.1 Zeros of the Denominator

Zeros of the denominator of (3.50) will lead to singularities in the spacetime. From condition IV of Theorem 3.1 it follows that these are just the points at which the matrix Φ cannot be normalized in the required way. This leads to the transcendental condition

$$\Theta_{pq}(\omega(\infty^-) + u + b) \neq 0 \,, \tag{4.1}$$

if one wants to exclude these poles. We will show in the next section how the zeros of the theta function in (4.1) can be found as the solution of a set of algebraic equations.

4.1.2 Essential Singularities

The integrals of the third kind occurring in Ω are nothing but a particular case of line integrals over contours with constant jump function $G(t)$. Therefore, we are left with the investigation of the integrals of the second kind at this point. Since the theta functions in (3.50) are regular as long as the Riemann surface $\mathcal{L}$ is, we are left with the exponent if (4.1) holds. For the behavior of the exponent, we get the following

Proposition 4.1. *The Ernst potential (3.50) has an essential singularity at the points where ξ coincides with the singularities of the integrals of the second kind on $\mathcal{L}$.*

Proof. The exponent has by construction an algebraic pole there and, therefore, the Ernst potential has an essential singularity.

□

An essential singularity of the real part of the Ernst potential corresponds to a line singularity of the metric function U. In the context of exterior solutions for bodies of revolution we are interested in, there seems to be no situation where such a line singularity in a spacetime might be interesting.

4.1.3 Contours

In case ξ lies on a contour Γ_i but not on an endpoint of Γ_i, on the axis, or on one of the branch points E_i or F_i, it can be easily seen that the integral in the exponent as well as the b-periods u_i are bounded since G is Hölder–continuous, finite and non-zero on Γ. At the endpoints of the contours Γ_i, singularities may occur. The value of these integrals at the remaining points will however not be the same in general if the contour is approached from one or the other side. This can be seen from the following fact: The point ξ is a branch point of $\mathcal{L}$. If it lies on the contour Γ, care has to be taken of the sign of the root whilst evaluating the integrals of the form

$$J_n = \int_\Gamma \ln G(\tau) \frac{\tau^n d\tau}{\mu} \ .$$

The decisive factor is $\mu_0^2 = (K - \xi)(K - \bar{\xi})$. We get for $K \in \Gamma$ with $K = K_1 + iK_2$, and $K_1, K_2 \in \mathbb{R}$ for the imaginary part of μ_0,

$$\Im\mu_0 = \pm\frac{1}{\sqrt{2}}\operatorname{sgn}\left((K_1 - \zeta)K_2\right)\sqrt{|\mu_0^2| - \Re(\mu_0^2)}, \qquad (4.2)$$

i.e. the sign of the imaginary part of μ depends on the sign of $K_1 - \zeta$. Thus the value of the integrals will in general not be the same if the contour is approached from the interior or the exterior region. This reasoning does not work for points $K = K_0$ not on the axis ($K_2 \neq 0$) with $K_1 = 0$. There the imaginary part of μ_0^2 is zero which means that μ_0 is either purely imaginary or real in the vicinity of $\xi = K_0$ depending on the sign of $K_2 - \zeta$. We conclude that the integrals over Γ with $\xi \in \Gamma$ have the form $J = J^1 + \operatorname{sgn}(\varepsilon)J^2$ (where the J^i are independent of ε which indicates if the contour is approached from the interior or the exterior) which implies that the limiting value of the Ernst potential calculated via (3.50) exists but depends on ε. Therefore, we have proven the following

Proposition 4.2. *Let $\xi \in \Gamma$ but not on the axis or be equal to one of the branch points E_i or F_i, or at an endpoint of Γ_i. Then $\mathcal{E}$ will, in general, have*

a jump at Γ. The limiting values of $\mathcal{E}$ will exist for both sides of the contour and be Hölder–continuous there. The Ernst potential may be singular at the endpoints of the Γ_i.

Thus the Ernst potential will be finite but possibly discontinuous at a contour Γ_ξ in the (ϱ, ζ)-plane given by $\xi \in \Gamma$ which means that the solution to the vacuum equations will not be regular at a surface in the (ϱ, ζ, ϕ)-space. If this surface is closed, it can possibly be identified with the surface of a rotating body. The interior of the body is supposed to be filled with matter. Therefore the vacuum solution is only considered in the exterior (that contains $\xi = \infty$); it is not regular at the boundary to the matter region.

4.1.4 Axis

The axis corresponds to a double point of the hyperelliptic curve since two branch points coincide. Whereas the algebraic curve is singular in this case, the corresponding Riemann surface is regular as will be shown below. In this case, all quantities may be considered on the Riemann surface Σ' given by

$$\mu'^2 = \prod_{i=1}^{g}(\tau - E_i)(\tau - F_i) \ ,$$

see e.g. Fay [128] and [196]. Let a prime denote here and in the following that the primed quantity is taken on Σ'. This surface is obtained from $\mathcal{L}$ by removing the cut $[\xi, \bar{\xi}]$. For the analysis of the axis, we will use a slightly different cut system than the one introduced in Chap. 3: we take a closed curve encircling $[\xi, \bar{\xi}]$ in the +-sheet as the cut a_g. All b-cuts shall begin at the cut $[E_1, F_1]$. The rest is unchanged with respect to the system of Fig. A.4. This implies for the characteristic of the Riemann theta function that it has the form

$$\begin{bmatrix} \boldsymbol{p}' & 1 \\ \boldsymbol{q}' & \varepsilon \end{bmatrix} \ , \tag{4.3}$$

where $\varepsilon = 0, 1$ and $p'_i = 0$.

Since the expansions of all characteristic quantities of the Riemann surface are smooth in ϱ except $\mathbf{B}_{gg}$ which is divergent as $\ln \varrho$ for $\varrho \to 0$, it follows that the Ernst potential has a regular expansion in ϱ. For points ξ not coinciding with real branch points or singularities of the exponent in (3.50), the Ernst potential is thus at least C^3. It follows from a theorem of Müller zum Hagen [197], which is based on a theorem by Morrey [198], that it is therefore analytic if there are no horizons. Consequently it is sufficient to calculate the limiting case. If this limit is well defined, the Ernst potential is regular at these points of the axis. The differentials of the first kind for $\varrho = 0$ have the limit

$$d\omega_i = d\omega'_i \ , \quad i = 1, \ldots, g-1 \ , \quad d\omega_g = -d\omega'_{\zeta+\zeta^-} \ , \tag{4.4}$$

where $d\omega'_{\zeta^+ + \zeta^-}$ is the normalized differential of the third kind on Σ' with poles in ζ^+ and ζ^-. This implies for the b-periods

$$\mathbf{B}_{ij} = \mathbf{B}'_{ij}, \quad i,j = 1,\ldots,g-1, \tag{4.5}$$

$$\mathbf{B}_{ig} = -\int_{\zeta^-}^{\zeta^+} d\omega'_i, \quad i = 1,\ldots,g-1, \tag{4.6}$$

$$\mathbf{B}_{gg} = 2\ln\varrho + \text{reg. terms}. \tag{4.7}$$

Since $\mathbf{B}_{gg}$ diverges, the theta function will break down to a sum of two theta series on Σ' (in the case of genus $g = 1$, the surface Σ' has genus 0; the formula below can however be used if one replaces the theta function Θ' simply by a factor 1 which means that the axis potential can be expressed in terms of elementary functions in this case). We introduce integrals $\omega'(P)$ of the first kind with the property $\omega'(E_1) = 0$. The differential $d\omega'_{\infty^+ + \infty^-}$ on the axis becomes $d\omega'_{\infty^+ + \infty^-}$. In the case of the contour integrals one has to observe that an additional factor $\text{sgn}(K_1 - \zeta)$ in the notation of (4.2) occurs for the same reasons as there. Since the Abelian integrals of the second kind can be obtained from the integrals of the third kind by a limiting procedure, the same holds for these integrals and their b-periods. With the above settings, we obtain for (3.50)

$$\mathcal{E} = \frac{\Theta'_{p'q'}\left(\omega'|_{\zeta^+}^{\infty^+} + u' + b'\right) + (-1)^\varepsilon e^{-(\omega'_g(\infty^+) + u_g + b_g)}\Theta'_{p'q'}\left(\omega'|_{\zeta^-}^{\infty^+} + u' + b'\right)}{\Theta'_{p'q'}\left(\omega'|_{\zeta^+}^{\infty^+} - u' - b'\right) + (-1)^\varepsilon e^{-(\omega'_g(\infty^+) - u_g - b_g)}\Theta'_{p'q'}\left(\omega'|_{\zeta^-}^{\infty^+} - u' - b'\right)}$$

$$\times \exp\left\{\Omega'|_{\infty^-}^{\infty^+} + \frac{1}{2\pi\mathrm{i}}\int_\Gamma \ln G(\tau)d\omega'_{\infty^+ + \infty^-}(\tau) + b_g + u_g\right\}. \tag{4.8}$$

It can be seen from the above formula that the limiting value of $\mathcal{E}$ exists even if u_g diverges, provided (4.1) holds ($\mathcal{E}$ will be Hölder–continuous if u_g diverges). The Ernst potential will however have an essential singularity at the real singularities of Ω. We can summarize the above results.

Proposition 4.3. *Let condition (4.1) hold. Then the Ernst potential is regular on the axis except at the points where ξ coincides with singularities of Ω, points of Γ, and branch points E_i, F_i.*

Remark 4.4. Though $\mathcal{E}$ is Hölder–continuous even if u_g diverges, it is interesting to note for the following when this will be the case. Obviously this can only happen at the real points of Γ. It can be seen however that u_g is always bounded at these points due to the reality condition, unless they are endpoints of Γ_i (this would lead to a conic singularity on the axis).

The above degeneration (see [128, 129]) implies that the Ernst potential has in the vicinity of the axis the form

$$\mathcal{E}(\varrho, \zeta) = \mathcal{E}_0(\zeta) + \varrho^2 \mathcal{E}_1(\zeta) + \mathcal{O}(\varrho^4) ; \tag{4.9}$$

here $\mathcal{E}_0$ and $\mathcal{E}_1$ are independent of ϱ, $\mathcal{E}_0$ is the axis potential. This means the Ernst potential is a function of ϱ^2 in the vicinity of a regular axis.

4.1.5 Asymptotic Behavior

Asymptotic flatness implies that the Ernst potential is of the form

$$\mathcal{E} = 1 - \frac{2m}{|\xi|} - \frac{2iJ\zeta}{|\xi|^3} + 0(1/|\xi|^3)$$

for $|\xi| \to \infty$ where m, the ADM mass, and J, the angular momentum, are real constants and where the mass has to be positive. A complex m is related to a so called NUT parameter that is comparable to a magnetic monopole, see e.g. [140].

Spatial infinity is a double point $\xi, \bar{\xi} \to \infty$ that can be treated as the axis. The asymptotic properties of the solutions (3.50) can be read off at the axis. Notice that the dw'_i are independent of ζ. For dw_g, we get

$$dw_g = dw'_{\infty^+\infty^-} \left(1 - \frac{1}{2\zeta} \sum_{i=1}^{g} (E_i + F_i)\right) + \frac{1}{\zeta} dw^{(1)}_{\infty^+}{}' + o(1/\zeta) , \tag{4.10}$$

where $dw^{(1)}_{\infty^+}{}'$ is the differential of the second kind with a pole of second order at ∞^+. Furthermore it can be seen that $\exp(-w_g(\infty^+))$ is proportional to $1/\zeta$ for $\zeta \to +\infty$. Thus we get

Proposition 4.5. *Let* $\lim_{\tau \to \infty} \tau \ln G(\tau) = 0$ *on all contours that go through* ∞^+ *or* ∞^- *and let*

$$\Theta'_{p'q'} (u' + b') \neq 0 .$$

Then $\mathcal{E}$ *has the form*

$$\mathcal{E} = 1 - \frac{2m}{\zeta} - \frac{2iJ}{\zeta^2} + 0(1/\zeta^3) , \tag{4.11}$$

for $\zeta \to +\infty$ *where* m, J *are complex constants which can be calculated in terms of elliptic theta functions.*

The proof of this proposition follows from (4.10) and (4.8).

4.1.6 Real Branch Points

If ξ coincides with a real branch point E_i or F_i, this will be a triple point on the hyperelliptic curve. We get the following

Proposition 4.6. *At points where* ξ *coincides with the real branch point* E_g, *the limiting value of* $\mathcal{E}$ *exists. The Ernst potential is in general not differentiable there.*

Proof. We use the same cut system and the same notation as on the axis. Put $\xi = E_g + x$ with $x = \delta e^{i\phi}$ and $\phi \in [0, 2\pi]$, $\delta \in \mathbb{R}^+$. In order to expand $\mathcal{E}$ in powers of x and $\bar{x}$, one has to consider the a-periods, in particular

$$
\oint_{a_g} \frac{d\tau}{\mu} = \frac{4}{\sqrt{\bar{x}(F_g - E_g)}} \int_1^{\frac{1}{k}} \frac{dt}{\sqrt{(1 - t^2)(1 - k^2 t^2)(F_g - E_g + xt^2)}\mu''(F_g + xt^2)}
$$

$$
= \frac{4}{\sqrt{\bar{x}(F_g - E_g)}\mu''(F_g)} (i\tilde{K}(k) + O(\delta)) , \tag{4.12}
$$

where $k = e^{i\phi}$, where $\tilde{K}(k) = K(\sqrt{1 - k^2})$ and $K(k)$ are the complete elliptic integrals of the first kind (see e.g. [199]), and where

$$
\mu''^2(\tau) = \prod_{i=1}^{g-1} (\tau - E_i)(\tau - F_i) .
$$

It can be seen from (4.12) that the a-period has an expansion in powers of $\sqrt{\delta}$. The coefficients of the expansion in $\sqrt{x}$, and $\sqrt{\bar{x}}$ are ϕ-dependent, since the module of the elliptic integrals is just $k = e^{i\phi}$. This implies for the differentials of the first kind

$$
d\omega_i = d\omega_i' + O(\sqrt{\delta})
$$

for $i = 1, \ldots, g - 1$, and

$$
d\omega_g = d\omega_{g-1} .
$$

Similarly

$$
d\omega_{\infty+\infty-} = d\omega_{\infty+\infty-}' + O\left(\sqrt{\delta}\right) .
$$

We get for the b-periods,

$$
\mathbf{B}_{gg} = -2\pi \frac{K(k)}{K_1(k)} \left(1 + O\left(\sqrt{\delta}\right)\right) , \tag{4.13}
$$

whereas $\mathbf{B}_{(g-1)g} = O(\sqrt{\delta})$ and $\mathbf{B}_{ij} = \mathbf{B}_{ij}'$ for $i, j = 1, \ldots, g - 1$ in the limit. Thus $\mathcal{E}$ can be expanded in $\sqrt{x}$ and $\sqrt{\bar{x}}$. Even in case that only integer powers in the expansion occur, the coefficients will be in general ϕ-dependent. Though the limiting value of $\mathcal{E}$ at $\xi = E_g$ exists, $\mathcal{E}$ will in general not be differentiable at this point.

$\square$

This implies that the real branch points can be singular points on the axis, possibly topological defects in the spacetime, see [93]. They should not occur in the context of exterior solutions for bodies of revolution we are interested in here.

4.1.7 Non-real Branch Points

If ξ coincides with a branch point $E_g = \bar{F}_g$, the points E_g and F_g will be double points on the algebraic curve. Thus the situation is similar to the one on the axis with the only exception that one ends up here with two double points. As on the axis, it is convenient to consider all quantities on a Riemann surface Σ'' given by

$$\mu''^2 = \prod_{i=1}^{g-1} (K - E_i)(K - F_i) \,, \tag{4.14}$$

where the double points are removed. All quantities with two primes are understood to be taken on this surface. We use the following cut system: let a_{g-1} be the circle around $[\xi, E_g]$, and a_g the circle around $[\bar{\xi}, F_g]$, both in the plus sheet. The remaining cuts are as on the axis, i. e. all b-cuts start at the cut $[E_1, F_1]$. We get

Proposition 4.7. Let (4.1) hold, and let $E_g = \bar{F}_g \notin \Gamma$. Then the Ernst potential is regular at the point $\xi = E_g$. For $E_g \in \Gamma$, $\mathcal{E}$ is in general Hölder-continuous at $\xi = E_g$.

The proof is similar to the one on the axis and basically uses again results of Fay [128].

Proof. The case $g = 1$ may be checked directly with the help of the standard theory of elliptic theta functions (see e. g. [199]). For $g > 1$ with the cut system in use and $\xi = E_g + x$, where x is chosen as in the case of the real branch points, the differentials of the first kind have a smooth expansion in x and $\bar{x}$. In contrast to the case of real branch points, the coefficients in the expansion are ϕ-independent. The differentials $d\omega_i$ become in leading order the differentials of the first kind $d\omega_i''$ on Σ''. The differential $d\omega_{g-1}$ becomes in the limit the differential $-d\omega''_{E_g^+ E_g^-}$, and similar for $d\omega_g$ at F_g. The differential of the third kind becomes

$$d\omega_{\infty^+ \infty^-} = d\omega''_{\infty^+ \infty^-} \,. \tag{4.15}$$

All these differentials have coefficients in the x and $\bar{x}$ expansion that contain Abelian integrals of the second kind with poles in $E_g^\pm$ and $F_g^\pm$ as may be checked by direct calculation. This implies for the b-periods that

$$\mathbf{B}_{ij} = \mathbf{B}''_{ij} \,, \tag{4.16}$$

for $i, j = 1, \ldots, g - 2$ and

$$\mathbf{B}_{(g-1)(g-1)} \sim \mathbf{B}_{gg} = 2\ln\delta + \ldots \,,$$
$$\mathbf{B}_{i(g-1)} = -2\omega''(E_g^+) \,,$$
$$\mathbf{B}_{ig} = -2\omega''(F_g^+) \,, \tag{4.17}$$

whereas $\mathbf{B}_{(g-1)g}$ is finite in the limit $\delta \to 0$. If $E_g \notin \Gamma$, the u_i as well as the Cauchy integral in the exponent have a smooth expansion in x and $\bar{x}$ with ϕ-independent finite coefficients. The Theorem of [197] then guarantees regularity in the absence of a horizon if the limiting value as calculated on the axis exists. The theta function on $\mathcal{L}$ breaks down to a sum of four theta functions on Σ'' times a multiplicative factor. If $E_g \in \Gamma$, the coefficients in the expansion of $\mathcal{E}$ in x and $\bar{x}$ will diverge which implies that $\mathcal{E}$ is possibly not differentiable there though the limiting value exists if (4.1) holds.

$\square$

4.2 Equatorial Symmetry

In Newtonian gravity it is known that isolated perfect fluid bodies in thermo-dynamical equilibrium lead to a spacetime with equatorial symmetry. There is the belief that the same holds in a general relativistic setting though a proof of this assertion is unknown. A reflection symmetry for the metric at the equatorial plane implies for the Ernst potential $\mathcal{E}(\varrho, -\zeta) = \bar{\mathcal{E}}(\varrho, \zeta)$ because of (2.45). It is therefore of special interest to single out equatorially symmetric solutions among those in (3.50). We get

Theorem 4.8. *Let $\mathcal{L}$ be a hyperelliptic surface of the form (3.35) with even genus $g = 2s$ and the property $\mu(-K, -\zeta) = \mu(K, \zeta)$. Let Γ be a piecewise smooth contour on $\mathcal{L}$ such that with $P = (K, \mu(K)) \in \Gamma$ also $\bar{P} \in \Gamma$ and $(-K, \mu(K)) \in \Gamma$. Let there be given a finite non-zero function G on Γ subject to $G(\bar{P}) = \bar{G}(P) = G((-K, \mu(K)))$. If $(p, \mu(p))$ is a singularity of Ω, the same should hold for $(-p, \mu(-p))$. Choose a cut system as in Fig. 4.1 in a way that the cuts a_i^1 $(i = 1, \ldots, s)$ encircle $[-F_i, -E_i]$ and a_i^2 encircle $[E_i, F_i]$ in the +-sheet (in the case of real branch points, the points are ordered in the way $E_i < F_i < E_{i+1} < \ldots$; points with the same real part are ordered in the way $\Im(E_i) < \Im(F_i) < \Im(E_{i+1}) < \ldots$ which implies that $E_i \neq \bar{F}_i$ in this special case).*

Then $\mathcal{E}$ is equatorially symmetric if the characteristics in the i-th position (any combination of these cases is allowed) have the form

$$\frac{1}{2}\begin{bmatrix} 0 & 0 \\ 1 & 1 \end{bmatrix}, \quad \frac{1}{2}\begin{bmatrix} 0 & 0 \\ 0 & 0 \end{bmatrix}, \quad \frac{1}{2}\begin{bmatrix} 1 & 1 \\ \frac{1}{2} & \frac{1}{2} \end{bmatrix}. \tag{4.18}$$

Proof. The property $\mu(\zeta, K) = \mu(-\zeta, -K)$ on $\mathcal{L}$ makes it possible to express quantities on a surface with $\zeta = -\zeta_0$ in terms of the corresponding quantities on the surface with $\zeta = \zeta_0$. We have $a_i^1(-\zeta) = \tau a_i^2(\zeta)$ and $b_i^1(-\zeta) = \tau b_i^2(\zeta)$ where ζ and $-\zeta$ denote the surface on which the quantity is considered, and where τ is the anti-holomorphic involution on $\mathcal{L}$. Together with the symmetry properties of the Abelian integrals in the exponent of (3.50), this implies that the transformation $\zeta \to -\zeta$ acts as the complex conjugation together with

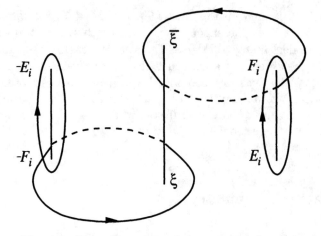

Fig. 4.1. Cut system adapted to equatorial symmetry

a change of the upper index. Thus we have for the characteristics in (4.18)
$\mathcal{E}(-\zeta) = \bar{\mathcal{E}}(\zeta)$.

$\square$

Remark 4.9. If the theta function contains only blocks of the form $\frac{1}{2}\begin{bmatrix} 0 & 0 \\ 1 & 1 \end{bmatrix}$, the resulting $\mathcal{E}$ is just the complex conjugate of the Ernst potential built with the Riemann theta function. In the case of a rotating body, complex conjugation of the Ernst potential implies that the angular velocity of the body changes its sign.

The above results suggest that it is possible to identify a whole subclass of solutions among (3.50) that are asymptotically flat, regular except at a closed contour and equatorially symmetric, i. e. solutions that might describe the exterior of a rotating body and might be helpful in the construction of solutions to boundary value problems for the Ernst equation. We get

Theorem 4.10. *Let $\mathcal{L}$ be a regular hyperelliptic surface of even genus $g = 2s$ of the form (3.35) without real branch points. Let Γ be a closed, smooth contour on $\mathcal{L}$ such that with $P = (K, \mu(K)) \in \Gamma$ also $\bar{P} = (\bar{K}, \mu(\bar{K})) \in \Gamma$ and $(-K, \mu(K)) \in \Gamma$ and $E_i \notin \Gamma$. Let there be given a finite nonzero function G on Γ subject to $G(\bar{P}) = \bar{G}(P) = G(-K, \mu(K))$. Choose the characteristic $[\boldsymbol{p}, \boldsymbol{q}]$ such that it consists of blocks of the form (4.18) as in Theorem 4.8. Then*

$$\mathcal{E}(\varrho, \zeta) = \frac{\Theta_{\boldsymbol{pq}}(\boldsymbol{\omega}(\infty^+) + \boldsymbol{u})}{\Theta_{\boldsymbol{pq}}(\boldsymbol{\omega}(\infty^+) - \boldsymbol{u})} \exp\left\{ \frac{1}{2\pi i} \int_\Gamma \ln G(\tau) \omega_{\infty^+ \infty^-}(\tau) \right\}, \quad (4.19)$$

is

(i)   a regular solution to the Ernst equation for $\xi \notin \Gamma$ if condition (4.1) holds,

(ii) in general discontinuous at Γ_ξ given by $\xi \in \Gamma$,

(iii) asymptotically ($|\xi| \to \infty$) given by

$$\mathcal{E} = 1 - 2m/|\xi| + \frac{2iJ\zeta}{|\xi|^3} \ ,$$

where m, J are finite real constants if

$$\Theta'_{p'q'}(u') \neq 0 \ ,$$

(iv) equatorially symmetric.

Proof. From (3.50) it can be seen that $\mathcal{E}$ is a solution to the Ernst equation. The regularity properties follow from the previous section. The asymptotic behavior and equatorial symmetry follow from above.

□

Remark 4.11. The choice of this class is mainly due to regularity requirements. If all singularities like real branch points or the singularities of the Abelian integrals of the second kind lie within the contour Γ where the solution is not considered since the region is assumed to be filled with matter, they would not affect the vacuum region. However this would not enlarge the degrees of freedom (one real-valued function and a set of complex parameters) if one wants to solve boundary value problems.

We will discuss the common properties of the solutions in this subclass in the following.

4.2.1 Reduction of the Ernst Potential

The explicit form of the solutions (4.19) in terms of theta functions makes it possible to identify physically interesting features directly as we will demonstrate in the following chapters. Since the theta functions as the trigonometric functions are transcendental functions, a numerical treatment will be necessary. As we will demonstrate in the following, the numerical treatment of theta functions is in general unproblematic since the exponential series converges rapidly due to the factor $\exp\left(\frac{1}{2}\mathbf{B}_{ij}n_in_j\right)$ where the real part of the Riemann matrix is negative definite. It is however obvious that the numerical evaluation becomes more and more tedious the larger the genus g of the Riemann surface $\mathcal{L}$ is. Therefore it is an important question whether the Riemann surface can be reduced in physically interesting cases to surfaces of lower genus. Loosely speaking this is possible if there exists a special relation between the branch points (see Weierstraß' discussion of the case $g = 2$ which is referred to in Chap. 7 of [1] and references given therein). Since the branch points ξ, $\bar{\xi}$ are parametrized by the physical coordinates and can thus take on arbitrary complex values, such a reduction will only be possible at special points of the spacetime which will in general not be of special physical interest. A general reduction of the Riemann surface is possible if there exist non-trivial automorphisms on the surface (i.e. automorphisms in addition to the hyperelliptic involution). For the class of equatorially symmetric solutions discussed here, this is the case in the equatorial plane and on the axis. There the surfaces $\mathcal{L}$ and Σ' have defining equations $\mu(K)$ and $\mu'(K)$ which both depend only on K^2. Thus on both surfaces there is the involution T defined by

$$T : (K, \mu(K)) \to (-K, \mu(-K)) \ .$$

For the sake of simplicity, we will only discuss the characteristic $p_i = q_i = 0$ and the case $E_i^1 = -\bar{E}_i^2$ (the general case can be inferred from the resulting relations without problems). We will concentrate on disks of radius ϱ_0 since they are an interesting model for galaxies which will be discussed in detail in the following. The Ernst potential simplifies in the equatorially symmetric case at the disk where the boundary data are prescribed. We recall that the first solution of such a boundary value problem was found for the rigidly rotating dust disk [109].

In the equatorial plane ($\zeta = 0$), the surface $\mathcal{L}$ is then given by

$$\mu^2(K) = (K^2 + \varrho^2) \prod_{i=1}^{s} (K^2 - E_i^2)(K^2 - \bar{E}_i^2) \ .$$

We cut the surface as before which implies

$$T a_i^1 = a_1^2 \ , \quad T b_i^1 = b_i^2 \ ,$$

and

$$d\omega_i^1(TP) = -d\omega_i^2(P) \ ,$$

with $P \in \mathcal{L}$. The Riemann surface $\Sigma_1 \doteq \mathcal{L}/T$ of genus s is then given by (the fixed points of the involution leads again to branch points of the surface $\mathcal{L}/T$)

$$\mu_1^2(x) = x(x + \varrho^2) \prod_{i=1}^{s} (x - E_i^2)(x - \bar{E}_i^2) \ . \tag{4.20}$$

The holomorphic differentials dv_i in Σ_1 dual to (a_i, b_i) (the projection of the cuts on $\mathcal{L}$ onto Σ_1) follow from

$$dv_i = d\omega_i^1 - d\omega_i^2 \ .$$

The so called Prym differentials dw_i are given by

$$dw_i = d\omega_i^1 + d\omega_i^2 \ .$$

They are holomorphic differentials on the Riemann surface Σ_2 of genus s with

$$\mu_2^2(y) = (y + \varrho^2) \prod_{i=1}^{s} (y - E_i^2)(y - \bar{E}_i^2) \,, \tag{4.21}$$

which implies that the Prym variety is a Jacobi variety in this case. The Riemann matrix on $\mathcal{L}$ has the form

$$\mathbf{B} = \frac{1}{2} \begin{pmatrix} \mathbf{B}^1 + \mathbf{B}^2 & \mathbf{B}^2 - \mathbf{B}^1 \\ \mathbf{B}^2 - \mathbf{B}^1 & \mathbf{B}^1 + \mathbf{B}^2 \end{pmatrix} \,, \tag{4.22}$$

where the $\mathbf{B}^i$ are the Riemann matrices on Σ^i respectively. The theta function on $\mathcal{L}$ thus factorizes into products of theta functions on the Σ_i,

$$\Theta(x_1 | x_2, \mathbf{B}) = \sum_\delta \Theta_{\delta 0}(x_1 + x_2; 2\mathbf{B}^2) \Theta_{\delta 0}(x_1 - x_2; 2\mathbf{B}^1) \,, \tag{4.23}$$

where each component of the s-dimensional vector δ takes the values $0, \frac{1}{2}$. Thus the theta function on the surface of genus $2s$ can be expressed in terms of theta functions on surfaces of genus s.

In the case of the Ernst potential (4.19), further simplifications follow from the fact that ∞ is a branch point of Σ_2. For the contour integrals u, we obtain for disks

$$u_v = \frac{1}{\pi i} \int_{\Gamma_v} \ln G dv \,, \qquad u_w = \mathrm{sgn} \zeta \frac{1}{\pi i} \int_{\Gamma_w} \ln G dw \,, \tag{4.24}$$

where Γ_v is the contour in the $+$-sheet of Σ_1 between 0 and $-\varrho^2$ along the real axis, and Γ_w is the part of the real axis in the upper sheet of Σ_2 between $-\infty$ and $-\varrho^2$. The formula for u_w shows that it does matter whether the equatorial plane is approached from the upper or the lower side (the Ernst potential is not regular at the disk). For $\varrho > \varrho_0$, we have $u_w = 0$ ($G = 1$ in the exterior of the disk). Similarly we get for the integral in the exponent of (4.19)

$$I_v \doteq \frac{1}{2\pi i} \int_{\Gamma} \ln G d\omega_{\infty^+ \infty^-} = \frac{1}{2\pi i} \int_{\Gamma_v} \ln G dv_{\infty^+ \infty^-} \,. \tag{4.25}$$

Summing up we can write the Ernst potential in the equatorial plane in the form

$$\mathcal{E} = \frac{\sum_\delta \Theta_{\delta 0}(v(\infty^+) + u_v; 2\mathbf{B}^1) \Theta_{\delta \beta}(u_w; 2\mathbf{B}^2)}{\sum_\delta \Theta_{\delta 0}(v(\infty^-) + u_v; 2\mathbf{B}^1) \Theta_{\delta \beta}(u_w; 2\mathbf{B}^2)} e^{I_v} \,, \tag{4.26}$$

where $\beta_i = \frac{1}{2}$, and where

$$v(P) = \int_{-\varrho^2}^{P} dv \,.$$

The reality properties of the above theta functions imply together with (4.24) the condition for equatorial symmetry. Thus the imaginary part of $\mathcal{E}$ jumps at the disk. For $\varrho > \varrho_0$ (where $u_w = 0$), only the terms with even characteristics in (4.26) will survive which leads to a real Ernst potential. This implies that the Ernst potential is regular in the equatorial plane in the exterior of the disk as it should be. The formula (4.26) can also be used to determine asymptotic quantities as the ADM mass in the limit of $\varrho \to \infty$ as was done previously on the axis.

A similar reduction as in the equatorial plane is possible on the axis. There the Riemann surface Σ' also has the involution T which makes it possible to factorize the surface into the surfaces Σ'_1 and Σ_2 where the Σ_i are as above and where Σ'_1 is Σ_1 with the cut $\left[0, -\varrho^2\right]$ removed. Thus the theta function Θ' on the surface Σ' of genus $2s - 1$ can be expressed via theta functions on surfaces of genus $s - 1$ and s respectively. In the case $g = 2$, this will not lower the genus of the Riemann surfaces under consideration on the axis. We will not give the Ernst potential on the axis since it will not be used here (the formula is helpful if one wants to calculate the multipole moments on the axis for higher genus).

4.3 Solitonic Limit

In Chap. 3 we had briefly discussed the formation of solitons in theta functional solutions of the KP equation. In the limit of a degenerate surface $\mathcal{L}$ given by (1.14) $E_i \to F_i$ for $i = 1, \ldots, g$ certain periods diverge and the theta functions reduce to elementary functions. In the cut system of Fig. A.4, the diagonal elements of the matrix of b-periods diverge whereas the remaining periods remain finite. Since the elliptic Ernst equation is not a wave equation it cannot have solitonic solutions in the sense of stable wave packets. But it is interesting to study to which solutions the 'solitonic' limit of the theta-functional solutions leads.

The result depends of course on the characteristic in (4.19). If the characteristic of the theta functions consists only of blocks of the first two possibilities in (4.18), the theta functions obviously tend to one in the solitonic limit. The resulting solutions are real and belong to the static Weyl class.

If there are blocks of the third form in (4.18), the limit will correspond to the multi-black hole solutions (see e.g. [35], [200,201] and [202]). We will discuss here the simplest non-trivial example that will lead to Kerr solution. One has

Proposition 4.12. *Let the characteristic of the theta functions in (3.77) be given by*

$$\left[\begin{matrix} \frac{1}{2} & \frac{1}{2} \\ -\frac{1}{4} + \frac{\ln a_1}{2\pi i} & \frac{1}{4} + \frac{\ln a_2}{2\pi i} \end{matrix} \right] , \tag{4.27}$$

where $a_1 = 1/a_2 = -\cot\frac{\varphi}{2}$ with $0 < \varphi < \pi/2$. Then in the limit $E_1 \to F_1 = -\alpha$, $E_2 \to F_2 = \alpha$ with $\alpha = m\cos\varphi$, the solution is identical to the Kerr solution and the Ernst potential is given by (1.8).

Proof. In the limit $E_i \rightarrow F_i$, $i = 1, 2$, all functions entering (3.77) are given on the Riemann surface $\mathcal{L}_0$. This implies that they can be expressed in terms of elementary functions. The holomorphic differentials $d\omega_i$ become the differentials of the third kind $d\omega_{E_i^- E_i^+}$. Since the b-periods $\mathbf{B}_{11}$ and $\mathbf{B}_{22}$ diverge, we get for the Ernst potential

$$\mathcal{E} = \frac{1 + a_1 a_2 e^{\omega_1(\infty^+) + \omega_2(\infty^+)} - ie^{-P}(a_1 e^{\omega_1(\infty^+)} - a_2 e^{\omega_2(\infty^+)})}{e^{\omega_1(\infty^+) + \omega_2(\infty^+)} + a_1 a_2 - ie^{-P}(a_1 e^{\omega_2(\infty^+)} - a_2 e^{\omega_1(\infty^+)})} ; \quad (4.28)$$

here

$$P = -\frac{1}{2}\mathbf{B}_{12} = \int_{\bar{\xi}}^{K_1} d\omega_2 = \ln \frac{(E_1 - E_2 + r_1 - r_2)(\bar{\xi} - E_2 + r_2)}{(E_1 - E_2 + r_1 + r_2)(\bar{\xi} - E_2 - r_2)}, \quad (4.29)$$

$r_i = \mu_0(E_i)$, and

$$\omega_i(\infty^+) = \ln \frac{E_i - \zeta + r_i}{-i\varrho}. \quad (4.30)$$

Writing the Ernst potential in the form

$$\mathcal{E} = \frac{\mathcal{G} - 1}{\mathcal{G} + 1}, \quad (4.31)$$

we get

$$\mathcal{G} = \frac{1 + a_1 a_2}{1 - a_1 a_2 - i(a_1 + a_2)} \frac{r_1 - r_2}{E_1 - E_2} + \frac{i(a_1 - a_2)}{1 - a_1 a_2 - i(a_1 + a_2)} \frac{r_1 + r_2}{E_1 - E_2}. \quad (4.32)$$

which is with the definition of the a_i the Kerr solution in the form (1.8). $\qquad \square$

In the used Weyl coordinates, the horizon is located on the axis between $\pm m \cos \varphi$. In the same way as above one gets for the metric functions af in (3.98) and e^{2k} in (3.102)

$$Z = -2\alpha \frac{(a_1 - a_2)(1 - X^2)(1 + Y^2) - (1 - Y^2)(a_1(1 + X)^2 - a_2(1 - X)^2)}{4Y^2 + (a_1 + a_2 + (a_1 - a_2)X)^2}, \quad (4.33)$$

and

$$e^{2k} = \frac{\sin^2 \varphi Y^2 + \cos^2 \varphi X^2 - 1}{(X^2 - Y^2)\cos^2 \varphi}, \quad (4.34)$$

where

$$X = \frac{r_1 + r_2}{E_2 - E_1}, \quad Y = \frac{r_1 - r_2}{E_2 - E_1}. \quad (4.35)$$

To compare with similar plots for the disk solutions to be discussed in the following chapters, we will now present plots of the metric functions for the Kerr solution for the case $m = 1$ and $\varphi = \frac{\pi}{4}$. The real part of the Ernst potential is shown in Fig. 4.2. It can be recognized that the function

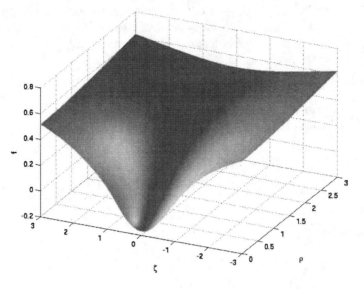

Fig. 4.2. Real part of the Ernst potential for the Kerr solution with $m = 1$ and $\varphi = \frac{\pi}{4}$

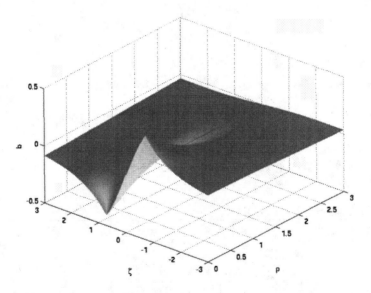

Fig. 4.3. Imaginary part of the Ernst potential for the Kerr solution with $m = 1$ and $\varphi = \frac{\pi}{4}$

f changes the sign at some contour, the ergosphere, which touches the axis at the horizon. This surface marks the limiting region where there can be no stationary observer with respect to infinity. The dragging due to the rotating black hole is too strong in this region. In locally corotating coordinates, the

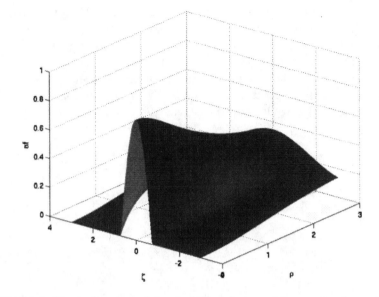

Fig. 4.4. Metric function af for the Kerr solution with $m = 1$ and $\varphi = \frac{\pi}{4}$

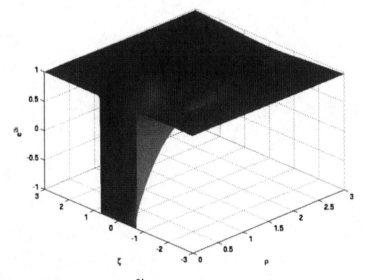

Fig. 4.5. Metric function e^{2k} for the Kerr solution with $m = 1$ and $\varphi = \frac{\pi}{4}$

metric function g'_{tt} is positive in the ergoregion up to the horizon where it vanishes. Asymptotically the function f tends to one, the value of the Minkowski spacetime. The function is obviously globally smooth and symmetric in ζ.

The imaginary part of the Ernst potential, see Fig. 4.3, is an odd function in ζ. It vanishes asymptotically and in the equatorial plane.

The metric function af as shown in Fig. 4.4 is equatorially symmetric. It vanishes asymptotically and on the regular part of the axis.

In contrast to the functions shown above, the metric function e^{2k} is not a Killing invariant. Nonetheless we will show it in Fig. 4.5 since it is related to the τ-function, an important concept in the theory of integrable systems, see [203] in this context. It is an equatorially symmetric function which is equal to one on the regular part of the axis and at infinity.

5 Boundary Value Problems and Solutions

In the previous chapters we have used the complete integrability of the stationary axisymmetric Einstein equations in vacuum to construct and to study rich classes of solutions which could describe the exterior gravitational field of stars and galaxies in thermodynamical equilibrium. In the present chapter we will use these methods to actually solve boundary value problems which are motivated by astrophysical models, in particular so-called dust disks.

Relativistic dust disks have been studied since the late sixties [204], the reasons for the interest in these configurations being both physical and mathematical. The physical motivation arises from the importance of disk-shaped matter distributions in certain galaxies and accretion disks. Whereas general relativistic effects do not play a role in the context of galaxies, they have to be taken into account in the case of disks around black-holes since black-holes are genuinely relativistic objects. Moreover disks can be considered as limiting configurations of fluid bodies for vanishing pressure (see e.g. [205]). From a more mathematical point of view, dust disks offer the opportunity to obtain global spacetimes containing matter distributions which can be physically interpreted. The Einstein equations for an ideal fluid do not seem to be integrable even in the stationary axisymmetric case. Infinitesimally thin disks provide a possibility to circumvent this problem because the matter is reduced to two spatial dimensions. This leads to ordinary differential equations inside the disk which can be integrated at least in principle. Consequently one has to solve a boundary value problem for the vacuum equations where the boundary data follow from the properties of the matter in the disk. Since dust disks have no radial pressures one can place the disks without loss of generality in the equatorial plane even in standard Weyl coordinates. Thus one avoids the complications of a free boundary value problem where the location of the disk has to be determined as part of the solution of the boundary value problem. The first solutions for relativistic dust disks were given by Morgan and Morgan [204]. They considered static spacetimes with disks which can be interpreted as being made up of two counter-rotating dust streams with vanishing total angular momentum. Bardeen and Wagoner [205] studied numerically and as a post-Newtonian expansion a uniformly rotating disk consisting of a single dust component. They compared this stationary solution to the Einstein equations to the static and the Newtonian case and gave

a detailed discussion of the physical features of the spacetime. Later Neugebauer and Meinel [109] gave an explicit solution for the Bardeen–Wagoner
disk in terms of Korotkin's solutions [52,94] on a Riemann surface of genus
2.

Since Newtonian dust disks are known to be unstable against fragmen-
tation and since numerical investigations (see e.g. [205]) indicate that the
same holds in the relativistic case, such solutions could be taken as exact
initial data for numerical collapse calculations: due to the inevitable numerical error such an unstable object will collapse if used as initial data. We will
investigate disks with counter-rotating dust streams which are discussed as
models for certain $S0$ and Sa galaxies (see [114] and references given therein
and [117,206]). These galaxies show counter-rotating matter components and
are believed to be the consequence of the merger of galaxies. Recent investigations have shown that there is a large number of galaxies (see [114,115],
the first was NGC 4550 in Virgo [116]) which show counter-rotating streams
in the disk with up to 50 % counter-rotation.

In the Newtonian case, dust disks can be treated in full generality (see
e.g. [106]) since the disks lead to boundary value problems for the Laplace
equations which can be solved explicitly. The unifying framework for both
the Laplace and the Ernst equation is provided by the previously discussed
Riemann–Hilbert problems: by solving the scalar problem for the Laplace
equation we obtain the Poisson integral for distributional densities. In this
sense the solutions to the Ernst equation constructed via Riemann–Hilbert
techniques can be viewed as a generalization of the Poisson integral to the
relativistic case.

Whereas the Poisson integral contains one free function which is sufficient
to solve boundary value problems for the scalar Newtonian potential, the
finite gap solutions contain one free function and a set of complex parameters,
the branch points of the Riemann surface. Thus one cannot hope to solve
general boundary value problems for the complex Ernst potential within this
class because this would imply the possibility to specify two free functions in
the solution according to the boundary data. This means that one can only
solve certain classes of boundary value problems on a given compact Riemann
surface. In this chapter we investigate the implications of the underlying
Riemann surface on the multipole moments and the boundary values taken
at a given boundary. The relations will be given for general genus of the
surface and will be discussed in detail in the case of genus 1 (elliptic surface)
and genus 2, which is the simplest case with generic equatorial symmetry. It is
shown that the solution of boundary value problems leads in general to non-
linear integral equations. However we can identify classes of boundary data
where only one linear integral equation has to be solved. Special attention will
be paid to counter-rotating dust disks which will lead us to the construction
of the solution for constant angular velocity and constant relative density
which was presented in [130]. It contains as limiting cases the static solutions

of Morgan and Morgan [204] and the disk with only one matter stream by Neugebauer and Meinel [111].

The chapter is organized as follows. In Sect. 5.1 we discuss Newtonian dust disks with Riemann–Hilbert methods and relate the corresponding boundary value problems to an Abelian integral equation. The boundary conditions for counter-rotating dust disks are summarized in Sect. 5.2. In Sect. 5.3, we establish relations for the corresponding Ernst potentials on the axis on a given Riemann surface of arbitrary genus. The found relation limits the possible choice of the multipole moments. We discuss in detail the elliptic and the genus 2 case with equatorial symmetry. This analysis is extended to the whole spacetime in Sect. 5.4 which leads to a set of differential and algebraic equations which is again discussed in detail for genus 1 and 2. The equations for genus 2 are used to study differentially counter-rotating dust disks in Sect. 5.5: First we discuss the Newtonian limit of disks of genus 2. Then we derive the class of counter-rotating dust disks with constant angular velocity and constant relative density of [113, 130] as an application of this constructive approach. We prove the regularity of the solution up to the ultrarelativistic limit in the whole spacetime except the disk. The found solution is thus a global solution to the boundary value problem.

5.1 Newtonian Dust Disks

To illustrate the basic concepts used in the following sections, we will briefly recall some facts on Newtonian dust disks. In Newtonian theory, gravitation is described by a scalar potential U which is a solution to the Laplace equation in the vacuum region. We place the disk made up of a pressureless two-dimensional perfect fluid with radius ϱ_0 in the equatorial plane $\zeta = 0$. In Newtonian theory stationary perfect fluid solutions and thus also the here considered disks are known to be equatorially symmetric.

Since we concentrate on dust disks, the only force to compensate gravitational attraction in the disk is the centrifugal force. This leads in the disk to

$$U_{,\varrho} = \Omega^2(\varrho)\varrho \,, \tag{5.1}$$

where $\Omega(\varrho)$ is the angular velocity of the dust at radius ϱ. Since all terms in (5.1) are quadratic in Ω there are no effects due to the sign of the angular velocity. The absence of these so-called gravitomagnetic effects in Newtonian theory implies that disks with counter-rotating components will behave with respect to gravity exactly as disks which are made up of only one component. We will therefore merely consider the case of one component in this section. Integrating (5.1) we get the boundary data $U(\varrho, 0)$ with an integration constant $U_0 = U(0, 0)$ which is related to the central redshift in the relativistic case.

To find the Newtonian solution for a given rotation law $\Omega(\varrho)$, we thus have to construct a solution to the Laplace equation which is everywhere regular except at the disk where it has to take on the boundary data (5.1). At the disk the normal derivatives of the potential will have a jump since the disk is a surface layer. Notice that one has to solve only the vacuum equations since the two-dimensional matter distribution merely leads to boundary conditions for the Laplace equation. In the Newtonian setting one thus has to determine the density for a given rotation law or vice versa, a well known problem (see e.g. [106] and references therein) for Newtonian dust disks.

The method we outline here has the advantage that it can be generalized to some extent to the relativistic case. We put $\varrho_0 = 1$ without loss of generality (we are only considering disks of finite non-zero radius) and obtain U as the solution of the Riemann–Hilbert problem in Chap. 4,

Theorem 5.1. *Let* $\ln G \in C^{1,\alpha}(\Gamma)$ *and* Γ *be the covering of the imaginary axis in the upper sheet of* $\mathcal{L}_0$ *between* $-i$ *and* i. *The function* G *has to be subject to the conditions* $G(\bar{\tau}) = \bar{G}(\tau)$ *and* $G(-\tau) = G(\tau)$. *Then*

$$U(\varrho, \zeta) = -\frac{1}{4\pi i} \int_\Gamma \frac{\ln G(\tau) \mathrm{d}\tau}{\sqrt{(\tau - \zeta)^2 + \varrho^2}} \tag{5.2}$$

is a real, equatorially symmetric solution to the Laplace equation which is everywhere regular except at the disk $\zeta = 0$, $\varrho \le 1$. *The function* $\ln G$ *is determined by the boundary data* $U(\varrho, 0)$ *or the energy density* σ *of the dust* $(2\pi\sigma = U_\zeta$ *in units where the velocity of light and the Newtonian gravitational constant are equal to 1) via*

$$\ln G(t) = 4\left(U_0 + t \int_0^t \frac{U_{,\varrho}(\varrho)\mathrm{d}\varrho}{\sqrt{t^2 - \varrho^2}}\right) \tag{5.3}$$

or

$$\ln G(t) = 4 \int_t^1 \frac{\varrho U_{,\zeta}}{\sqrt{\varrho^2 - t^2}} \mathrm{d}\varrho \tag{5.4}$$

respectively where $t = -i\tau$.

We briefly outline the

Proof. It may be checked by direct calculation that U in (5.2) is a solution to the Laplace equation except at the disk. The reality condition on G leads to a real potential, whereas the symmetry condition with respect to the involution $\tau \to -\tau$ leads to equatorial symmetry. At the disk the potential takes due to the equatorial symmetry the boundary values

$$U(\varrho, 0) = -\frac{1}{2\pi} \int_0^\varrho \frac{\ln G(t)}{\sqrt{\varrho^2 - t^2}} \mathrm{d}t \tag{5.5}$$

and

$$U_{,\zeta}(\varrho, 0) = -\frac{1}{2\pi} \int_\varrho^1 \frac{\partial_t(\ln G(t))}{\sqrt{t^2 - \varrho^2}} \mathrm{d}t . \tag{5.6}$$

Both equations constitute integral equations for the jump data $\ln G$ of the Riemann–Hilbert problem if the respective left-hand side is known. The equations (5.5) and (5.6) are both Abelian integral equations and can be solved in terms of quadratures, i.e. (5.3) and (5.4). To show the regularity of the potential U we prove that the integral (5.2) is identical to the Poisson integral for a distributional density which reads at the disk

$$U(\varrho) = -2 \int_0^1 \sigma(\varrho')\varrho'd\varrho' \int_0^{2\pi} \frac{d\phi}{\sqrt{(\varrho + \varrho')^2 - 4\varrho\varrho' \cos\phi}}$$

$$= -4 \int_0^1 \sigma(\varrho')\varrho'd\varrho' \frac{K(k(\varrho, \varrho'))}{\varrho + \varrho'} , \qquad (5.7)$$

where $k(\varrho, \varrho') = 2\sqrt{\varrho\varrho'}/(\varrho + \varrho')$ and where K is the complete elliptic integral of the first kind (see the appendix and references given therein). Eliminating $\ln G$ in (5.5) via (5.4) we obtain after interchange of the order of integration

$$U = -\frac{2}{\pi} \left(\int_0^\varrho U_{,\zeta} \frac{\varrho'}{\varrho} K\left(\frac{\varrho'}{\varrho}\right) d\varrho' + \int_\varrho^1 U_{,\zeta} K\left(\frac{\varrho}{\varrho'}\right) d\varrho' \right) \qquad (5.8)$$

which is identical to (5.7) since $K(2\sqrt{k}/(1 + k)) = (1 + k)K(k)$. Thus the integral (5.2) has the properties known from the Poisson integral: it is a solution to the Laplace equation which is everywhere regular except at the disk where the normal derivatives are discontinuous.

$\square$

Remark 5.2. We note that it is possible in the Newtonian case to solve the boundary value problem purely locally at the disk. The regularity properties of the Poisson integral then ensure global regularity of the solution except at the disk. Such a purely local treatment will not be possible in the relativistic case.

The above considerations make it clear that one cannot prescribe both U at the disk (and thus the rotation law) and the density independently. This just reflects the fact that the Laplace equation is an elliptic equation for which Cauchy problems are ill-posed. If $\ln G$ is determined by either (5.3) or (5.4) for given rotation law or density, expression (5.2) gives the analytic continuation of the boundary data to the whole spacetime. In case we prescribe the angular velocity, the constant U_0 is determined by the condition $\ln G(i) = 0$ which excludes a ring singularity at the rim of the disk. For rigid rotation ($\Omega = const$), we get e.g.

$$\ln G(\tau) = 4\Omega^2(\tau^2 + 1) \qquad (5.9)$$

which leads with (5.2) to the well-known Maclaurin disk, the disk limit of the Maclaurin ellipsoids.

5.2 Boundary Conditions for Counter-rotating Dust Disks

In the stationary relativistic case the Euler–Darboux equation of Newtonian theory is replaced by the Ernst equation. In static situations the Ernst potential is real and belongs to the Weyl class, the Ernst equation reduces again to the Euler–Darboux equation. Hence static disks like the counter-rotating disks of Morgan and Morgan [204] can be treated in the same way as the Newtonian disks in the previous section.

In this section we want to study the boundary values at a stationary infinitesimally thin disk made up of two components of pressureless matter which are counter-rotating. These models are simple enough that explicit solutions can be constructed, and they show typical features of general boundary value problems one might consider in the context of the Ernst equation. It is also possible to study explicitly the transition from a stationary to a static spacetime with a matter source of finite extension for these models. Counter-rotating disks of infinite extension but finite mass were treated in [206–208], disks producing the Kerr metric in [117].

To obtain the boundary conditions at a relativistic dust disk, it seems best to use Israel's invariant junction conditions for matching spacetimes across non-null hypersurfaces [209]. Again we place the disk in the equatorial plane and match the regions $V^{\pm}$ ($\pm\zeta > 0$) at the equatorial plane. This is possible with the used coordinates since we are only considering dust i.e. vanishing radial stresses in the disk. The jump $\gamma_{\alpha\beta} = K^+_{\alpha\beta} - K^-_{\alpha\beta}$ in the extrinsic curvature $K_{\alpha\beta}$ of the hypersurface $\zeta = 0$ with respect to its embeddings into $V^{\pm} = \{\pm\zeta > 0\}$ is due to the energy-momentum tensor $S_{\alpha\beta}$ of the disk via

$$-8\pi S_{\alpha\beta} = \gamma_{\alpha\beta} - h_{\alpha\beta}\gamma^\varepsilon_\varepsilon \,, \tag{5.10}$$

where h is the metric on the hypersurface (greek indices take here the values 0, 1, 3 corresponding to the coordinates t, ϱ, ϕ). As a consequence of the field equations the energy-momentum tensor is divergence free, $S^{\alpha\beta}_{;\beta} = 0$ where the semicolon denotes the covariant derivative with respect to h.

The energy-momentum tensor of the disk is written in the form

$$S^{\mu\nu} = \sigma_+ u^\mu_+ u^\nu_+ + \sigma_- u^\mu_- u^\nu_- \,, \tag{5.11}$$

where the vectors $u^\alpha_\pm$ are a linear combination of the Killing vectors, $(u^\alpha_\pm) = (1, 0, \pm\Omega(\varrho))$. This has to be considered as an algebraic definition of the tensor components. Since the vectors $u_\pm$ are not normalized, the quantities $\sigma_\pm$ have no direct physical significance, they are just used to parametrize $S^{\mu\nu}$. The energy-momentum tensor was chosen in a way to interpolate continuously between the static case and the one-component case with constant angular velocity. An energy-momentum tensor $S^{\mu\nu}$ with three independent components can always be written as

$$S^{\mu\nu} = \sigma_p^* v^\mu v^\nu + p_p^* w^\mu w^\nu , \tag{5.12}$$

where v and w are the unit timelike respectively spacelike vectors $(v^\mu) = N_1(1, 0, \omega_\phi)$ and $(w^\mu) = N_2(\kappa, 0, 1)$. This corresponds to the introduction of observers (called ϕ-isotropic observers (FIOs) in [117]) for which the energy-momentum tensor is diagonal. The condition $w_\mu v^\mu = 0$ determines κ in terms of ω_ϕ and the metric,

$$\kappa = -\frac{g_{03} + \omega_\phi g_{33}}{g_{00} + \omega_\phi g_{03}} . \tag{5.13}$$

If $p_p^*/\sigma_p^* < 1$ the matter in the disk can be interpreted as in [204] either as having a purely azimuthal pressure or as being made up of two counter-rotating streams of pressureless matter with proper surface energy density $\sigma_p^*/2$ which are counter-rotating with the same angular velocity $\sqrt{p_p^*/\sigma_p^*}$,

$$S^{\mu\nu} = \frac{1}{2}\sigma^*(U_+^\mu U_+^\nu + U_-^\mu U_-^\nu) , \tag{5.14}$$

where $(U_\pm^\mu) = U^*(v^\mu \pm \sqrt{p_p^*/\sigma_p^*}w^\mu)$ is a unit timelike vector. We will always adopt the latter interpretation if the condition $p_p^*/\sigma_p^* < 1$ is satisfied which is the case in the example we will discuss in more detail in Sect. 5.5. The energy-momentum tensor (5.14) is just the sum of two energy-momentum tensors for dust. Furthermore it can be shown that the vectors $U_\pm$ are geodesic vectors with respect to the inner geometry of the disk: this is a consequence of the equation $S_{;\nu}^{\mu\nu} = 0$ together with the fact that $U_\pm$ is a linear combination of the Killing vectors. In the discussion of the physical properties of the disk we will refer only to the measurable quantities ω_ϕ, σ_p^* and p_p^* which are obtained by the introduction of the FIOs whereas $\sigma_\pm$ and Ω are just used to generate a sufficiently general energy-momentum tensor.

Equation $S_{;\beta}^{\alpha\beta} = 0$ leads to the condition

$$U_{,\varrho}\left(1 + 2\gamma\Omega a + \Omega^2 a^2\right) + \Omega a_{,\varrho}(\gamma + \Omega a) + \Omega^2 \varrho(\varrho U_{,\varrho} - 1)e^{-4U} = 0 , \tag{5.15}$$

where

$$\gamma(\varrho) = \frac{\sigma_+(\varrho) - \sigma_-(\varrho)}{\sigma_+(\varrho) + \sigma_-(\varrho)} . \tag{5.16}$$

The function $\gamma(\varrho)$ is a measure for the relative energy density of the counter-rotating matter streams. For $\gamma \equiv 1$, there is only one component of matter, for $\gamma \equiv 0$, the matter streams have identical density which leads to a static spacetime of the Morgan and Morgan class.

As in the Newtonian case, one cannot prescribe both the proper energy densities $\sigma_\pm$ and the rotation law Ω at the disk since the Ernst equation is an elliptic equation. For the matter model (5.11), we get at the disk

Theorem 5.3. Let $\tilde{\sigma}(\varrho) = \sigma_+(\varrho) + \sigma_-(\varrho)$ and let $R(\varrho)$ and $\delta(\varrho)$ be given by

$$R = \left(a + \frac{\gamma}{\Omega}\right) e^{2U} , \tag{5.17}$$

and

$$\delta(\varrho) = \frac{1 - \gamma^2(\varrho)}{\Omega^2(\varrho)} . \tag{5.18}$$

Then for prescribed $\Omega(\varrho)$ and $\delta(\varrho)$, the boundary data at the disk take the form

$$\mathcal{E}_{,\zeta} = -\mathrm{i}\frac{R^2 + \varrho^2 + \delta e^{4U}}{2R\varrho}\mathcal{E}_{,\varrho} + \frac{\mathrm{i}}{R}e^{2U} . \tag{5.19}$$

Let σ be given by $\sigma = \tilde{\sigma}e^{k-U}$. Then for given density σ and γ, the boundary data read,

$$\left(\varrho^2 + \delta e^{4U}\right)\left(\left(e^{2U}\right)_{,\varrho}\left(e^{2U}\right)_{,\zeta} + b_{,\varrho}b_{,\zeta}\right)^2$$
$$-2\varrho e^{2U}\left(e^{2U}\right)_{,\zeta}\left(\left(e^{2U}\right)_{,\varrho}\left(e^{2U}\right)_{,\zeta} + b_{,\varrho}b_{,\zeta}\right) + b_{,\varrho}^2 e^{4U} = 0 , \tag{5.20}$$

and

$$\left(b_{,\varrho} - a\left(\left(e^{2U}\right)_{,\varrho}\left(e^{2U}\right)_{,\zeta} + b_{,\varrho}b_{,\zeta}\right)\right)^2$$
$$+8\pi\varrho\sigma e^{2U}\gamma^2\left(\left(e^{2U}\right)_{,\varrho}\left(e^{2U}\right)_{,\zeta} + b_{,\varrho}b_{,\zeta}\right) = 0 . \tag{5.21}$$

Proof. The relations (5.10) lead to

$$-4\pi e^{(k-U)}S_{00} = (k_{,\zeta} - 2U_{,\zeta}) e^{2U} ,$$
$$-4\pi e^{(k-U)}(S_{03} - aS_{00}) = -\frac{1}{2}a_{,\zeta}e^{2U} ,$$
$$-4\pi e^{(k-U)}(S_{33} - 2aS_{03} + a^2 S_{00}) = -k_{,\zeta}\varrho^2 e^{-2U} , \tag{5.22}$$

where

$$S_{00} = \tilde{\sigma}e^{4U}\left(1 + \Omega^2 a^2 + 2\Omega a\gamma\right) ,$$
$$S_{03} - aS_{00} = -\tilde{\sigma}\varrho^2\Omega\left(\Omega a + \gamma\right) ,$$
$$S_{33} - 2aS_{03} + a^2 S_{00} = \tilde{\sigma}\Omega^2\varrho^4 e^{-4U} . \tag{5.23}$$

One can substitute one of the above equations by (5.15) in the same way as one replaces one of the field equations by the covariant conservation of the energy-momentum tensor in the case of three-dimensional perfect fluids. This makes it possible to eliminate $k_{,\zeta}$ from (5.22) and to treat the boundary value problem purely on the level of the Ernst equation. The function k will then be determined via (2.42) with the found solution of the Ernst equation. It is straight forward to check the consistency of this approach with the help of (2.42).

If Ω and γ (and thus δ) are given, one has to eliminate $\tilde{\sigma}$ from (5.22) and (5.23). This can be combined with (5.15) and (2.45) to (5.19).

If the functions γ and σ are prescribed (this makes it possible to treat the problem completely on the level of the Ernst equation), one has to eliminate Ω from (5.15), (5.22) and (5.23) which leads to (5.20) and (5.21).

□

Remark 5.4. For given $\Omega(\varrho)$ and $\delta(\varrho)$, equation (5.15) is an ordinary nonlinear differential equation for e^{2U},

$$(R^2 - \varrho^2)_{,\varrho} e^{2U} - 2Re^{4U} \left(\frac{\gamma}{\Omega}\right)_{,\varrho} = (R^2 - \varrho^2 - \delta e^{4U})\left(e^{2U}\right)_{,\varrho} . \tag{5.24}$$

For constant Ω and γ we get

$$R^2 - \varrho^2 + \delta e^{4U} = \frac{2}{\lambda}e^{2U} , \tag{5.25}$$

where

$$\lambda = 2\Omega^2 e^{-2U_0} . \tag{5.26}$$

For given boundary values as in Theorem 5.3, the task is to to find a solution to the Ernst equation which is regular in the whole spacetime except at the disk where it has to satisfy two real boundary conditions. In the following we will concentrate on the case where the angular velocity Ω and the relative density γ are prescribed.

Remark 5.5. To solve boundary value problems with the class of solutions (4.19), one has two kinds of freedom: the function G as before and the branch points E_i of the Riemann surface as a discrete degree of freedom. Since one would need to specify two free functions to solve a general boundary value problem for the Ernst equation, it is obvious that one can only solve a restricted class of problems on a given surface, and that one cannot expect to solve general problems on a surface of finite genus. But once one has constructed a solution which takes the imposed boundary data at the disk, one has to check the condition $\Theta(\omega(\infty^-) + u) \neq 0$ in the whole spacetime to actually prove that one has found the desired solution: a solution that is everywhere regular except at the disk where it has to take the imposed boundary conditions.

There are in principle two ways of generalizing the approach used for the Newtonian case: One can eliminate Ω from the two real equations (5.19) and enter the resulting equation with a solution (4.19) on a chosen Riemann surface. This will lead for given γ to a non-linear integral equation for $\ln G$. In general there is little hope to get explicit solutions to this equation (for a numerical treatment of differentially rotating disks along this line in the genus 2 case see [210]). Once a function G is found, one can read off the rotation law Ω on a given Riemann surface from (4.19). Another approach is to establish the relations between the real and the imaginary part of the Ernst potential which exist on a given Riemann surface for arbitrary G. The simplest example for such a relation is provided by the function $w = e^{i\psi}$ which is a function on a Riemann surface of genus 0, where we have obviously $|w| = 1$. As we will point out in the following, similar relations also exist for an Ernst potential of the form (4.19), but they will lead to a system of differential equations. Once one has established these relations for a given Riemann surface, one can determine in principle which boundary value problems can be solved there (in our example which classes of functions Ω, γ can occur) by the condition that one of the boundary conditions must be identically satisfied. The second equation will then be used to determine G as the solution of an integral equation which is possibly non-linear. Following the second approach, we want to study the implications of the hyperelliptic Riemann surface for the physical properties of the solutions.

5.3 Axis Relations

In order to establish relations between the real and the imaginary part of the Ernst potential, we will first consider the axis of symmetry ($\varrho = 0$) where the situation simplifies decisively. In addition the axis is of interest since the asymptotically defined multipole moments [211, 212] can be read off there as in [213].

In Chap. 4 we have shown that the Ernst potential can be expressed on the axis in terms of functions defined on the Riemann surface Σ' given by $\mu'^2 = \prod_{i=1}^{g} (K - E_i)(K - \bar{E}_i)$, i.e. the Riemann surface obtained from $\mathcal{L}$ by removing the cut $[\xi, \bar{\xi}]$ which just collapses on the axis. We use the notation of Sect. 4.1.4 and let a prime denote that the corresponding quantity is defined on the surface Σ'. We get with (4.8)

$$\mathcal{E}(0, \zeta) = \frac{\vartheta\left(\int_{\zeta+}^{\infty^+} d\omega' + u'\right) - e^{-(\omega_g'(\infty^+) + u_g)}\vartheta\left(\int_{\zeta-}^{\infty^+} d\omega' + u'\right)}{\vartheta\left(\int_{\zeta+}^{\infty^+} d\omega' - u'\right) - e^{-(\omega_g'(\infty^+) - u_g)}\vartheta\left(\int_{\zeta-}^{\infty^+} d\omega' - u'\right)} e^{I + u_g} ,$$

$$(5.27)$$

where ϑ is the theta function on Σ' with the characteristic $p_i' = 0$, $q_i' = 1$ for $i = 1, \ldots, g - 1$.

Notice that the u_i' and I are constant with respect to ζ. The only term dependent both on G and on ζ is u_g. To establish a relation on the axis between the real and the imaginary part of the Ernst potential independent of G, the first step must be thus to eliminate u_g. We can state the following

Theorem 5.6. *The Ernst potential (4.8) satisfies for $g > 1$ the relation*

$$\mathcal{P}_1(\zeta)\mathcal{E}\bar{\mathcal{E}} + \mathcal{P}_2(\zeta)b + \mathcal{P}_3(\zeta) = 0 ,$$

$$(5.28)$$

where the $\mathcal{P}_i$ are real polynomials in ζ with coefficients depending on the branch points E_i and the g real constants $\int_\Gamma \ln G \tau^i \mathrm{d}\tau/\mu'(\tau)$ with $i = 0, \ldots, g-1$. The degree of the polynomials $\mathcal{P}_1$ and $\mathcal{P}_3$ is $2g-3$ or less, the degree of $\mathcal{P}_2$ is $2g-2$ or less.

To prove this theorem we need the fact that one can express integrals of the third kind via theta functions with odd characteristic denoted by $\vartheta_\star$, see (3.83),

$$\exp(-\omega_g(\infty^+)) = -\frac{\vartheta_\star(\boldsymbol{\omega}'(\infty^+) - \boldsymbol{\omega}'(\zeta^+))}{\vartheta_\star(\boldsymbol{\omega}'(\infty^+) - \boldsymbol{\omega}'(\zeta^-))} . \tag{5.29}$$

Proof. The first step is to establish the relation

$$A\mathcal{E}\bar{\mathcal{E}} + Bib + 1 = 0 , \tag{5.30}$$

where

$$Ae^{2I} = -\frac{\vartheta\left(\boldsymbol{u}' + \int_{\zeta^-}^{\infty^-} \mathrm{d}\boldsymbol{\omega}'\right) \vartheta\left(\boldsymbol{u}' + \int_{\zeta^+}^{\infty^-} \mathrm{d}\boldsymbol{\omega}'\right)}{\vartheta\left(\boldsymbol{u}' + \int_{\zeta^-}^{\infty^+} \mathrm{d}\boldsymbol{\omega}'\right) \vartheta\left(\boldsymbol{u}' + \int_{\zeta^+}^{\infty^+} \mathrm{d}\boldsymbol{\omega}'\right)} \tag{5.31}$$

and

$$Be^I = e^{-\omega_g(\infty^+)}\frac{\vartheta\left(\boldsymbol{u}' + \int_{\zeta^+}^{\infty^-} \mathrm{d}\boldsymbol{\omega}'\right)}{\vartheta\left(\boldsymbol{u}' + \int_{\zeta^+}^{\infty^+} \mathrm{d}\boldsymbol{\omega}'\right)} + e^{\omega_g(\infty^+)}\frac{\vartheta\left(\boldsymbol{u}' + \int_{\zeta^-}^{\infty^-} \mathrm{d}\boldsymbol{\omega}'\right)}{\vartheta\left(\boldsymbol{u}' + \int_{\zeta^-}^{\infty^+} \mathrm{d}\boldsymbol{\omega}'\right)} \tag{5.32}$$

which may be checked with (4.8) by direct calculation. The reality properties of the Riemann surface Σ' and the function G imply that A is real and that B is purely imaginary. We use the addition theorem (A.65) with $[m_1] = \ldots = [m_4]$ equal to the characteristic of ϑ in (5.31) to get

$$Ae^{2I} = -\frac{\displaystyle\sum_{2a\in(Z_2)^{2g}} \exp(-4\pi\mathrm{i}\langle m_1^1, a^2\rangle)\vartheta^2[a](\boldsymbol{u}' + \boldsymbol{\omega}'(\infty^-))\vartheta^2[a](\boldsymbol{\omega}'(\zeta^+))}{\displaystyle\sum_{2a\in(Z_2)^{2g}} \exp(-4\pi\mathrm{i}\langle m_1^1, a^2\rangle)\vartheta^2[a](\boldsymbol{u}' + \boldsymbol{\omega}'(\infty^+))\vartheta^2[a](\boldsymbol{\omega}'(\zeta^+))} . \tag{5.33}$$

This term is already in the desired form. Using the relation for root functions (A.84), one can directly see that the right-hand side is a quotient of polynomials of order $g-1$ or lower in ζ. For (5.32) we use (5.29) with $[\tilde{m}_1] = [\tilde{m}_2] = [\mathcal{K}]$ where $\mathcal{K}$ is the Riemann vector as the characteristic of the odd theta function $\vartheta_\star$ and let $[\tilde{m}_3] = [\tilde{m}_4]$ be equal to the characteristic of ϑ. The addition theorem A.65 then leads to

$$Be^I = -\frac{\displaystyle\sum_{2a\in(Z_2)^{2g}} \exp(-4\pi\mathrm{i}\langle \tilde{m}_1^1, a^2\rangle)\Theta'^2[n+a](\boldsymbol{u}' + \boldsymbol{\omega}'(\infty^-))\vartheta^2[a](\boldsymbol{\omega}'(\zeta^+))}{\displaystyle\sum_{2a\in(Z_2)^{2g}} \exp(-4\pi\mathrm{i}\langle m_1^1, a^2\rangle)\vartheta^2[a](\boldsymbol{u}' + \boldsymbol{\omega}'(\infty^+))\vartheta^2[a](\boldsymbol{\omega}'(\zeta^+))}$$

$$\times \sum_{2a\in(Z_2)^{2g}} \exp(-4\pi i\langle m_1^1, a^2\rangle)\frac{\vartheta^2[a](u')}{\vartheta^2(0)}$$

$$\times \left(\frac{\vartheta^2[a](\omega'(\infty^+)-\omega'(\zeta^+))}{\vartheta_*^2(\omega'(\infty^+)-\omega'(\zeta^+))} + \frac{\vartheta^2[a](\omega'(\infty^+)-\omega'(\zeta^-))}{\vartheta_*^2(\omega'(\infty^+)-\omega'(\zeta^-))}\right) , \quad (5.34)$$

where n follows from $\tilde{m}$ as in Theorem A.65. The first fraction in (5.34) is again the quotient of polynomials of degree $g-1$ in ζ for the same reasons as above. But since the quotient must vanish for $\zeta \to \infty$, the leading terms in the numerator just cancel. It is thus a quotient of polynomials of degree $g-2$ or less in the numerator and $g-1$ or less in the denominator. To deal with the quotients $\vartheta^2[a](\omega'(\infty^+)-\omega'(\zeta^\pm))/\vartheta_*^2(\omega'(\infty^+)-\omega'(\zeta^\pm))$, we define the divisors $\mathfrak{T}^\pm = T_1^\pm +\ldots+ T_{g-1}^\pm$ as the solutions of the Jacobi inversion problems $\omega'(\mathfrak{T}^\pm)-\omega'(\mathfrak{Y}) = \omega'(\infty^+)+\omega'(\zeta^\pm)$ where $\mathfrak{Y}$ is the divisor $\mathfrak{Y} = E_1 +\ldots+ E_{g-1}$. Abel's theorem then implies for arbitrary $K \in C$

$$\prod_{i=1}^{g-1}(K-T_i^\pm)(K-\zeta) = (K-A^\pm)^2 \prod_{i=1}^{g-1}(K-E_i)-(K-E_g)\prod_{i=1}^{g}(K-\bar{E}_i) , \quad (5.35)$$

where

$$\zeta - A^\pm = \pm\frac{\mu'(\zeta)}{\prod_{i=1}^{g-1}(\zeta - E_i)} . \quad (5.36)$$

Let Q_j be given by the condition $[Q_j + \mathcal{K}] = [a]$, i.e. Q_j is a branch point of Σ'. Then we get for the quotient

$$\frac{\vartheta^2[a](\omega'(\infty^+) - \omega'(\zeta^\pm))}{\vartheta_*^2(\omega'(\infty^+) - \omega'(\zeta^\pm))} = const \prod_{i=1}^{g-1} \frac{T_i^\pm - Q_j}{T_i^\pm - E_1} , \quad (5.37)$$

where *const* is independent with respect to ζ. With the help of (5.35), it is straight forward to see that for $Q_j \in \mathfrak{Y}$, the theta quotient is just proportional to $(\zeta - E_1)/(\zeta - Q_j)$ whereas for $Q_j \notin \mathfrak{Y}$, the term is proportional to $(\zeta - E_1)(Q_j - A^\pm)^2/(\zeta - Q_j)$. Using (5.36) one recognizes that the terms containing roots just cancel in (5.34). The remaining terms are just quotients of polynomials in ζ with maximal degree g in the numerator and $g-2$ in denominator.

$\square$

Remark 5.7. The remaining dependence on G via u' and I can only be eliminated by differentiating relation (5.27) g times with respect to ζ. If we prescribe e.g. the function b on a given Riemann surface (this just reflects the fact that the function G can be freely chosen in (5.27)), we can read off e^{2U} from (5.28). To fix the constants related to G in (5.28) one needs to know the Ernst potential and $g-1$ derivatives at some point on the axis where the Ernst potential is regular, e.g. at the origin or at infinity, where one has to prescribe the multipole moments. If the Ernst potential were known on some regular part of the axis, one could use (5.28) to read off the Riemann surface (genus and branch points). Equation (5.27) is then an integral equation for G for known sources. This just reflects a result of [214] that the Ernst potential for known sources can be constructed via Riemann–Hilbert techniques if it is known on some regular part of the axis.

In practice it is difficult to express the coefficients in the polynomials $\mathcal{P}_i$ via the constants u'_i and I, and it will be difficult to get explicit expressions. We will therefore concentrate on the general structure of the relation (5.28), its implications on the multipoles and some instructive examples. Let us first consider the case genus 1 which is not generically equatorially symmetric. In this case the Riemann surface Σ' is of genus 0. One can use formula (5.27) for the axis potential if one replaces the theta functions by 1. Putting $E_1 = \alpha_1 + i\beta_1$ we thus end up with

$$\mathcal{E}\bar{\mathcal{E}} - 2b\frac{\zeta - \alpha_1}{\beta_1}e^I = e^{2I} . \tag{5.38}$$

Here the only remaining G-dependence is in I. If $\mathcal{E}_0 = \mathcal{E}(0,0)$ is given, e^I follows from $\mathcal{E}_0\bar{\mathcal{E}}_0 + 2b_0\alpha_1 e^I/\beta_1 = e^{2I}$, if the in general non-real mass M is known, the constant e^I follows from $1 + 4\mathrm{Im}Me^I/\beta_1 = e^{2I}$. In the latter case the imaginary part of the ADM mass (this corresponds to a NUT parameter) will be sufficient. Differentiating (5.38) once will lead to a differential relation between the real and the imaginary part of the Ernst potential which holds for all G, which means it reflects only the impact of the underlying Riemann surface on the structure of the solution.

Remark 5.8. For equatorially symmetric solutions, one has on the positive axis the relation $\mathcal{E}(-\zeta)\bar{\mathcal{E}}(\zeta) = 1$ (see [215, 216]). This is to be understood in the following way: the function $|\zeta|$ is even in ζ, but restricted to positive ζ it seems to be an odd function, and it is exactly this behavior which is addressed by the above formula. This leads to the conditions

$$\mathcal{P}_1(-\zeta) = -\mathcal{P}_3(\zeta) , \quad \mathcal{P}_2(-\zeta) = P_2(\zeta) . \tag{5.39}$$

The coefficients in the polynomials depend on the $g/2$ integrals $\int_\Gamma \ln G\tau^{2i}\frac{d\tau}{\mu'(\tau)}$ $(i = 0, \ldots, g/2 - 1)$ and the branch points.

The simplest interesting example is genus 2, where we get with $E_1^2 = \alpha + i\beta$

$$\mathcal{E}\bar{\mathcal{E}}(\zeta - C_1) + \frac{\sqrt{2}}{C_2}(\zeta^2 - \alpha - C_2^2)b = \zeta + C_1 , \tag{5.40}$$

i.e. a relation which contains two real constants C_1, C_2 related to G. In case the Ernst potential at the origin is known, one can express these constants via $\mathcal{E}_0$. A relation of this type, which is as shown typical for the whole class of solutions, was observed in [109] for the rigidly rotating dust disk.

5.4 Differential Relations in the Whole Spacetime

The above considerations on the axis have shown that it is possible there to obtain relations between the real and the imaginary part of the Ernst potential which are independent of the function G and thus reflect only properties of the underlying Riemann surface. The found algebraic relations contain however g real constants related to the function G, which means that one has to differentiate g times to get a differential relation which is completely free of the function G. These constants were just the integrals u' and I which are only constant with respect to the physical coordinates on the axis where the Riemann surface Σ degenerates. Thus one cannot hope to get an algebraic relation in the whole spacetime as on the axis. Instead one has to deal with integral equations or to look directly for a differential relation.

To avoid the differentiation of theta functions with respect to a branch point of the Riemann surface, we use the algebraic formulation of the hyperelliptic solutions which we have discussed in Sect. 3.4. With this form of the solutions it can also be seen how one could get a relation independent of G without differentiation: one can consider the equations (3.56) and (3.57) as integral equations for G. In principle one could try to eliminate G and X from these equations and (3.63). We will not investigate this approach but try to establish a differential relation. To this end it proves helpful to define the symmetric (in the K_n) functions S_i via

$$\prod_{i=1}^{g}(K - K_i) =: K^g - S_{g-1}K^{g-1} + \dots + S_0 , \qquad (5.41)$$

i.e. $S_0 = K_1 K_2 \dots K_g, \ \dots, \ S_{g-1} = K_1 + \dots + K_g$. The equations (3.63) are bilinear in the real and imaginary parts of the S_i which will be denoted by R_i and I_i respectively. With this notation we get

Proposition 5.9. *The x_i and the Ernst potential $\mathcal{E}$ are subject to the system of differential equations*

$$0 = (R_0 - \xi R_1 + \dots + \xi^g(-1)^g) \, x_{,\xi} + \frac{1}{2}Q_2(\xi)$$

$$-\frac{i}{2}(1 - x^2)(\ln \mathcal{E}\bar{\mathcal{E}})_{,\xi} \left(I_0 - \xi I_1 + \dots + (-1)^{g-1}I_{g-1}\xi^{g-1}\right) \quad (5.42)$$

and for $g > 1$

$$x_{j,\xi} = x_{,\xi} \left((-1)^{j+1}R_{j+1} + \dots + \xi^{g-j-1}\right) + (x_{j+1} + \dots + x\xi^{g-j-2})$$

$$-\frac{i}{2}(1 - x^2)(\ln \mathcal{E}\bar{\mathcal{E}})_{,\xi}((-1)^{j+1}I_{j+1} + \dots - \xi^{g-j-2}I_{g-1}) . \quad (5.43)$$

Proof. Differentiating (3.63) with respect to ξ and eliminating the derivatives $K_{i,\xi}$ via the Picard–Fuchs relations (3.61), we end up with a linear system of equations for the derivatives of the x_i and x which can be solved in standard manner. The Vandemonde-type determinants can be expressed via the symmetric functions. For $x_{,\xi}$ one gets (5.42). The equations for the $x_{j,\xi}$ are bilinear in the symmetric functions. They can be combined with (5.42) to (5.43).

□

Remark 5.10. If one can solve (3.63) for the K_i, the equations (5.42) and (5.43) will be a non-linear differential system in ξ (and $\bar{\xi}$ which follows from the reality properties) for x_i, x and $\mathcal{E}$ which only contains the branch points of the Riemann surface as parameters.

For the metric function af, we get with (3.51)

Proposition 5.11. *The metric function a is related to the functions x_i and S_i via*

$$Z = \frac{ix_{g-2}}{1-x^2} - I_{g-1} - \frac{ix\zeta}{1-x^2} \qquad (5.44)$$

for $g > 1$ and

$$Z = -I_0 + \frac{ix(\alpha_1 - \zeta)}{1-x^2} \qquad (5.45)$$

for $g = 1$.

Proof. To express the function Z via the divisor $\mathfrak{K}$ in (3.56), we define the divisor $\mathfrak{T} = T_1 + \ldots + T_g$ as the solution of the Jacobi inversion problem $\omega(\mathfrak{T}) = \omega(\mathfrak{K}) + \omega(P)$ where P is in the vicinity of ∞^- (only terms of first order in the local parameter near ∞^- are needed). Using the formula for root functions (A.85), we get for the quantity Z in (3.51)

$$Z = \frac{i}{2}D_\infty - \ln \prod_{i=1}^{g} \frac{T_i - \bar{\xi}}{T_i - \xi} . \qquad (5.46)$$

Applying Abel's theorem to the definition of $\mathfrak{T}$ and expanding in the local parameter near ∞^-, we end up with (5.44) for general $g > 1$ and with (5.45) for $g = 1$.

□

Remark 5.12. For $g > 1$ equation (5.44) can be used to replace the relation for $x_{g-2,\xi}$ in (5.43) since the latter is identically fulfilled with (5.44) and (2.45).

Remark 5.13. An interesting limiting case is $G \approx 1$ where $\mathcal{E} \approx 1$, i.e. the limit where the solution is close to Minkowski spacetime. By the definition (3.56), the divisor $\mathfrak{K}$ is in this case approximately equal to $\mathfrak{D}$. Thus the symmetric functions in (5.43) and (5.42) can be considered as being constant and given by the branch points E_i. Relation (5.44) implies that the quantity

Z is approximately equal to I_{g-2} in this limit, i.e. it is mainly equal to the constant a_0 in lowest order. Since the differential system (5.42) and (5.43) is linear in this limit, it is straight forward to establish two real differential equations of order g for the real and the imaginary part of the Ernst potential. In principle this works also in the non-linear case, where sign ambiguities in the solution of (3.56) can be fixed by the Minkowskian limit.

To illustrate the above equations we will first consider the elliptic case. This is the only case where an algebraic relation between Z and b independent of G could be established. Equations (3.63) lead to

$$(1 - x^2)R_0 = \alpha_1 - \zeta x^2 ,$$
$$(1 - x^2)S_0 \bar{S}_0 = E_1 \bar{E}_1 - P_0 \bar{P}_0 x^2 . \qquad (5.47)$$

Formula (5.45) takes with (5.47) (the sign of I_0 is fixed by the condition that $I_0 = -\beta_1$ for $x = 0$) the form

$$(1 - x^2)Z = ix(\alpha_1 - \zeta) + \sqrt{(1 - x^2)(\beta_1^2 - \varrho^2 x^2) - x^2(\alpha_1 - \zeta)^2} . \qquad (5.48)$$

This relation holds in the whole spacetime for all elliptic potentials, i.e. for all possible choices of G in (4.19). This implies that one can only solve boundary value problems on elliptic surfaces where the boundary data at some given contour Γ_ξ satisfy condition (5.48).

In the case genus 2, we get for (3.63)

$$(1 - x^2)R_1 = \alpha_1 + \alpha_2 - \zeta x^2 + xx_0 ,$$
$$(1 - x^2)(R_1^2 + I_1^2 + 2R_0) = (\alpha_1 + \alpha_2)^2 + 2\alpha_1\alpha_2 + \beta_1^2 + \beta_2^2$$
$$- x_0^2 - x^2(\varrho^2 + \zeta^2) + 4\zeta xx_0 ,$$
$$(1 - x^2)(R_1 R_0 + I_1 I_0) = \alpha_1 \alpha_2 (\alpha_1 + \alpha_2) + \alpha_1 \beta_2^2 + \alpha_2 \beta_1^2$$
$$- \zeta x_0^2 + (\varrho^2 + \zeta^2)xx_0 ,$$
$$(1 - x^2)(R_0^2 + I_0^2) = (\alpha_1^2 + \beta_1^2)(\alpha_2^2 + \beta_2^2) - (\varrho^2 + \zeta^2)x_0^2 . \qquad (5.49)$$

The aim is to determine the S_i and x_0 from (5.49) and

$$(1 - x^2)(Z + I_1) = ix_0 - \zeta ix , \qquad (5.50)$$

and to eliminate these quantities in

$$(R_0 - \xi R_1 + \xi^2)x_{,\xi} = -\frac{1}{2}(x_0 + \xi x) + \frac{i}{2}(1 - x^2)(\ln \mathcal{E}\bar{\mathcal{E}})_{,\xi}(I_0 - \xi I_1) \qquad (5.51)$$

which follows from (5.42).

Remark 5.14. Since the above relations will hold in the whole spacetime, it is possible to extend them to an arbitrary smooth boundary Γ_ξ, where the Ernst potential may be singular (a jump discontinuity) and where one wants to prescribe boundary data (combinations of $\mathcal{E}$, $\mathcal{E}_\xi$). If these data are of sufficient differentiability (at least $C^{g,\alpha}(\Gamma_\xi)$), we can check the solvability of the problem on a given surface with the above formulas. The conditions on the differentiability of the boundary data can be relaxed by working directly with the equations (3.56) and (3.57) which can be considered as integral equations for $\ln G$. The latter is not very convenient if one wants to construct explicit solutions, but it makes it possible to treat boundary value problems where the boundary data are Hölder continuous. We will only work with the differential relations and consider merely the derivatives tangential to Γ_ξ in (5.43) to establish the desired differential relations between a, b and U. One ends up with two differential equations which involve only U, b and derivatives. The aim is to construct the spacetime which corresponds to the prescribed boundary data from these relations. To this end one has to integrate the differential relations using the boundary conditions. Integrating one of these equations, one gets g real integration constants which cannot be freely chosen since they arise from applying the tangential derivatives in (5.43). Thus they have to be fixed in a way that the integrals on the right-hand side of (3.56) are in fact the b-periods of the second integral on the right-hand side of (3.56) and that (3.57) holds. The second differential equation arises from the use of normal derivatives of the Ernst potential in (5.42). To satisfy the b-period condition (3.56), one has to fix a free function in the integrated form of the corresponding differential equation. Thus one has to complement the two differential equations following from (5.42) with an integral equation which is obtained by eliminating G from e.g. $\tilde{u}_1$ and $\tilde{u}_2$ in (3.56). For given boundary data, the system following from (3.56) may in principle be integrated to give e^{2U} and b in dependence of the boundary data. Then the in general non-linear integral equation will establish whether the boundary data are compatible with the considered Riemann surface. This is typically a rather tedious procedure. As shown in the next section, there is however a class of problems where it is unnecessary to use this integral equation. If the differential equations hold for an arbitrary function e^{2U}, the integral equation will only be used to determine this metric function, but the boundary value problem will be always solvable (locally). This offers a constructive approach to solve boundary value problems without having to consider non-linear integral equations.

5.5 Counter-rotating Disks of Genus 2

Since it is not very instructive to establish the differential relations for genus 2 in the general case, we will concentrate in this section on the form these equations take in the equatorially symmetric case for counter-rotating dust disks. In this case, the solutions are parametrized by $E_1^2 = \alpha + i\beta$. We will always assume in the following that the boundary data are at least $C^2(\Gamma_\xi)$ (the normal derivatives of the metric functions have a jump at the disk, but the tangential derivatives are supposed to exist up to at least second order).

Putting $s = be^{-2U}$, we get for (5.51) for $\zeta = 0$, $\varrho \leq 1$

$$ix_0 = \left(R_0 - \varrho^2 - sI_0\right)\frac{b_{,\zeta}}{f} - \varrho\left(R_1 - sI_1\right)\frac{b_{,\varrho}}{f}$$
$$- \left(s(R_0 - \varrho^2) + I_0\right)\frac{f_{,\zeta}}{f} + \varrho\left(sR_1 + I_1\right)\frac{f_{,\varrho}}{f}\,,$$

$$\varrho s = \left(R_0 - \varrho^2 - sI_0\right)\frac{b_{,\varrho}}{f} + \varrho\left(R_1 - sI_1\right)\frac{b_{,\zeta}}{f}$$
$$- \left(s(R_0 - \varrho^2) + I_0\right)\frac{f_{,\varrho}}{f} - \varrho\left(sR_1 + I_1\right)\frac{f_{,\zeta}}{f}\,, \tag{5.52}$$

where the S_i and ix_0 are taken from (5.49) and (5.50). Since counter-rotating dust disks are subject to the boundary conditions (5.19), we can replace the normal derivatives in (5.52) via (5.19) which leads to a differential system where only tangential derivatives at the disk occur. With (5.49) and (5.50) we get

$$ix_0 + (Z - ix_0)\frac{R_0 - \varrho^2}{I_1 R} = \left(\varrho - \frac{R^2 + \varrho^2 + \delta f^2}{2R\varrho}\frac{R_0 - \varrho^2}{I_1}\right)$$
$$\times \left((-Z + ix_0)\frac{f_{,\varrho}}{f} - sZ\frac{b_{,\varrho}}{f}\right)\,, \tag{5.53}$$

$$\varrho s\left(1 - \frac{Z}{R}\right) = \left(\frac{R_0 - \varrho^2}{I_1} - \frac{R^2 + \varrho^2 + \delta f^2}{2R}\right)$$
$$\times \left(sZ\frac{f_{,\varrho}}{f} + (-Z + ix_0)\frac{b_{,\varrho}}{f}\right)\,. \tag{5.54}$$

With $I_1 = ix_0/(1 - x^2) - Z$ and

$$R_0 = \frac{ix_0 Z - \alpha - \frac{\varrho^2}{2}}{1 - x^2} - \frac{Z^2 - \varrho^2}{2}\,, \tag{5.55}$$

the function ix_0 follows from

$$R_0^2 + \frac{(R_0 - \varrho^2)^2}{I_1^2}\frac{x^2 x_0^2}{(1 - x^2)^2} = \frac{\alpha^2 + \beta^2 - \varrho^2 x_0^2}{1 - x^2}\,, \tag{5.56}$$

i.e. an algebraic equation of fourth order for ix_0 which can be uniquely solved by respecting the Minkowskian limit. Thus (5.53) and (5.54) are in fact a differential system which determines b and f in dependence of the angular velocity Ω.

5.5.1 Newtonian Limit

For illustration we will first study the Newtonian limit of the equations (5.53) to (5.56) (where counter-rotation does not play a role). This means we are looking for dust disks with an angular velocity of the form $\Omega = \omega q(\varrho)$ where $|q(\varrho)| \leq 1$ for $\varrho \leq 1$, and where the dimensionless constant $\omega \ll 1$. Since we have put the radius ϱ_0 of the disk equal to 1, $\omega = \omega \varrho_0$ is the upper limit for the velocity in the disk. The condition $\omega \ll 1$ just means that the maximal velocity in the disk is much smaller than the velocity of light which is equal to 1 in the used units. An expansion in ω is thus equivalent to a standard post-Newtonian expansion. Of course there may be dust disks of genus 2 which do not have such a limit, but we will study in the following which constraints are imposed by the Riemann surface on the Newtonian limit of the disks where such a limit exists.

The invariance of the metric (2.40) under the transformation $t \rightarrow -t$ and $\Omega \rightarrow -\Omega$ implies that U is an even function in ω whereas b has to be odd. Since we have chosen an asymptotically non-rotating frame, we can make the ansatz $f = 1 + \omega^2 f_2 + \ldots$, $b = \omega^3 b_3 + \ldots$, and $a = \omega^3 a_3 + \ldots$. The boundary conditions (5.19) imply in lowest order $f_{2,\varrho} = 2q^2 \varrho$, the well-known Newtonian limit. Since the Ernst equation reduces to the Laplace equation for f_2 in order ω^2, we can use the methods of Sect. 5.1 to construct the corresponding solution. In order ω^3, the boundary conditions (5.19) lead to

$$b_{3,\varrho} = 2\varrho q f_{2,\zeta} , \qquad (5.57)$$

whereas equation (2.43) leads to the Laplace equation for b_3. Again we can use the methods of Sect. 5.1, but this time we have to construct a solution which is odd in ζ because of the equatorial symmetry. In principle one can extend this perturbative approach to higher order, where the field equations (2.43) lead to Poisson equations with terms of lower order acting as source terms, and where the boundary conditions can also be obtained iteratively from (5.19).

With this notation we get

Proposition 5.15. *Dust disks of genus 2 which have a Newtonian limit, i.e. a limit in which $\Omega = \omega q(\varrho)$ where $|q(\varrho)| \leq 1$ for $\varrho \leq 1$, are either rigidly rotating ($q = 1$) or q is a solution to the integro-differential equation*

$$b_3 = \left((R_0^0 - \varrho^2) 2q - \kappa \right) f_{2,\zeta} , \qquad (5.58)$$

where in the first case $I_1^0/R_0^0 = 2\omega$ and in the second $I_1 = \kappa\omega$ with R_0^0 and κ being ω independent constants.

We note that one can show with the techniques of [217], an integrated version of the above Picard–Fuchs system, that only constant Ω is possible in this limit.

Proof. Since the right hand side of (3.56) vanishes, we have $K_i = E_i$ for $\omega \rightarrow 0$, and thus $a_0 = I_1$ up to at least order ω^3. Keeping only terms in lowest order and denoting the corresponding terms of the symmetric functions by S_i^0, we obtain for (5.54)

$$\omega^3 b_3 = f_{2,\zeta}\left(2q(R_0^0 - \varrho^2)\omega^3 - \omega^2 I_1^0\right) .\tag{5.59}$$

The second equation (5.54) involves $b_{3,\zeta}$ and is thus of higher order. If (5.59) holds, this equation will be automatically fulfilled.

The ω-dependence in (5.59) implies that R_0^0, I_1^0 and thus the branch points must depend on ω. Since $f_{2,\zeta}$ is proportional to the density in the Newtonian case, it must not vanish identically. The possible cases following from equation (5.59) are constant Ω or (5.58). Using (5.6) and (5.3), one can express $U_{,\zeta}$ directly via Ω which leads to

$$f_{2,\zeta} = \frac{4}{\pi}\int_0^1 \frac{d\varrho'}{\varrho + \varrho'}\partial_{\varrho'}(q^2\varrho'^2)K(k)\tag{5.60}$$

with $k = 2\sqrt{\varrho\varrho'}/(\varrho + \varrho')$ and K being the complete elliptic integral of the first kind. Thus (5.58) is in fact an integro-differential equation for q.

$\square$

5.5.2 Explicit Solution for Constant Angular Velocity and Constant Relative Density

The simplifications of the Newtonian equation (5.59) for constant Ω give rise to the hope that rigid rotation could be generalized to the relativistic case which is what we will check in the following. Constant γ/Ω makes it in fact possible to avoid the solution of a differential equation and leads thus to the simplest example. We restrict ourselves to the case of constant relative density, $\gamma = const$. The structure of equation (5.54) suggests that it is sensible to choose the constant a_0 as $a_0 = -\gamma/\Omega$ since in this case $Z = R$. This is the only freedom in the choice of the parameters α and β on the Riemann surface one has for $g = 2$ since one of the parameters will be fixed as in the Newtonian case by the condition that the disk has to be regular at its rim. The second parameter will be determined as an integration constant of the Picard–Fuchs system.

We get

Theorem 5.16. *The boundary conditions (5.19) and (5.25) for the counter-rotating dust disk with constant Ω and constant γ are satisfied by an Ernst potential of the form (4.19) on a hyperelliptic Riemann surface of genus 2 with the branch points specified by*

$$\alpha = -1 + \frac{\delta}{2}, \quad \beta = \sqrt{\frac{1}{\lambda^2} + \delta - \frac{\delta^2}{4}} .\tag{5.61}$$

The parameter δ varies between $\delta = 0$ (only one component) and $\delta = \delta_s$,

$$\delta_s = 2\left(1 + \sqrt{1 + \frac{1}{\lambda^2}}\right) ,\tag{5.62}$$

the static limit. The function G is given by

$$G(\tau) = \frac{\sqrt{(\tau^2 - \alpha)^2 + \beta^2} + \tau^2 + 1}{\sqrt{(\tau^2 - \alpha)^2 + \beta^2} - (\tau^2 + 1)} \, . \tag{5.63}$$

Proof. The proof of the theorem is performed in several steps.
1. Since the second factor on the right-hand side in (5.54) must not vanish in the Newtonian limit, we find that for $Z = R$

$$\frac{R_0 - \varrho^2}{I_1} = \frac{Z^2 + \varrho^2 + \delta f^2}{2Z} \, . \tag{5.64}$$

With this relation it is possible to solve (5.55) and (5.50),

$$ix_0 = \frac{Z(\varrho^2 + 2\alpha - \delta f^2(1 - x^2))}{Z^2 - \varrho^2 - \delta f^2} \, , \tag{5.65}$$

$$\frac{\delta^2 f^2}{2}(1 - x^2) = -\frac{1}{\lambda}\left(\frac{1}{\lambda} - \delta f\right) + \delta\left(\alpha + \frac{\varrho^2}{2}\right)$$

$$+ \frac{\frac{1}{\lambda} - \delta f}{\sqrt{\frac{1}{\lambda^2} + \delta\varrho^2}}\sqrt{\left(\frac{1}{\lambda^2} - \alpha\delta\right)^2 + \delta^2\beta^2} \, .$$

One may easily check that (5.53) is identically fulfilled in this case. Thus the two differential equations (5.53) and (5.54) are satisfied for an unspecified f which implies that the boundary value problem for the rigidly rotating dust disk can be solved on a Riemann surface of genus 2 (the remaining integral equation which we will discuss below determines then f).

2. To establish the integral equations which determine the function G and the metric potential e^{2U}, we use equations (3.56). Since we have expressed above the K_i as a function of e^{2U} alone, the left-hand sides of (3.56) are known in dependence of e^{2U}. It proves helpful to make explicit use of the equatorial symmetry at the disk. By construction the Riemann surface $\mathcal{L}$ is for $\zeta = 0$ invariant under the involution $K \to -K$. This implies that the theta functions factorize and can be expressed via theta functions on the covered surface Σ_1 given by $\mu_1^2(\tau) = \tau(\tau + \varrho^2)((\tau - \alpha)^2 + \beta^2)$ and the Prym variety Σ_2 (which is here also a Riemann surface) given by $\mu_2^2(\tau) = (\tau + \varrho^2)((\tau - \alpha)^2 + \beta^2)$ (see Sect. 4.2.1). On these surfaces we define the divisors V and W respectively as the solution of the Jacobi inversion problem

$$u_v = \frac{1}{i\pi}\int_0^{-\varrho^2} \frac{\ln G(\sqrt{\tau})d\tau}{\mu_1(\tau)} \doteq \int_0^V \frac{d\tau}{\mu_1} \, ,$$

$$u_w = \frac{1}{i\pi}\int_{-\varrho^2}^{-1} \frac{\ln G(\sqrt{\tau})d\tau}{\mu_2(\tau)} \doteq \int_\infty^W \frac{d\tau}{\mu_2} \, . \tag{5.66}$$

For the Ernst potential we get

$$\ln \mathcal{E}\bar{\mathcal{E}} = -\ln\left(1 - \frac{2ix_0}{Z(1-x^2)}\right) + \int_0^V \frac{\tau d\tau}{\mu_1} - I_v . \tag{5.67}$$

where

$$I_v = \frac{1}{2\pi i}\int_0^{-\varrho^2} \frac{\ln G(\sqrt{\tau})\tau d\tau}{\mu_1(\tau)} . \tag{5.68}$$

3. Using Abel's theorem and (3.56), we can express V and W in terms of the divisor $\mathfrak{K}$ which leads to

$$V = -\frac{\varrho^2 x_0^2}{Z^2(1-x^2) - 2Zix_0} \tag{5.69}$$

and

$$W + \varrho^2 = -\frac{1}{x^2}\left(Z^2(1-x^2) - 2Zix_0 - x_0^2\right) . \tag{5.70}$$

4. Since V and I_v vanish for $\varrho = 0$, we can use (5.67) for $\varrho = 0$ to determine the integration constant of the Picard–Fuchs system. We get with (5.65)

$$\beta^2 = \frac{1}{\lambda^2} - \delta\alpha + \frac{\delta^2}{4} . \tag{5.71}$$

5. Since V in (5.69) is with (5.65) a rational function of ϱ, α and β and does not depend on the metric function e^{2U}, we can use the first equation in (5.66) to determine G as the solution of an Abelian integral which is obviously linear. With G determined in this way, the second equation in (5.66) can then be used to calculate e^{2U} at the disk which leads to elliptic theta functions (see also [129]). (In the general case, one would have to eliminate e^{2U} in the relations for u_v and u_w to end up with a non-linear integral equation for G.)

The integral equation following from (5.66),

$$\int_0^V \frac{d\tau}{\mu_1(\tau)} = \frac{1}{i\pi}\int_0^{-\varrho^2} \frac{\ln G}{\mu_1(\tau)} d\tau \tag{5.72}$$

is an Abelian equation and can be solved in standard manner by integrating both sides of the equation with a factor $1/\sqrt{K-r}$ with respect to r from 0 to r where $r = -\varrho^2$. With (5.69) we get for what is essentially an integral over a rational function

$$G(K) = \frac{\sqrt{(K-\alpha)^2 + \beta^2} + K - \alpha + \frac{\delta}{2}}{\sqrt{(K-\alpha)^2 + \beta^2} - (K - \alpha + \frac{\delta}{2})} . \tag{5.73}$$

6. The condition $G(-1) = 1$ excludes ring singularities at the rim of the disk and leads to a continuous potential and density there. It determines the last degree of freedom in (5.73) to

$$\alpha = -1 + \frac{\delta}{2} . \tag{5.74}$$

7. The static limit of the counter-rotating disks is reached for $\beta = 0$, i.e. the value δ_s.

$\square$

Remark 5.17. It is interesting to note that there exist algebraic relations between a, b and e^{2U} though they are expressed in terms of theta functions, i.e. transcendental functions, also at the disk.

5.5.3 Global Regularity

In Theorem 5.16 it was shown that one can identify an Ernst potential on a genus 2 surface which takes the required boundary data at the disk. One has to notice however that this is only a local statement which does not ensure one has found the desired global solution which has to be regular in the whole spacetime except at the disk. It was shown in [118, 129] that this is the case if $\Theta(\omega(\infty^-) + u) \neq 0$. In the Newtonian theory (see Sect. 5.1), the boundary value problem could be treated at the disk alone because of the regularity properties of the Poisson integral. Thus one knows that the above condition will hold in the Newtonian limit of the hyperelliptic solutions if the latter exists. For physical reasons it is however clear that this will not be the case for arbitrary values of the physical parameters: if more and more energy is concentrated in a region of spacetime, a black-hole or a singularity is expected to form (see e.g. the hoop conjecture [218]). The black-hole limit will be a stability limit for the above disk solutions. Thus one expects that additional singularities will occur in the spacetime if one goes beyond the black-hole limit. The task is to find the range of the physical parameters, here λ and δ, where the solution is regular except at the disk.

We can state

Theorem 5.18. *Let Σ' be the Riemann surface given by $\mu'^2 = (K^2 - E)(K^2 - \bar{E})$ and let a prime denote that the primed quantity is defined on Σ'. Let $\lambda_c(\delta)$ be the smallest positive value λ for which $\Theta'(u') = 0$. Then $\Theta(\omega(\infty^-) + u) \neq 0$ for all ϱ, ζ and $0 < \lambda < \lambda_c(\delta)$ and $0 \leq \delta \leq \delta_s$.*

This defines the range of the physical parameters where the Ernst potential of Theorem 5.16 is regular in the whole spacetime except at the disk. Since it was shown in [118, 129] that $\Theta'(u') = 0$ defines the limit in which the solution can be interpreted as the extreme Kerr solution, the disk solution is regular up to the black-hole limit if this limit is reached. This ultrarelativistic limit will be discussed in more detail in Chap. 7.

Proof. 1. Using the divisor $\mathfrak{K}$ of (3.56) and the vanishing condition for the Riemann theta function, we find that $\Theta(\omega(\infty^-) + u) = 0$ is equivalent to the condition that ∞^+ is in $\mathfrak{K}$. The reality of the $\tilde{u}_i$ implies that $\mathfrak{K} = \infty^+ + \xi$. Equation (3.56) thus leads to

$$\int_{E_1}^{\infty^+} \frac{d\tau}{\mu} + \int_{E_2}^{\xi} \frac{d\tau}{\mu} - \frac{1}{2\pi i} \int_{\Gamma} \frac{\ln G d\tau}{\mu} \equiv 0 \, ,$$

$$\int_{E_1}^{\infty^+} \frac{\tau d\tau}{\mu} + \int_{E_2}^{\xi} \frac{\tau d\tau}{\mu} - \frac{1}{2\pi i} \int_{\Gamma} \frac{\ln G \tau d\tau}{\mu} \equiv 0 \, , \tag{5.75}$$

where $\equiv$ denotes equality up to periods. The reality and the symmetry with respect to ζ of the above expressions limits the possible choices of the periods. It is straight forward to show that $\Theta(\omega(\infty^-) + u) = 0$ if and only if the functions F_i defined by

$$F_1 \doteq \int_{E_1}^{\infty^+} \frac{d\tau}{\mu} + \int_{E_2}^{\xi} \frac{d\tau}{\mu} - n_1 \left(2 \oint_{b_1} \frac{d\tau}{\mu} + 2 \oint_{b_2} \frac{d\tau}{\mu} + \oint_{a_1} \frac{d\tau}{\mu} + \oint_{a_2} \frac{d\tau}{\mu} \right)$$
$$- \frac{1}{2\pi i} \int_{\Gamma} \frac{\ln G d\tau}{\mu} \, ,$$

$$F_2 \doteq \int_{E_1}^{\infty^+} \frac{\tau d\tau}{\mu} + \int_{E_2}^{\xi} \frac{\tau d\tau}{\mu} - n_2 \left(2 \oint_{b_1} \frac{\tau d\tau}{\mu} + 2 \oint_{b_2} \frac{\tau d\tau}{\mu} \right.$$
$$\left. + \oint_{a_1} \frac{\tau d\tau}{\mu} + \oint_{a_2} \frac{\tau d\tau}{\mu} \right) - \frac{1}{2\pi i} \int_{\Gamma} \frac{\ln G \tau d\tau}{\mu} \tag{5.76}$$

with the cut system of Fig. A.4 and with $n_{1,2} \in \mathbb{Z}$ vanish for the same values of ϱ, ζ, λ, δ. The functions F_i are both real, F_1 is even in ζ whereas F_2 is odd. Thus F_2 is identically zero in the equatorial plane outside the disk.

2. In the Newtonian limit $\lambda \approx 0$, the above expressions take in leading order of λ the form

$$F_1 = \lambda \left((-8n_1 + 1)c_1(\varrho, \zeta) \ln \lambda - d_1(\varrho, \zeta)\lambda \right) \, , \tag{5.77}$$

and

$$F_2 = \sqrt{\lambda} \left((-8n_2 + 1)c_2(\varrho, \zeta) \ln \lambda - d_2(\varrho, \zeta)\lambda^{\frac{3}{2}} \right) \, , \tag{5.78}$$

where we have used the same approach as in the calculation of the axis potential in (4.8) (see [129] and references given therein); the functions c_1, d_1 are non-negative whereas c_2/d_2 is positive in $\mathbb{C}/\{\zeta = 0\}$. Thus the F_i are zero for $\lambda = 0$ which is Minkowski spacetime ($\mathcal{E} = 1$), but they are not simultaneously zero for small enough λ, i.e. $\mathcal{E}$ is regular in the Newtonian regime in accordance with the regularity properties of the Poisson integral. The F_i may vanish however at some value λ_s for given ϱ, ζ and δ. Since we are looking for zeros of the F_i in the vicinity of the Newtonian regime, we may put $n_{1,2} = 1$ here.

3. Let $\mathcal{G}$ be the open domain $\mathbb{C}/\{\zeta = 0, \varrho \leq 1 \vee \varrho = 0\}$. It is straight forward to check that the F_i are a solution to the Laplace equation $\Delta F_i = 0$ for $\xi, \bar{\xi} \in \mathcal{G}$. Thus by the maximum principle the F_i do not have an extremum in $\mathcal{G}$.

4. At the axis for $\zeta > 0$, the $\tilde{u}_i$ are finite whereas the F_i diverge proportional to $-\ln \varrho$ for all λ, δ. Thus $\mathcal{E}$ is always regular at the axis.

5. Relation (5.42) at the disk can be written in the form $(f + A)^2 + b^2 = B^2$ where A and B are finite real quantities. Thus the Ernst potential is always regular at the disk. Due to symmetry reasons $F_2 \equiv \tilde{u}_2$ which is non-zero except at the rim of the disk. For F_1 one gets at the disk

$$F_1 = \int_{-\varrho^2}^{\infty^+} \frac{d\tau}{\mu_1(\tau)} + \int_0^E \frac{d\tau}{\mu_1(\tau)} + \int_0^{\bar{E}} \frac{d\tau}{\mu_1(\tau)} - u_v . \tag{5.79}$$

With (5.72) one can see that F_1 is always positive at the disk.

6. Since F_1 is strictly positive on the axis and the disk and a solution to the Laplace equation in $\mathcal{G}$, it is positive in $\bar{\mathbb{C}}$ if it is positive at infinity. F_1 is regular for $|\xi| \to \infty$ and can be expanded as $F_1 = F_{11}/|\xi| + o(1/|\xi|)$ where F_{11} can be expressed via quantities on Σ'. We get

$$F_{11} = \frac{1}{2} \oint_{b_1'} \frac{d\tau}{\mu'} - \frac{1}{2\pi i} \int_{-i}^i \frac{\ln G d\tau}{\mu'} . \tag{5.80}$$

The quantity $F_{11} \equiv 0$ iff $\Theta'(u') = 0$. The condition $F_{11} > 0$ is thus equivalent to the condition that $\lambda < \lambda_c(\delta)$ where $\lambda_c(\delta)$ is the first positive zero of $\Theta'(u')$.

□

Remark 5.19. Since $F_2(\varrho, 0) = 0$ for $\varrho \geq 1$, the reasoning in 6. of the above proof shows that there will be a zero of $\Theta(\omega(\infty^-) + u)$ and thus a pole of the Ernst potential in the equatorial plane for $\lambda > \lambda_c(\delta)$ if the theta function in the numerator does not vanish at the same point. In the equatorial plane the Ernst potential can be expressed via elliptic theta functions (see [129]) which have first order zeros. Thus F_{11} will be negative for $\lambda > \lambda_c$ in the vicinity of λ_c, and consequently the same holds for F_1 in the equatorial plane at some value $\varrho > 1$. It will be shown in Chap. 7 that the spacetime has a singular ring in the equatorial plane in this case. The disk is however still regular and the imposed boundary conditions are still satisfied. This provides a striking example that one cannot treat boundary value problems locally at the disk alone in the relativistic case. Instead one has to identify the range of the physical parameters where the solution is regular except at the disk.

6 Hyperelliptic Theta Functions and Spectral Methods

The solutions to the Ernst equation discussed in the previous chapters are given in terms of multi-dimensional theta functions. Though theta-functional solutions to integrable equations are known since the beginning of the seventies for equations like KdV, the work with these solutions admittedly has not reached the importance of solitons.

The main reason for the more widespread use of solitons is that they are given in terms of algebraic or exponential functions. On the other hand the parameterization of theta functions by the underlying Riemann surface is very implicit. The main parameters, typically the branch points of the Riemann surface, enter the solutions as parameters in integrals on the Riemann surface. A full understanding of the functional dependence on these parameters seems to be only possible numerically. In recent years algorithms have been developed to establish such relations for rather general Riemann surfaces as in [219] or via Schottky uniformization (see Chap. 5 of [1]), which have been incorporated successively in numerical and symbolic codes, see [220–224] and references therein (the last two references are distributed along with Maple 6, respectively Maple 8, and in an improved version as a Java implementation at [225]). For an approach to express periods of hyperelliptic Riemann surfaces via theta constants see [226].

These codes are convenient to study theta-functional solutions of equations of KdV-type where the considered Riemann surfaces are 'static', i.e., independent of the physical coordinates. In these cases the characteristic quantities of the Riemann surface have to be calculated once, just the (for low genus) comparatively fast summation in the approximation of the theta series via a finite sum as e.g. in [224] has to be carried out in dependence of the spacetime coordinates.

In the case of the theta-functional solutions to the Ernst equation, the branch points of the underlying hyperelliptic Riemann surface are parameterized by the physical coordinates, the spectral curve of the Ernst equation is in this sense 'dynamical'. This implies that the time-consuming calculation of the periods of the Riemann surface has to be carried out for each point in the spacetime. This includes limiting cases where the surface is almost degenerate. In addition the theta-functional solutions should be calculated to high precision in order to be able to test numerical solutions for rapidly rotating neutron stars such as provided e.g. by the spectral code **LORENE** [132]. This requires a very efficient code of high precision.

In [131] a numerical code for hyperelliptic surfaces was presented where the integrals entering the solution are calculated by expanding the integrands with a Fast Cosine Transformation in MATLAB. The precision of the numerical evaluation is tested by checking identities for periods on Riemann surfaces and by comparison with exact solutions. The code was originally optimized for the solution discussed in Chap. 5, but was generalized in [178] to the case of arbitrary hyperelliptic surfaces. It is shown for the solution of the counter-rotating dust disk that an accuracy of the order of machine precision ($\sim 10^{-14}$) can be achieved at a spacetime point in general position with 32 polynomials and with at most 256 polynomials in the case of almost degenerate surfaces which occurs e.g., when the point approaches the symmetry axis. Global tests of the numerical accuracy of the solutions to the Ernst equation are provided by integral identities for the Ernst potential and its derivatives: the equality of the ADM mass and the Komar mass (see [146, 227]) and a generalization of the Newtonian virial theorem as derived in [228]. The so determined numerical data for the theta functions can be used to provide 'exact' boundary values on a sphere for the program library **LORENE** [132] which was developed for a numerical treatment of rapidly rotating neutron stars. **LORENE** solves the boundary value problem for the stationary axisymmetric Einstein equations with spectral methods. It was shown in [131] that the theta-functional solution is reproduced to the order of 10^{-11} and better.

The chapter is organized as follows: in Sect. 6.1 we summarize basic features of spectral methods and explain our implementation of various quantities. The calculation of the periods of the hyperelliptic surface and the non-Abelian line integrals entering the solution is performed together with tests of the precision of the numerics. In Sect. 6.2 we check integral identities for the Ernst potential. The test of the spectral code **LORENE** is presented in Sect. 6.3.

6.1 Numerical Implementations

The numerical task in this work is to approximate and evaluate analytically defined functions as accurately and efficiently as possible. To this end it is advantageous to use (pseudo-)spectral methods which are distinguished by their excellent approximation properties when applied to smooth functions. Here the functions are known to be analytic except for isolated points. In this section we explain the basic ideas behind the use of spectral methods and describe in detail how the theta functions and the Ernst potential can be obtained to a high degree of accuracy.

6.1.1 Spectral Approximation

The basic idea of spectral methods is to approximate a given function f globally on its domain of definition by a linear combination

$$f \approx \sum_{k=0}^{N} a_k \phi_k ,$$

where the functions ϕ_k are taken from some class of functions which is chosen appropriately for the problem at hand.

The coefficients a_k are determined by requiring that the linear combination should be 'close' to f. Thus, one could require that $\|f - \sum_{k=0}^{N} a_k \phi_k\|$ should be minimal for some norm. Another possibility is to require that $\left\langle f - \sum_{k=0}^{N} a_k \phi_k, \chi_l \right\rangle = 0$ for $l = 0 : N$ with an appropriate inner product and associated orthonormal basis χ_l. This is called the Galerkin method. Finally, one can demand that $f(x_l) = \sum_{k=0}^{N} a_k \phi_k(x_l)$ at selected points $(x_l)_{l=0:N}$. This is the so-called collocation method which is the one we will use here. In this case the function values $f_l = f(x_l)$ and the coefficients a_k are related by the matrix $\Phi_{lk} = \phi_k(x_l)$.

The choice of the expansion basis depends to a large extent on the specific problem. For periodic functions there is the obvious choice of trigonometric functions $\phi_k(x) = \exp(2\pi i k/N)$ while for functions defined on a finite interval the most used $\phi_k(x)$ are orthogonal polynomials, in particular Chebyshev and Legendre polynomials. While the latter are important because of their relationship with the spherical harmonics on the sphere, the former are used because they have very good approximation properties and because one can use fast transform methods when computing the expansion coefficients from the function values provided one chooses the collocation points $x_l = \cos(\pi l/N)$ (see [229] and references therein). We will use here collocation with Chebyshev polynomials.

Let us briefly summarize their basic properties. The Chebyshev polynomials $T_n(x)$ are defined on the interval $I = [-1, 1]$ by the relation

$$T_n(\cos(t)) = \cos(nt) , \text{ where } x = \cos(t) , \qquad t \in [0, \pi] .$$

They satisfy the differential equation

$$(1 - x^2)\,\phi''(x) - x\phi'(x) + n^2\phi(x) = 0 . \tag{6.1}$$

The addition theorems for sine and cosine imply the recursion relations

$$T_{n+1}(x) - 2x\,T_n(x) + T_{n-1}(x) = 0 , \tag{6.2}$$

for the polynomials T_n and

$$\frac{T'_{n+1}(x)}{n+1} - \frac{T'_{n-1}(x)}{n-1} = 2T_n(x) \tag{6.3}$$

for their derivatives. The Chebyshev polynomials are orthogonal on I with respect to the hermitian inner product

$$\langle f, g \rangle = \int_{-1}^{1} f(x)\bar{g}(x) \frac{dx}{\sqrt{1 - x^2}} .$$

We have

$$\langle T_m, T_n \rangle = c_m \frac{\pi}{2} \delta_{mn} \tag{6.4}$$

where $c_0 = 2$ and $c_l = 1$ otherwise.

Now suppose that a function f on I is sampled at the points $x_l = \cos(\pi l/N)$ and that $\sum_{n=0}^{N} a_n T_n$ is the interpolating polynomial. Defining $c_0 = c_N = 2$, $c_n = 1$ for $0 < n < N$ in the discrete case and the numbers $F_n = c_n a_n$ we have

$$f_l = \sum_{n=0}^{N} a_n T_n(x_l) = \sum_{n=0}^{N} a_n T_n(\cos(\pi l/N))$$

$$= \sum_{n=0}^{N} a_n \cos(\pi n l/N) = \sum_{n=0}^{N} \frac{F_n}{c_n} \cos(\pi n l/N) .$$

This looks very much like a discrete cosine series and in fact one can show [230] that the coefficients F_n are related to the values f_l of the function by an inverse discrete cosine transform (DCT)

$$F_n = \frac{2}{N} \sum_{l=0}^{N} \frac{f_l}{c_l} \cos(\pi n l/N) .$$

Note, that up to a numerical factor the DCT is idempotent, i.e., it is its own inverse. This relationship between the Chebyshev polynomials and the DCT is the basis for the efficient computations because the DCT can be performed numerically by using the fast Fourier transform (FFT) and pre- and postprocessing of the coefficients [229]. The fast transform allows us to switch easily between the representations of the function in terms of its sampled values and in terms of the expansion coefficients a_n (or F_n).

The fact that f is approximated globally by a finite sum of polynomials allows us to express any operation applied to f approximately in terms of the coefficients. Let us illustrate this in the case of integration. So we assume that $f = p_N = \sum_{n=0}^{N} a_n T_n$ and we want to find an approximation of the integral for p_N, i.e., the function

$$F(x) = \int_{-1}^{x} f(s) \, ds ,$$

so that $F'(x) = f(x)$. We make the ansatz $F(x) = \sum_{n=0}^{N} b_n T_n(x)$ and obtain the equation

$$F' = \sum_{n=0}^{N} b_n T'_n = \sum_{n=0}^{N} a_n T_n = f .$$

Expressing T_n in terms of the T'_n using (6.3) and comparing coefficients implies the equations

$$b_1 = \frac{2a_0 - a_2}{2} , \qquad b_n = \frac{a_{n-1} - a_{n+1}}{2n} \quad \text{for } 0 < n < N , \qquad b_N = \frac{a_{N-1}}{2N} .$$

between the coefficients which determine all b_l in terms of the a_n except for b_0. This free constant is determined by the requirement that $F(-1) = 0$ which implies (because $T_n(-1) = (-1)^n$)

$$b_0 = -\sum_{n=1}^{N} (-1)^n b_n .$$

These coefficients b_n determine a polynomial q_N of degree N which approximates the indefinite integral $F(x)$ of the N-th degree polynomial f. The exact function is a polynomial of degree $N + 1$ whose highest coefficient is proportional to the highest coefficient a_N of f. Thus, ignoring this term we make an error whose magnitude is of the order of $|a_N|$ so that the approximation will be the better the smaller $|a_N|$ is. The same is true when a smooth function f is approximated by a polynomial p_N. Then, again, the indefinite integral will be approximated well by the polynomial q_N whose coefficients are determined as above provided the highest coefficients in the approximating polynomial p_N are small.

From the coefficients b_n we can also find an approximation to the definite integral $\int_{-1}^{1} f(s) \, ds = F(1)$ by evaluating

$$q_N(1) = \sum_{n=0}^{N} b_n = 2 \sum_{l=0}^{\lfloor N/2 \rfloor} b_{2l+1} .$$

Thus, to find an approximation of the integral of a function f we proceed as described above, first computing the coefficients a_n of f, computing the b_n and then calculating the sum of the odd coefficients.

6.1.2 Implementation of the Square-root

The Riemann surface $\mathcal{L}$ is defined by an algebraic curve of the form

$$\mu^2 = (K - \xi)(K - \bar{\xi}) \prod_{i=1}^{g} (K - E_i)(K - \bar{E}_i) ,$$

where we have $g = 2$ for the counter-rotating disk. In order to compute the periods and the theta functions related to this Riemann surface it is necessary

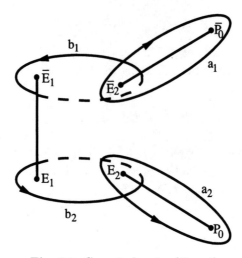

Fig. 6.1. Canonical cycles $(P_0 = \xi)$

to evaluate the square-root $\sqrt{\mu^2(K)}$ for arbitrary complex numbers K. In order to make this a well defined problem we introduce the cut-system as indicated in Fig. 6.1. On the cut surface the square-root $\mu(K)$ is defined as in [231] as the product of square-roots of monomials

$$\mu = \sqrt{K - \xi}\,\sqrt{K - \bar{\xi}}\prod_{i=1}^{g}\sqrt{K - E_i}\sqrt{K - \bar{E}_i}\,. \tag{6.5}$$

The square-root routines such as the one available in MATLAB usually have their branch-cut along the negative real axis. The expression (6.5) is holomorphic on the cut surface so that we cannot simply take the built-in square-root when computing $\sqrt{\mu^2(K)}$. Instead we need to use the information provided by the cut-system to define adapted square-roots.

Let $\arg(z)$ be the argument of a complex number z with values in $]-\pi,\pi[$ and consider two factors in (6.5) such as

$$\sqrt{K - P_1}\sqrt{K - P_2}\,,$$

where P_1 and P_2 are two branch-points connected by a branch-cut. Let $\alpha = \arg(P_2 - P_1)$ be the argument at the line from P_1 to P_2. Now we define the square-root $\sqrt[(\alpha)]{\cdot}$ with branch-cut along the ray with argument α by computing for each $z \in \mathbb{C}$ the square-root $s \doteq \sqrt{z}$ with the available MATLAB routine and then putting

$$\sqrt[(\alpha)]{z} = \begin{cases} s & \alpha/2 < \arg(s) < \alpha/2 + \pi \\ -s & \text{otherwise} \end{cases}\,.$$

With this square-root we compute the two factors

$$\sqrt[(\alpha)]{K - P_1} \, \sqrt[(\alpha)]{K - P_2} \, .$$

It is easy to see that this expression changes sign exactly when the branch-cut between P_1 and P_2 is crossed. We compute the expression (6.5) by multiplying the pairs of factors which correspond to the branch-cuts.

This procedure is not possible in the case of the non-linear transformations we are using to evaluate the periods in certain limiting cases. In these cases the root is chosen in a way that the integrand is a continuous function on the path of integration.

6.1.3 Numerical Treatment of the Periods

The quantities entering formula (4.19) for the Ernst potential are the periods of the Riemann surface and the line integrals u and I. The value of the theta function is then approximated by a finite sum.

The periods of a hyperelliptic Riemann surface can be expressed as integrals between branch points. Since we need in our example the periods of the holomorphic differentials and the differential of the third kind with poles at $\infty^{\pm}$, we have to consider integrals of the form

$$\int_{P_i}^{P_j} \frac{K^n \mathrm{d}K}{\mu(K)} \, , \quad n = 0, 1, 2 \, , \tag{6.6}$$

where the P_i, $i, j = 1, \dots, 6$ denote the branch points of $\mathcal{L}$.

In general position we use a linear transformation of the form $K = ct + d$ to transform the integral (6.6) to the normal form

$$\int_{-1}^{1} \frac{\alpha_0 + \alpha_1 t + \alpha_2 t^2}{\sqrt{1 - t^2}} \, H(t) \, \mathrm{d}t \, , \tag{6.7}$$

where the α_i are complex constants and where $H(t)$ is a continuous (in fact, analytic) complex valued function on the interval $[-1, 1]$. This form of the integral suggests to express the powers t^n in the numerator in terms of the first three Chebyshev polynomials $T_0(t) = 1$, $T_1(t) = t$ and $T_2(t) = 2t^2 - 1$ and to approximate the function $H(t)$ by a linear combination of Chebyshev polynomials

$$H(t) = \sum_{n \geq 0} h_n T_n(t) \, .$$

The integral is then calculated with the help of the orthogonality relation (6.4) of the Chebyshev polynomials.

Since the Ernst potential has to be calculated for all $\varrho, \zeta \in \mathbb{R}_0^+$, it is convenient to use the cut-system (6.1). In this system the moving cut does not cross the immovable cut as in the one in Fig. A.4 used for the analytical calculations. In addition the system is adapted to the symmetries and reality properties of $\mathcal{L}$. Thus the periods a_2 and b_2 are related to a_1 and b_1 via complex conjugation. For the analytical calculations of the Ernst potential in the limit of collapsing cuts, we have chosen in Sect. 4.1 cut systems adapted to the respective situation. In the limit $\xi \to \bar{\xi}$ we were using for instance a system where a_2 is the cycle around the cut $[\xi, \bar{\xi}]$. This has the effect that only the b-period b_2 diverges logarithmically in this case whereas the remaining periods stay finite as ϱ tends to 0. In the cut system of Fig. 6.1, all periods diverge as $\ln \varrho$. Since the divergence is only logarithmical this does not pose a problem for values of $\varrho > 10^{-5}$. In addition the integrals which have to be calculated in the evaluation of the periods are the same in both cut-system. Thus there is no advantage in using different cut systems for the numerical work.

To test the numerics we use the fact that the integral of any holomorphic differential along a contour surrounding the cut $[E_1, F_1]$ in positive direction is equal to minus the sum of all a-periods of this integral. Since this condition is not implemented in the code it provides a strong test for the numerics. It can be seen in Fig. 6.2 that 16 to 32 polynomials are sufficient in general position to achieve optimal accuracy. Since MATLAB works with 16 digits, machine precision is in general limited to 14 digits due to rounding errors. These rounding errors are also the reason why the accuracy drops slightly when a higher number of polynomials is used. The use of a low number of polynomials consequently does not only require less computational resources but has the additional benefit of reducing the rounding errors. It is therefore worthwhile to reformulate a problem if a high number of polynomials would be necessary to obtain optimal accuracy. These situations occur in the calculation of the periods when the moving branch points almost coincide which happens on the axis of symmetry in the spacetime or at spatial infinity. As can be seen from Fig. 6.2, for $\varrho = 10^{-3}$ and $\zeta = 10^3$ not even 2048 polynomials (this is the limit due to memory on the low end computers we were using) produce sufficient accuracy. The reason for these problems is that the function H in (6.7) behaves like $1/\sqrt{t + \varrho}$ near $t = 0$. For small ϱ this behavior is only satisfactorily approximated by a large number of polynomials. We therefore split the integral in two integrals between F_2 and $(F_2 + \bar{\xi})/2$ and between $(F_2 + \bar{\xi})/2$ and $\bar{\xi}$. The first integral is calculated with the Chebyshev integration routine after the substitution $t = \sqrt{K - F_2}$. This substitution leads to a regular integrand also at the branch point F_2. The second integral is calculated with the Chebyshev integration routine after the substitution $K - \zeta = \varrho \sinh(t)$. This takes care of the almost collapsing cut $[\xi, \bar{\xi}]$. It can be seen in Fig. 6.2 that 128 polynomials are sufficient to obtain machine precision even in almost degenerate situations.

The cut-system in Fig. 6.1 is adapted to the limit $\bar{\xi} \to F_2$ in what concerns the a-periods, since the cut which collapses in this limit is encircled by an a-cycle. However there will be similar problems as above in the determination of the b-periods. For $\bar{\xi} \sim F_2$ we split the integrals for the b-periods as above in two integrals between F_1 and 0, and 0 and F_2. For the first integral we use the

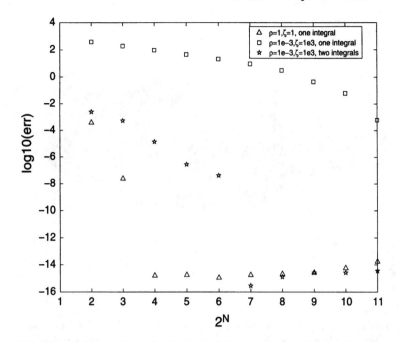

Fig. 6.2. Test of the numerics for the a-periods at several points in the spacetime. The error is shown in dependence of the number N of Chebychev polynomials

integration variable $t = \sqrt{K - F_1}$, for the second $K = \Re F_2 - i\Im F_2 \sinh t$. Since the Riemann matrix is symmetric, the error in the numerical evaluation of the b-periods can be estimated via the asymmetry of the calculated Riemann matrix. We define the function $err(\varrho, \zeta)$ as the maximum of the norm of the difference in the a-periods and the integral of a closed cycle around $[\xi, \bar{\xi}]$ discussed above and the difference of the off-diagonal elements of the Riemann matrix. This error is presented for a whole spacetime in Fig. 6.3. The values for ϱ and ζ vary between 10^{-4} and 10^4. On the axis and at the disk we give the error for the elliptic integrals (only the error in the evaluation of the a-periods, since the Riemann matrix has just one component). For $\xi \to \infty$ the asymptotic formulas for the Ernst potential are used. The calculation is performed with 128 polynomials, and up to 256 for $|\xi| > 10^3$. It can be seen that the error is in this case globally below 10^{-13}.

It is possible to address the above problems for arbitrary genus as long as the branch points coincide at most pairwise as in the solitonic limit. The idea of [178] is to use substitutions in the integrals (6.6) leading to a regular integrand. To determine the a-periods in the case $\tilde{E}_i \to \tilde{F}_i$, where the a-cycle is the closed contour around the cut $[\tilde{E}_i, \tilde{F}_i]$, we use

$$K = \frac{\tilde{E}_i + \tilde{F}_i}{2} + \frac{\tilde{F}_i - \tilde{E}_i}{2} \cosh x \ . \tag{6.8}$$

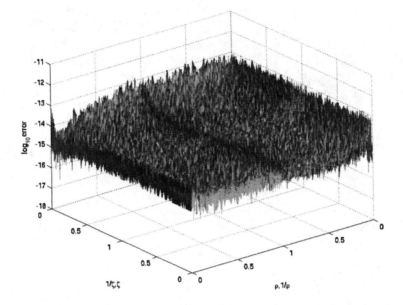

Fig. 6.3. A measure for the error in the determination of the periods in dependence of the physical coordinates. For $\varrho, \zeta > 1$ we use $1/\varrho, 1/\zeta$ as coordinates

After a linear transformation which transforms the integration path to the interval $[-1, 1]$, the integral is computed with the Chebyshev integration routine sketched above. This also works in situations close to the solitonic limit. To treat the b-periods in this case, we split the integral from $\tilde{F}_i$ to $\tilde{E}_{i+1}$ in two integrals from $\tilde{F}_i$ to $(\tilde{F}_i + \tilde{E}_{i+1})/2$ and from $(\tilde{F}_i + \tilde{E}_{i+1})/2$ to $\tilde{E}_{i+1}$. In the former case we use the substitution (6.8) and

$$K = \frac{\tilde{E}_{i+1} + \tilde{F}_{i+1}}{2} + \frac{\tilde{F}_{i+1} - \tilde{E}_{i+1}}{2} \cosh x \qquad (6.9)$$

in the latter. These substitutions lead to a regular integrand even in situations close to the solitonic limit. After a linear transformation, the integrals are computed with the Chebyshev integration routine.

The above procedure to determine the periods can be directly implemented within MATLAB for arbitrary genus. Due to the efficient vectorization algorithms of MATLAB, the calculation of the periods with optimal accuracy takes only a second even for genus 10 on the used low-end computers. The limiting factor is here whether the matrix A of a-periods is ill conditioned. This is the case if a considerable number of the entries of A are of the order of the rounding error (10^{-16}). Thus due to the limited number of digits, the inversion of the matrix A which is necessary to determine the Riemann matrix can only be carried out with reduced accuracy which is independent of the number of polynomials used in the spectral expansion. For large genus, this becomes the limiting factor in the determination of the

Riemann matrix. For instance in the case of the genus 20 surface with branch points $[-21, -20, \ldots, 20]$, the identity for a-periods is satisfied with 32 polynomials up to 10^{-15}, the symmetry of the Riemann matrix only up to 10^{-6} since the matrix A is badly scaled. The calculation of the periods takes less than a second in this case on the used computers.

6.1.4 Numerical Treatment of the Line Integrals

The line integrals $\boldsymbol{u}$ and I in (4.19) are linear combinations of integrals of the form

$$\int_{-i}^{i} \frac{\ln G(K) K^{l} dK}{\mu(K)} \ , \qquad l = 0, 1, 2 \ . \tag{6.10}$$

In general position, i.e. not close to the disk and λ small enough, the integrals can be directly calculated after the transformation $K = it$ with the Chebyshev integration routine. To test the numerics we consider the Newtonian limit $(\lambda \to 0)$ where the function $\ln G$ is proportional to $1 + K^2$, i.e. we calculate the test integral

$$\int_{-i}^{i} \frac{(1 + K^2)\, dK}{\sqrt{(K - \zeta)^2 + \varrho^2}} \ . \tag{6.11}$$

We compare the numerical with the analytical result in Fig. 6.4. In general position machine precision is reached with 32 polynomials.

When the moving cut approaches the path Γ, i.e., when the spacetime point comes close to the disk, the integrand in (6.11) develops cusps near the points ξ and $\bar{\xi}$. In this case a satisfactory approximation becomes difficult even with a large number of polynomials. Therefore we split the integration path in $[-i, -i\varrho]$, $[-i\varrho, i\varrho]$ and $[i\varrho, i]$. Using the reality properties of the integrands, we only calculate the integrals between 0 and $i\varrho$, and between $i\varrho$ and i. In the first case we use the transformation $K = \zeta + \varrho \sinh t$ to evaluate the integral with the Chebyshev integration routine, in the second case we use the transformation $t = \sqrt{K - \bar{\xi}}$. It can be seen in Fig. 6.4 that machine precision can be reached even at the disk with 64 to 128 polynomials. The values at the disk are, however, determined in terms of elliptic functions which is more efficient than the hyperelliptic formulae.

To treat the case where $\delta\lambda^2$ is not small, it is convenient to rewrite the function G in (5.63) in the form

$$\ln G(K) = 2 \ln \left(\sqrt{(K^2 - \alpha)^2 + \beta^2} + K^2 + 1 \right) - \ln \left(\frac{1}{\lambda^2} - \delta K^2 \right) \ . \tag{6.12}$$

In the limit $\delta\lambda^2 \to \infty$ with δ finite, the second term in (6.12) becomes singular for $K = 0$. Even for $\delta\lambda^2$ large but finite, the approximation of the integrand by Chebyshev polynomials requires a huge number of coefficients as can be seen from Fig. 6.5. It is therefore sensible to 'regularize' the integrand near

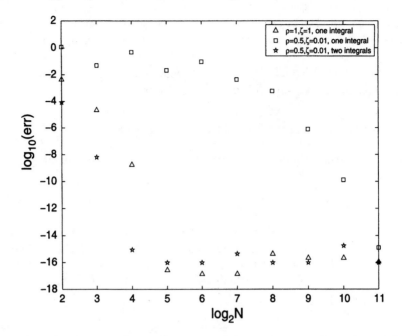

Fig. 6.4. Error in the integrals for the Maclaurin solution in dependence of the number N of Chebychev polynomials

$K = 0$. We consider instead of the function $\ln(\frac{1}{\lambda^2} - \delta K^2)F(K)$ where $F(K)$ is a C^∞ function near $K = 0$, the function

$$\ln\left(\frac{1}{\lambda^2} - \delta K^2\right)\left(F(K) - F(0) - F'(0)K - \ldots - \frac{1}{n!}F^{(n)}(0)K^n\right) . \quad (6.13)$$

The parameter n is chosen such that the spectral coefficients of (6.13) are of the order of 10^{-14} for a given number of polynomials, see Fig. 6.5. There we consider the integral

$$\int_{-i}^{i} \frac{\ln G(K)dK}{\sqrt{(K^2 - \alpha)^2 + \beta^2}} , \quad (6.14)$$

which has to be calculated on the axis. We show the absolute values of the coefficients a_k in an expansion of the integrand in Chebyshev polynomials, $\sum_{k=1}^{N} a_k T_k$. It can be seen that one has to include values of $n = 6$ in (6.13). The integral $\int_\Gamma \ln G(K)F(K)$ is then calculated numerically as the integral of the function (6.13), the subtracted terms are integrated analytically. In this way one can ensure that the line integrals are calculated in the whole spacetime with machine precision: close to the Newtonian limit, we use an analytically known test function to check the integration routine, for general situations we check the quality of the approximation of the integrand by Chebyshev polynomials via the spectral coefficients which have to become smaller than 10^{-14}.

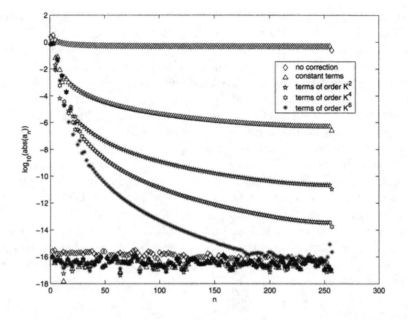

Fig. 6.5. Spectral coefficients for the integral (6.14) for $\delta = 1$ and $\lambda = 10^{16}$ in dependence of the number of Chebychev polynomials

6.1.5 Theta Functions

The theta series (A.53) for the Riemann theta function (the theta function in (A.53) with zero characteristic, theta functions with characteristic follow from (A.54)) is approximated as the sum

$$\Theta(\boldsymbol{x}|\mathbf{B}) = \sum_{n_1=-N}^{N} \sum_{n_2=-N}^{N} \exp\left\{ \frac{1}{2}n_1^2 \mathbf{B}_{11} + n_1 n_2 \mathbf{B}_{12} + \frac{1}{2}n_2^2 \mathbf{B}_{22} \right.$$
$$\left. + n_1 x_1 + n_2 x_2 \right\} . \tag{6.15}$$

The value of N is determined in an a priori estimate by the condition that terms in the series (A.53) for $n > N$ are strictly smaller than some threshold value ε_0 which is taken to be of the order of 10^{-16}. To this end we determine the eigenvalues of $\mathbf{B}$ and demand that

$$N > -\frac{1}{B_{\max}} \left(\|\boldsymbol{x}\| + \sqrt{\|\boldsymbol{x}\|^2 + 2B_{\max} \ln \varepsilon_0} \right) , \tag{6.16}$$

where $B_{\max}$ is the real part of the eigenvalue with maximal real part (the real part of $\mathbf{B}$ is negative definite). For a more sophisticated analysis of theta summations see [224]. In general position we find values of N between 4 and 8. For very large values of ζ close to the axis, N can become larger than 40 which however did not lead to any computational problems. To treat more extreme cases it could be helpful to take care of the fact that the eigenvalues of **B** can differ by more than an order of magnitude in our example. In these cases a summation over an ellipse rather than over a sphere in the plane (n_1, n_2), i.e. different limiting values for n_1 and n_2 as in [224] could be more efficient. The summation over the hypercube has, however, the advantage that it can be implemented in MATLAB for arbitrary genus. In addition it makes full use of MATLAB's vectorization algorithms outlined below. Thus it is questionable whether a summation over an ellipse would be more efficient in terms of computation time in this setting.

Here we made use of MATLAB's efficient way to handle matrices. We generate a $2N + 1$-dimensional array containing all possible index combinations and thus all components in the sum (6.15) which is then summed. To illustrate this we consider the simple example of genus 2 with $N = 2$. The summation indices are written as $(2N + 1) \times (2N + 1)$-matrices since $g = 2$. Each of these matrices contains $2N + 1$ copies of the vector with integers $-(2N + 1), \ldots, 2N + 1$. N_2 is the transposed matrix of N_1. Explicitly, we have

$$
N_1 = \begin{pmatrix} 2 & 2 & 2 & 2 & 2 \\ 1 & 1 & 1 & 1 & 1 \\ 0 & 0 & 0 & 0 & 0 \\ -1 & -1 & -1 & -1 & -1 \\ -2 & -2 & -2 & -2 & -2 \end{pmatrix}, \quad N_2 = \begin{pmatrix} 2 & 1 & 0 & -1 & -2 \\ 2 & 1 & 0 & -1 & -2 \\ 2 & 1 & 0 & -1 & -2 \\ 2 & 1 & 0 & -1 & -2 \\ 2 & 1 & 0 & -1 & -2 \end{pmatrix}. \quad (6.17)
$$

The terms in the sum (6.15) can thus be written in matrix form

$$
\exp\left(\frac{1}{2}N_1 \star N_1 \mathbf{B}_{11} + N_1 \star N_2 \mathbf{B}_{12} + \frac{1}{2}N_2 \star N_2 \mathbf{B}_{22} + N_1 \star x_1 + N_2 \star x_2\right),
$$
(6.18)

where the operation $N_1 \star N_2$ denotes that each element of N_1 is multiplied with the corresponding element of N_2. Thus, the argument of exp is a $(2N + 1) \times (2N + 1)$-dimensional matrix. Furthermore, the exponential function is understood to act not on the matrix but on each of its elements individually, producing a matrix of the same size. The approximate value of the theta function is then obtained by summing up all the elements in (6.18).

The most time consuming operations are the determination of the bilinear terms involving the Riemann matrix in (6.18). If one wants to calculate solutions to e.g. the KP equation, these terms only have to be determined once for a given surface. The integer N is in this case fixed for the largest $\|x\|$ in the plot. We note that the summation is very fast even though the used determination of N is rather crude. For instance in the case of genus 2 with $N = 100$, the calculation of the theta function takes 0.1s on the used computers.

The limiting factor is here the available memory since arrays of the order $(2N + 1)^g$ have to be multiplied with each other. On the used low-end computers we could deal with rather general genus 4 situations, but the limit was reached for genus 6 and $N = 5$. The summation is still very efficient, the calculation of the bilinear terms and the determination of the coefficients took 16s in the latter case, the subsequent calculation of the linear terms in (6.18) and the summation, which have to be carried out for each value of x and t, took roughly 4s. Thus the limitations we had to face were not due to the computing time but due to missing memory.

In case of genus 2 solutions to the Ernst equation the computation of the integrals entering the theta functions was however always the most time consuming. The theta summation always took less than 10 % of the calculation time for a value of the Ernst potential. Between 50 and 70 % of the processor time are used for the determination of the periods. On the used low-end PCs, the calculation time varied between 0.4 and 1.2s depending on the used number of polynomials.

We show a plot of the real part of the Ernst potential for $\lambda = 10$ and $\delta = 1$ in Fig. 6.6. For $\varrho, \zeta > 1$, we use $1/\varrho, 1/\zeta$ as coordinates which makes it possible to plot the whole spacetime in Weyl coordinates. The non-smoothness of the coordinates across $\varrho = 1 = 1/\varrho$ and $\zeta = 1 = 1/\zeta$ is noticeable in the plot. Asymptotically the potential is equal to 1. The disk is situated in the equatorial plane between $\varrho = 0$ and $\varrho = 1$. At the disk, the normal derivatives of e^{2U} are discontinuous.

The coordinate ϱ can take all non-negative real values, the coordinate ζ all real values. Since the example we are studying here has an equatorial symmetry, it is sufficient to consider only non-negative values of ζ. The case $\varrho = 0$ corresponds to the axis of symmetry where the branch cut $[\xi, \bar{\xi}]$ degenerates to a point, and where the Ernst potential is given in terms of elliptic theta functions (4.8). Formula (4.9) could be used to calculate the Ernst potential in the vicinity of the axis. However we considered only values of ϱ greater than 10^{-5} and did not experience any numerical problems. Consequently we did not use formula (4.9). For large values of $r = |\xi|$, the Ernst potential has the asymptotic expansion (4.11). The mass and the angular momentum are calculated in terms of elliptic functions on the axis. Formula (4.11) is used for values of $r > 10^6$.

6.2 Integral Identities

In the previous section we have tested the accuracy of the numerics locally, i.e. at single points in the spacetime. Integral identities have the advantage that they provide some sort of global test of the numerical precision since they sum up the errors. In addition they require the calculation of the potentials in extended regions of the spacetime which allows to explore the numerics for rather general values of the physical coordinates.

The identities we are considering in the following are the well known equivalence of a mass calculated at the disk (the Komar mass) and the ADM mass

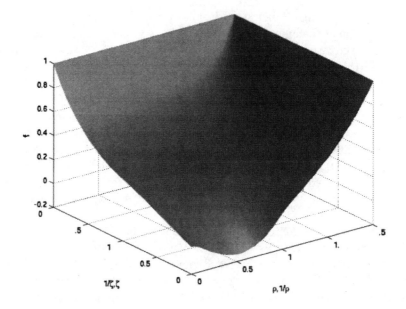

Fig. 6.6. The real part of the Ernst potential for $\lambda = 10$ and $\delta = 1$ in dependence of the physical coordinates. For $\varrho, \zeta > 1$ we use $1/\varrho, 1/\zeta$ as coordinates

determined at infinity, see [146, 227], and a generalization of the Newtonian virial identity, see [228] and [131]. The derivatives of the Ernst potential occurring in the integrands can be related to derivatives of theta functions as shown in Chap. 3. Since we are interested here in the numerical treatment of theta functions with spectral methods, we determine the derivatives with spectral methods, too (see Sect. 6.1). The integrals are again calculated with the Chebyshev integration routine. The main problem in this context is the singular behavior of the integrands e.g. at the disk which is a singularity for the spacetime. As before this will lead to problems in the approximation of these terms via Chebyshev polynomials. This could cause a drop in accuracy which is mainly due to numerical errors in the evaluation of the integrand and not of the potentials which we want to test. An important point is therefore the use of integration variables which are adapted to the possible singularities.

6.2.1 Mass Equalities

The equality between the ADM mass and the Komar mass provides a test of the numerical treatment of the elliptic theta functions at the disk by means of the elliptic theta functions on the axis. Since this equality is not implemented in the code, it provides a strong test.

The Komar mass at the disk is given by

$$2 \int_{disk} \mathrm{d}V \left(T_{ab} - \frac{1}{2} g_{ab} T_c^c \right) n^a \xi^b =: m_K , \qquad (6.19)$$

where the integration is carried out over the disk, where n_a is the normal at the disk, and where T_{ab} is the energy momentum tensor of the disk of Chap. 5. In the example we are considering here, the normal derivatives at the disk can be expressed via tangential derivatives (see (5.19)) which makes a calculation of the derivatives solely within the disk possible. We implement the Komar mass in the form

$$m_K = \int_0^1 \mathrm{d}\varrho \frac{b_{,\varrho}}{4\Omega^2 \sqrt{\varrho^2 - \delta \mathrm{e}^{4U} + 2\mathrm{e}^{2U}/\lambda}} \left(\mathrm{e}^{2U} + \Omega^2 \mathrm{e}^{-2U} (\varrho^2 - a^2 \mathrm{e}^{4U}) \right) .$$
$$(6.20)$$

The integrand is known to vanish as $\sqrt{1 - \varrho^2}$ at the rim of the disk, which is the typical behavior for such disk solutions. Since $\sqrt{1 - \varrho^2}$ is not analytic in ϱ, an expansion of the integrand (6.20) in Chebyshev polynomials in ϱ would not be efficient. We will thus use $t = \sqrt{1 - \varrho^2}$ as the integration variable. This takes care of the behavior at the rim of the disk. Since in general the integrand in (6.20) depends on ϱ^2, this variable can be used in the whole disk. In the ultra-relativistic limit for $\delta \neq 0$, the function e^{2U} vanishes as ϱ. In such cases it is convenient either to take two domains of integration or to use a different variable of integration. We chose the second approach with $\varrho = \sin x$ (this corresponds to the disk coordinates (6.22)). Yet, strongly relativistic situations still lead to problems since e^{2U} vanishes in this case at the center of the disk as does $b_{,\varrho}$ which leads to a '0/0' limit. In Fig. 6.7 one can see that the masses are in general equal to the order of 10^{-14}. In these calculations 128 up to 256 polynomials were used. We show the dependence for $\gamma = 0.7$ and several values of the parameter[1] ε, as well as for $\varepsilon = 0.8$ and several values of γ. The accuracy drops in the strongly relativistic, almost static situations (ε close to 1, γ close to zero) since the Riemann surface is almost degenerate in this case ($\beta \to 0$). In the ultra-relativistic limit for $\delta = 0$, the situation is no longer asymptotically flat which implies that the masses formally diverge. For $\varepsilon = 0.95$, the masses are still equal to the order of 10^{-13}. Not surprisingly the accuracy drops for $\varepsilon = 0.9996$ to the order of 10^{-4}.

---

[1] For physical reasons it is convenient to discuss the solution in dependence of the two real parameters ε and γ, where γ is defined in (5.16). The parameter $\varepsilon = z_R/(1 + z_R)$ is related to the redshift z_R of photons emitted at the center of the disk and detected at infinity. It varies between 0 in the Newtonian limit, and 1 in the ultra-relativistic limit, where photons cannot escape to infinity. Thus, ε is a measure of how relativistic the situation is. For the functional relations between ε, γ and λ, δ see Chap. 7.

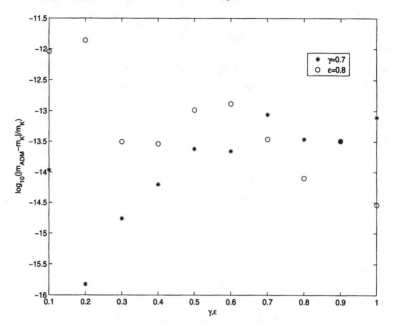

Fig. 6.7. The relative difference of the ADM mass and the Komar mass for $\gamma = 0.7$ and several values of ε, and for $\varepsilon = 0.8$ and several values of γ

6.2.2 Virial-type Identities

Generalizations of the Newtonian virial theorem are used in numerics (see [228]) as a test of the quality of the numerical solution of the Einstein equations. Since they involve integrals over the whole spacetime, they test the numerics globally and thus provide a valid criterion for the entire range of the physical coordinates.

The identity which is checked here is a variant of the one given in [228] which is adapted to possible problems at the zeros of the real part of the Ernst potential, the ergosphere,

$$\frac{3}{2}\int_0^\infty \int_0^\infty d\varrho d\zeta \varrho^2 (\mathcal{E}_{,\varrho}(\bar{\mathcal{E}}_{,\varrho}^2 + \bar{\mathcal{E}}_{,\zeta}^2) + \bar{\mathcal{E}}_{,\varrho}(\mathcal{E}_{,\varrho}^2 + \mathcal{E}_{,\zeta}^2))$$
$$= \int_0^1 d\varrho \varrho^2 e^{2U}(\mathcal{E}_{,\varrho}\bar{\mathcal{E}}_{,\zeta} + \mathcal{E}_{,\zeta}\bar{\mathcal{E}}_{,\varrho}) + 2\int_0^\infty \int_0^\infty d\varrho d\zeta \varrho e^{2U}\mathcal{E}_{,\zeta}\bar{\mathcal{E}}_{,\zeta} , \quad (6.21)$$

see [131] for a derivation. Eq. (6.21) relates integrals of the Ernst potential and its derivatives over the whole spacetime to corresponding integrals at the disk. Since the numerics at the disk has been tested above, this provides a global test of the evaluation of the Ernst potential. As before, derivatives and integrals will be calculated via spectral methods.

The problem one faces when integrating over the whole spacetime is the singular behavior of the fields on the disk which represents a discontinuity of the Ernst potential. The Weyl coordinates in which the solution is given are not optimal to describe the geometry near the disk. Hence a huge number of polynomials is necessary to approximate the integrands in (6.21). Even with 512 polynomials for each coordinate, the coefficients of an expansion in Chebyshev polynomials did not drop below 10^{-6} in more relativistic situations. Though the computational limits are reached, the identity (6.21) is only satisfied to the order of 10^{-8} which is clearly related to the bad choice of coordinates.

We therefore use for this calculation so-called disk coordinates η, θ (see [106]) which are related to Weyl coordinates via

$$\varrho + i\zeta = \cosh(\eta + i\theta) \,. \tag{6.22}$$

The coordinate η varies between $\eta = 0$, the disk, and infinity, the coordinate θ between $-\pi/2$ and $\pi/2$. The axis is given by $\theta = \pm\pi/2$, the equatorial plane in the exterior of the disk by $\theta = 0$ and $\eta \neq 0$. Because of the equatorial symmetry, we consider only positive values of θ. The surfaces of constant η are confocal ellipsoids which approach the disk for small η. For large η, the coordinates are close to spherical coordinates.

To evaluate the integrals in (6.21), we perform the η-integration up to a value η_0 as well as the θ-integration with the Chebyshev integration routine. The parameter η_0 is chosen in a way that the deviation from spherical coordinates becomes negligible, typically $\eta_0 = 15$. The integral from η_0 to infinity is then carried out analytically with the asymptotic formula (4.11). It turns out that an expansion in 64 to 128 polynomials for each coordinate is sufficient to provide a numerically optimal approximation within the used precision. This illustrates the convenience of the disk coordinates in this context. The virial identity is then satisfied to the order of 10^{-12}. We plot the deviation of the sum of the integrals in (6.21) from zero for several values of ε and γ in Fig. 6.8. The drop in accuracy for strongly relativistic almost static situations (γ small and ε close to 1) is again due to the almost degenerate Riemann surface. The lower accuracy in the case of strongly relativistic situations for $\gamma = 1$ reflects the fact that the disk is shrinking to a point in this limit, see the discussion in Chap. 7 of the ultrarelativistic limit of the counter-rotating disk. To maintain the needed resolution one would have to use more polynomials in the evaluation of the virial-type identity which was not possible on the used computers.

6.3 Testing LORENE

One purpose of exact solutions of the Einstein equations is to provide testbeds for numerical codes to check the quality of the numerical approximation. In the previous sections we have established that the theta-functional solutions can be numerically evaluated to the order of machine precision which implies they can be used in this respect.

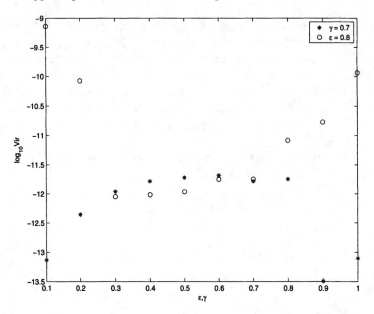

Fig. 6.8. The deviation from zero of the virial-type identity for $\gamma = 0.7$ and several values of ε, and for $\varepsilon = 0.8$ and several values of γ

The code we are considering here is a C++-library called LORENE [132] which was constructed to treat problems from relativistic astrophysics such as rapidly rotating neutron stars. The main idea is to solve Poisson-type equations iteratively via spectral methods. To this end an equation as the Ernst equation (1.4) is written in the form

$$\Delta \mathcal{F} = \mathcal{G}(\mathcal{F}, r, \theta, \phi) \,, \tag{6.23}$$

where spherical coordinates r, θ, ϕ are used, where Δ is the Laplace operator in these coordinates, and where $\mathcal{G}$ is some possibly non-linear functional of $\mathcal{F}$ and the coordinates. The system (6.23) is to be solved for $\mathcal{F}$ which can be a vector. In an iterative approach, the equation is rewritten as

$$\Delta \mathcal{F}_{n+1} = \mathcal{G}(\mathcal{F}_n, r, \theta, \phi) \,, \quad n = 1, 2, \ldots . \tag{6.24}$$

Starting from some initial function $\mathcal{F}_0$, in each step of the iteration a Poisson equation is solved for a known right-hand side. For the stationary axisymmetric Einstein equations which we are considering here, it was proven in [107, 108] that this iteration will converge exponentially for small enough boundary data if the initial values are close to the solution of the equation in some Banach space norm. It turns out that one can always start the iteration with Minkowski data, but it is necessary to use a relaxation: instead of the solution $\mathcal{F}_{n+1}$ of (6.24), it is better to take a combination $\tilde{\mathcal{F}}_{n+1} = \mathcal{F}_{n+1} + \kappa \mathcal{F}_n$ with $\kappa \in]0, 1[$ (typically $\kappa = 0.5$) as a new value in the source $\mathcal{G}_{n+1}$ to provide numerical stability. The iteration is in general stopped if $\|\mathcal{F}_{n+1} - \mathcal{F}_n\| < 10^{-10}$.

The Ernst equation (1.4) is already in the form (6.23), but it has the disadvantage that the equation is no longer strongly elliptic at the ergosphere where $\Re \mathcal{E} = 0$. In physical terms, this apparent singularity is just a coordinate singularity, and the theta-functional solutions are analytic there. The Ernst equation in the form (6.23) has a right-hand side of the form '0/0' for $\Re \mathcal{E} = 0$ which causes numerical problems especially in the iteration process since the zeros of the numerator and the denominator will only coincide for the exact solution. The disk solutions we are studying here have ergospheres in the shape of cusped toroids as will be shown in Chap. 7. Therefore it is difficult to take care of the limit 0/0 by using adapted coordinates. Consequently the use of the Ernst picture is restricted to weakly relativistic situations without ergospheres in this framework.

To be able to treat strongly relativistic situations, we use a different form of the stationary axisymmetric vacuum Einstein equations which is derived from the standard $3+1$-decomposition, see [232]. We introduce the functions ν and N_ϕ via

$$e^{2\nu} = \frac{\varrho^2 e^{2U}}{\varrho^2 - a^2 e^{4U}} , \quad N_\phi = \frac{\varrho a e^{4U}}{\varrho^2 - a^2 e^{4U}} . \tag{6.25}$$

The vacuum Einstein equations for the functions (6.25) read in cylindrical coordinates

$$\Delta \nu = \frac{1}{2} \varrho^2 e^{-4\nu} (N_{\phi,\varrho}^2 + N_{\phi,\zeta}^2) , \tag{6.26}$$

$$\Delta N_\phi - \frac{1}{\varrho^2} N_\phi = 4\varrho(N_{\phi,\varrho}(e^{2\nu})_\varrho + N_{\phi,\zeta}(e^{2\nu})_\zeta) . \tag{6.27}$$

By putting $V = N_\phi \cos\phi$ we obtain the flat 3-dimensional Laplacian acting on V on the left-hand side,

$$\Delta V = 4\varrho(V_\varrho(e^{2\nu})_\varrho + V_\zeta(e^{2\nu})_\zeta) . \tag{6.28}$$

Since the function $e^{2\nu}$ can only vanish at a horizon, it is globally non-zero in the examples we are considering here. Thus the system of equations (6.26) and (6.28) is strongly elliptic, even at an ergosphere.

The disadvantage of this regular system is the non-linear dependence of the potentials ν and N_ϕ on the Ernst potential and a via (6.25). Thus we loose accuracy due to rounding errors of roughly an order of magnitude. Though we have shown in the previous sections that we can guarantee the numerical accuracy of the data for e^{2U} and ae^{2U} to the order of 10^{-14}, the values for ν and V are only reliable to the order of 10^{-13}.

To test the spectral methods implemented in LORENE, we provide boundary data for the disk solutions discussed above on a sphere around the disk. For these solutions it would have been more appropriate to prescribe data at the disk, but LORENE was developed to treat objects of spherical topology such as stars which suggests the use of spherical coordinates. It would be possible to include coordinates like the disk coordinates of the previous section in LORENE, but the code should be tested in its present form. Instead we want to use the Poisson–Dirichlet routine which solves a Dirichlet boundary value problem for the Poisson equation for data prescribed at a sphere. We prescribe the data for ν and N_ϕ on a sphere of radius R and solve the system (6.26) and (6.28) iteratively in the exterior of the sphere. If the iteration converges, we compare the numerical solution in the exterior of the sphere with the exact solution.

Since spherical coordinates are not adapted to the disk geometry, a huge number of spherical harmonics would be necessary to approximate the potentials if R is close to the disk radius. The limited memory on the used computers imposes an upper limit of 64 to 128 harmonics. We choose the radius R and the number of harmonics in a way that the Fourier coefficients in θ drop below 10^{-14} to make sure that the provided boundary data contain the related information to the order of machine precision. The exterior of the sphere where the boundary data are prescribed is divided in two domains, one from R to $2R$ and one from $2R$ to infinity. In the second domain $1/r$ is used as a coordinate. For the ϕ dependence which is needed only for the operator in (6.28), 4 harmonics in ϕ are sufficient.

Since LORENE is adapted to the solution of the Poisson equation, it is to be expected that it reproduces the exact solution best for nearly static situations, since the static solutions solve the Laplace equation. The most significant deviations from the exact solution are therefore expected for $\delta = 0$. For the case $\lambda = 3$, we consider 32 harmonics in θ on a sphere of radius $R = 1.5$. The iteration is stopped if $||\mathcal{F}_{n+1} - \mathcal{F}_n|| < 5 * 10^{-10}$ which is the case in this example after 90 steps. The exact solution is reproduced to the order of 10^{-11}. The absolute value of the difference between the exact and the numerical solution on a sphere of radius 3 is plotted in Fig. 6.9 in dependence of θ. There is no significant dependence of the error on θ. The maximal deviation is typically found on or near the axis. As can be seen from Fig. 6.10 which gives the dependence on r on the axis, the error decreases almost linearly with $1/r$ except for some small oscillations near infinity.

We have plotted the maximal difference between the numerical and the exact solution for a range of the physical parameters λ and δ in Fig. 6.11. As can be seen, the expectation is met that the deviation from the exact solution increases if the solution becomes more relativistic (larger ε). As already mentioned, the solution can be considered as exactly reproduced if the deviation is below 10^{-13}. Increasing the value of γ for fixed ε leads to less significant effects though the solutions become less static with increasing γ.

For $\delta = 0$, the ultra-relativistic limit $\lambda \to 4.629\ldots$ corresponds to a spacetime with a singular axis which is not asymptotically flat, see [119] and the discussion in Chap. 7. Since LORENE expands all functions in a Galerkin basis with regular axis in an asymptotically flat setting, solutions close to

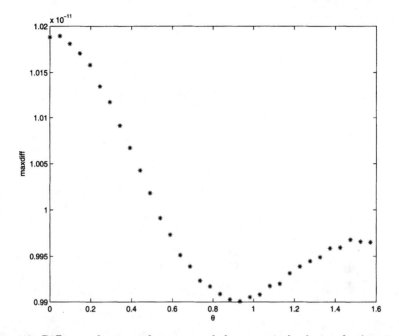

Fig. 6.9. Difference between the exact and the numerical solution for $\lambda = 3$ and $\delta = 0$ for $r = 3$ in dependence on θ

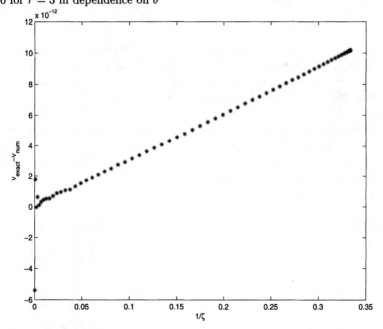

Fig. 6.10. Difference between the exact and the numerical solution for $\lambda = 3$ and $\delta = 0$ on the axis in dependence on ζ

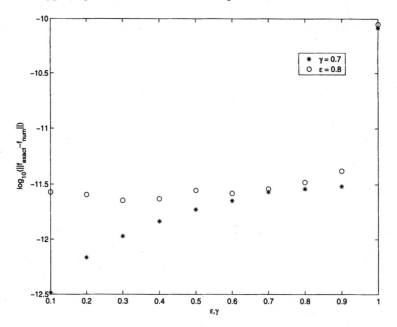

Fig. 6.11. Difference between the exact and the numerical solution for $\gamma = 0.7$ and several values of ε, and for $\varepsilon = 0.8$ and several values of γ

this singular limit cannot be approximated. Convergence gets much slower and can only be achieved with considerable relaxation. For $\lambda = 4$ and $\delta = 0$ we needed nearly 2000 iterations with a relaxation parameter of $\kappa = 0.9$. The approximation is rather crude (to the order of one percent). For higher values of λ no convergence could be obtained.

This is however due to the singular behavior of the solution in the ultra-relativistic limit. In all other cases, LORENE is able to reproduce the solution to the order of 10^{-11} and better. More static and less relativistic cases are reproduced with the provided accuracy.

7 Physical Properties

In Chap. 5 we studied solutions describing stationary counter-rotating dust disks in terms of hyperelliptic functions. As an example of this approach we gave an explicit solution on a Riemann surface of genus 2 in Theorem 5.16 where the two counter-rotating dust streams have constant angular velocity and constant relative density. In the present chapter we discuss the physical features of the class of hyperelliptic solutions (4.19) which are a subclass of Korotkin's finite gap solutions [52, 94] for the example of this disk. We demonstrate how one can extract physically interesting quantities from the hyperelliptic functions in terms of which the metric is given. The metric depends on two physical parameters: $\varepsilon = z_R/(1 + z_R)$ is related to the redshift z_R of photons emitted from the center of the disk and detected at infinity; γ is the relative density of the counter-rotating streams in the disk. In the Newtonian limit ε is approximately 0 whereas it tends to 1 in the ultrarelativistic limit where the central redshift diverges. The limit of a single component disk is reached for $\gamma = 1$ (we will only consider positive values of γ), the static limit for $\gamma = 0$.

We give analytic expressions for the mass and the angular momentum as an expansion of the metric functions at infinity. As in [117] and Chap. 5 we discuss the matter in the disk using observers which rotate in a way that the energy-momentum tensor is diagonal for them. We study the angular velocity of these observers with respect to the locally non-rotating frames, and the angular velocities and the energy densities of the dust components which these observers measure. In the limit of diverging central redshift the spacetime is no longer asymptotically flat in the case of a one component disk, and the axis is no longer elementary flat. This behavior can be related as in [205] to the vanishing of the radius ϱ_0 of the disk which was used as a length scale. If one carries out the limit $\varrho_0 \to 0$ for $\varrho \neq 0$, the metric becomes the extreme Kerr metric. In this limit the disk vanishes behind the horizon of the extreme Kerr solution. In the case of two counter-rotating dust components the radius of the disk remains finite even in the limit where the central redshift diverges. In the ultrarelativistic limit of the static disks, the matter in the disk moves at the speed of light, the energy density diverges at the center of the disk but the mass remains finite.

We follow closely the discussion in the pioneering paper [205], but this time for a class of solutions which depend on two parameters which continuously interpolate between the Newtonian and the ultrarelativistic regime, and the static and the Bardeen–Wagoner case respectively. The chapter is organized is follows: In Sect. 7.1 we write down the complete metric corresponding to the Ernst potential of Theorem 5.16 in terms of theta functions. In Sect. 7.2 we discuss various physical properties of the solutions: We relate the physical parameters ε and γ to the parameters on which the analytic solution depends and discuss mass and angular momentum. The angular velocity Ω is discussed as a function of ε and γ. We study the energy-momentum tensor at the disk as in [117] as well as the occurrence of ergospheres. In Sect. 7.3 we discuss the ultrarelativistic limit of the solutions. We briefly study the over-extreme case for the one-component solution where the boundary value problem at the disk is still solved but where a ring singularity exists in the spacetime since the parameters of the solution are beyond the ultrarelativistic limit.

7.1 Metric Functions

In Theorem 5.16 we have constructed the solution for the counter-rotating dust disk with constant γ and Ω, which takes in the cut-system of Fig. 6.1 the form

$$\mathcal{E}(\varrho, \zeta) = \frac{\Theta[m](\boldsymbol{\omega}(\infty^+) + \boldsymbol{u})}{\Theta[m](\boldsymbol{\omega}(\infty^+) - \boldsymbol{u})} e^I , \qquad (7.1)$$

where $\Theta[m]$ is the theta function on $\mathcal{L}$ with half-integer characteristic $[m]$ where $[m] = \frac{1}{2}\begin{bmatrix} 1 & 0 \\ 1 & 0 \end{bmatrix}$.

As shown in Sect. 3.6, the complete metric (2.40) can be expressed via theta functions. Carrying out the partial degeneration of the Riemann surface used there, we get with the characteristics

$$[n_1] = \frac{1}{2}\begin{bmatrix} 1 & 1 \\ 1 & 1 \end{bmatrix} , \quad [n_2] = \frac{1}{2}\begin{bmatrix} 0 & 0 \\ 1 & 1 \end{bmatrix} , \quad [n_3] = \frac{1}{2}\begin{bmatrix} 1 & 0 \\ 1 & 0 \end{bmatrix} , \quad [n_4] = \frac{1}{2}\begin{bmatrix} 0 & 1 \\ 1 & 0 \end{bmatrix} , \qquad (7.2)$$

for the function e^{2U} with (3.79)

$$e^{2U} = \frac{\Theta[n_1](\boldsymbol{u})\Theta[n_2](\boldsymbol{u})}{\Theta[n_1](0)\Theta[n_2](0)} \frac{\Theta[n_3](\boldsymbol{\omega}(\infty^-))\Theta[n_4](\boldsymbol{\omega}(\infty^-))}{\Theta[n_3](\boldsymbol{\omega}(\infty^-) + \boldsymbol{u})\Theta[n_4](\boldsymbol{\omega}(\infty^-) + \boldsymbol{u})} e^I . \qquad (7.3)$$

This form is especially adapted for determining ergospheres which are just the zeros of e^{2U}. It can be seen that the real part of the Ernst potential can only vanish if

$$\Theta[n_1](\boldsymbol{u})\Theta[n_2](\boldsymbol{u}) = 0 \qquad (7.4)$$

which provides a necessary condition for the occurrence of ergospheres (the sufficient condition is that the denominator in (7.3) is non-zero in this case).

For the metric function a we obtain with (3.98)

$$(a - a_0) e^{2U} = \tag{7.5}$$
$$-\varrho \left(\frac{\Theta[n_1](0)\Theta[n_2](0)}{\Theta[n_3](\omega(\infty^-))\Theta[n_4](\omega(\infty^-))} \frac{\Theta[n_1](u)\Theta[n_2](u + 2\omega(\infty^-))}{\Theta[n_3](u + \omega(\infty^-))\Theta[n_4](u + \omega(\infty^-))} - 1 \right) ,$$

where the constant $a_0 = -\gamma/\Omega$. The constant can be expressed via theta functions on the elliptic surface Σ' given by $\mu'^2 = (K - E_1^2)(K - \bar{E}_1^2)$. As before we denote quantities defined on Σ' by a prime and get

$$a_0 = \frac{\beta_1}{\alpha_1} \sqrt{\alpha_1^2 + \beta_1^2} \left(\frac{\vartheta_4^2(0)}{\vartheta_3(\omega'(\infty^-))\vartheta_4(\omega'(\infty^-))} \right)^2 \frac{\vartheta_4(u' + 2\omega'(\infty^-))}{\vartheta_4(u')} e^{-I'} , \tag{7.6}$$

where $d\omega_1 = d\omega'$, $d\omega_2 = d\omega'_{\zeta^- \zeta^+}$, $u_i = \frac{1}{2\pi i} \int_\Gamma \ln G d\omega_i$, and where $I' = \frac{1}{2\pi i} \int_\Gamma \ln G d\omega'_{\infty^+ \infty^-}$. For a definition of Jacobi's elliptic theta functions see the appendix or references given therein.

For the metric function k we get with (3.102)

$$e^{2k} = C \frac{\Theta[n_1](u)\Theta[n_2](u)}{\Theta[n_1](0)\Theta[n_2](0)} \tag{7.7}$$
$$\times \exp \left(\frac{2}{(4\pi i)^2} \int_\Gamma \int_\Gamma dK_1 dK_2 h(K_1)h(K_2) \ln \frac{\Theta_\star(\omega(K_1) - \omega(K_2))}{K_1 - K_2} \right) ,$$

where $\Theta_\star$ is a theta function with an odd non-singular half-integer characteristic, where $h(\tau) = \partial_\tau \ln G(\tau)$, and where C is a constant which is determined by the condition that k vanishes on the regular part of the axis and at infinity. It reads

$$1/C = \frac{\vartheta_4^2(u')}{\vartheta_4^2(0)} \tag{7.8}$$
$$\times \exp \left(\frac{2}{(4\pi i)^2} \int_\Gamma \int_\Gamma dK_1 dK_2 h(K_1)h(K_2) \ln \frac{\vartheta_1(\omega'(K_1) - \omega'(K_2))}{K_1 - K_2} \right) .$$

In an ergoregion, the function $\Theta[n_1](u)\Theta[n_2](u)$ becomes negative. Since the remaining terms in (7.7) cannot change sign, the function e^{2k} is always negative where e^{2U} is negative. The metric function $g_{11} = g_{22} = e^{2(k-U)}$ is consequently non-negative.

Since we can concentrate on positive values of ζ because of the equatorial symmetry of the solution, the Riemann surface can only become singular if ξ coincides with $\bar{\xi}$, i.e. on the axis, or if it coincides with E_2. Coinciding branch points imply that some of the periods diverge. To avoid numerical problems in the immediate vicinity of the axis, we substitute the analytic expression (4.8). The real part of the Ernst potential can be written in the form

$$e^{2U} = \frac{\vartheta_4^2(u')}{\vartheta_4^2(0)} \tag{7.9}$$

$$\times \frac{\vartheta_4^2 \left(\int_{\zeta^+}^{\infty^-} d\omega'\right) - \exp\left(-2\omega_2(\infty^-)\right) \vartheta_4^2 \left(\int_{\zeta^-}^{\infty^-} d\omega'\right)}{\vartheta_4^2 \left(u' + \int_{\zeta^+}^{\infty^-} d\omega'\right) - \exp\left(-2\omega_2(\infty^-) - 2u_2\right) \vartheta_4^2 \left(u' + \int_{\zeta^-}^{\infty^-} d\omega'\right)} \;.$$

For finite ζ it can only vanish for

$$\vartheta_4(u') = 0 \;. \tag{7.10}$$

Since this condition is independent of ζ, the real part of the Ernst potential will vanish on the whole axis in this case which means that the axis is an ergosphere. The spacetime is not asymptotically flat in this case. For an interpretation of this case as an ultrarelativistic limit see Sect. 7.3.3.

If ξ coincides with E_2, the Ernst potential and the metric functions can be expressed in terms of quantities defined on the Riemann surface Σ'' of genus 0 given by $\mu''^2(\tau) = (\tau - E_1)(\tau - \bar{E}_1)$ i.e., via elementary functions as shown in Sect. 4.1. For $\xi = E_2$ the differentials on $\mathcal{L}$ reduce to differentials on Σ'', $d\omega_1 = d\omega''_{E_2^- E_2^+}$, $d\omega_2 = d\omega''_{\bar{E}_2^- \bar{E}_2^+}$ and $I = I'' = \frac{1}{2\pi i} \int_\Gamma \ln G d\omega''_{\infty^+ \infty^-}$ where a double prime denotes that the quantity is defined on Σ''. The Ernst potential reads

$$\mathcal{E} = \frac{\sinh \frac{\omega_1(\infty^+) + u_1}{2}}{\sinh \frac{\omega_1(\infty^+) - u_1}{2}} e^{I''} \;, \tag{7.11}$$

the function a follows from

$$(a - a_0)e^{2U} + \varrho$$

$$= \frac{\varrho \sinh \frac{B_{12}}{4}}{2 \sinh \frac{\omega_1(\infty^+)}{2} \sinh \frac{\omega_2(\infty^+)}{2} \sinh \frac{u_1 - \omega_1(\infty^+)}{2} \sinh \frac{u_2 - \omega_2(\infty^+)}{2}} \times$$

$$\left(\exp\left(\frac{B_{12}}{4}\right) \cosh \frac{u_1 + u_2 + 2\omega_1(\infty^+) + 2\omega_2(\infty^+)}{2}\right.$$

$$\left. - \exp\left(-\frac{B_{12}}{4}\right) \cosh \frac{u_1 - u_2 + 2\omega_1(\infty^+) - 2\omega_2(\infty^+)}{2}\right) \;, \tag{7.12}$$

and the function e^{2k} is given by

$$e^{2k} = C \frac{\exp\left(\frac{B_{12}}{4}\right) \cosh \frac{u_1 + u_2}{2} - \exp\left(-\frac{B_{12}}{4}\right) \cosh \frac{u_1 - u_2}{2}}{2 \sinh \frac{B_{12}}{4}} \tag{7.13}$$

$$\exp\left(\frac{1}{(4\pi i)^2} \int_\Gamma \int_\Gamma \frac{dK_1 dK_2}{(K_1 - K_2)^2} \ln G(K_1) \ln G(K_2)\right.$$

$$\left. \times \left(\sqrt{\frac{(K_1 - E_1)(K_2 - \bar{E}_1)}{(K_1 - \bar{E}_1)(K_2 - E_1)}} + \sqrt{\frac{(K_1 - \bar{E}_1)(K_2 - E_1)}{(K_1 - E_1)(K_2 - \bar{E}_1)}} - 2\right)\right) \;.$$

At the disk the branch points ξ, $\bar{\xi}$ lie on the contour Γ which implies that care has to be taken in the evaluation of the path integrals. The situation is however simplified by the equatorial symmetry of the solution which is reflected by the additional involution $K \to -K$ of the Riemann surface $\mathcal{L}$ for $\zeta = 0$. This makes it possible to express the metric functions in terms of elliptic theta functions. In Sect. 4.2 we gave especially efficient formulas for the functions needed to calculate the energy-momentum tensor at the disk. With the notation of Sect. 4.2.1 the real part of the Ernst potential at the disk can be written as

$$
e^{2U} = \frac{1}{Y - \delta} \left(-\frac{1}{\lambda} - \frac{Y}{\delta} \left(\frac{\frac{1}{\lambda^2} + \delta}{\sqrt{\frac{1}{\lambda^2} + \delta \varrho^2}} - \frac{1}{\lambda} \right) \right.
$$
$$
\left. + \sqrt{\frac{Y^2((\varrho^2 + \alpha)^2 + \beta^2)}{\frac{1}{\lambda^2} + \delta \varrho^2} - 2Y(\varrho^2 + \alpha) + \frac{1}{\lambda^2} + \delta \varrho^2} \right), \quad (7.14)
$$

where

$$
Y = \frac{\frac{1}{\lambda^2} + \delta \varrho^2}{\sqrt{(\varrho^2 + \alpha)^2 + \beta^2}} \frac{\vartheta_3^2(u_w)}{\vartheta_1^2(u_w)}. \quad (7.15)
$$

In Theorem 5.16 it was shown that there exist algebraic relations between the real and imaginary parts of the Ernst potential, and the function Z defined in (3.10). Consequently these functions can be expressed in terms of the metric function e^{2U}.

At the rim of the disk ($\varrho = 1$ and $\zeta = 0$) the value of the metric function e^{2U} thus has the form

$$
e^{2U(1,0)} = 1 - \frac{1}{\delta} \left(\sqrt{\frac{1}{\lambda^2} + \delta} - \frac{1}{\lambda} \right). \quad (7.16)
$$

The imaginary part of the Ernst potential vanishes for $\gamma \neq 0$ at the rim of the disk as $(1 - \varrho^2)^{\frac{3}{2}}$. These explicit relations at the rim of the disk can be used as a test for the numerics.

To illustrate the metric functions we show plots for $\varepsilon = 0.85$ and $\gamma = 0.99$ ($\lambda = 10.12$ and $\delta = 0.856$), i.e., a disk in a strongly relativistic situation. The metric function e^{2U} (see Fig. 7.1) tends to 1 for large distances from the disk. At the disk it is continuous but its normal derivatives have a jump. In the vicinity of the disk, the function is negative which indicates the presence of an ergosphere. In the exterior of the disk, e^{2U} is completely smooth and does not take a local extremum in the whole physical range of the parameters. The function thus shows the same analytic properties as a solution to the Laplace equation.

The imaginary part of the Ernst potential (see Fig. 7.2) is an odd function in ζ. Thus it vanishes in the equatorial plane in the exterior of the disk. For large distances from the disk it tends to zero because of the asymptotic

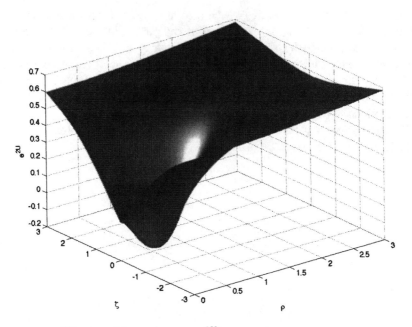

Fig. 7.1. Metric function e^{2U} for $\varepsilon = 0.85$ and $\gamma = 0.99$

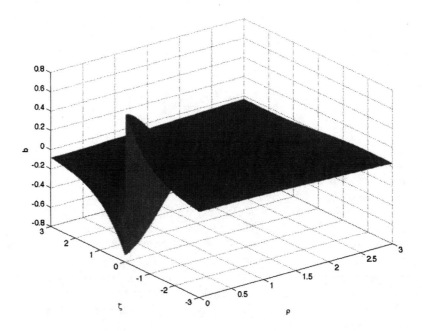

Fig. 7.2. Imaginary part of the Ernst potential for $\varepsilon = 0.85$ and $\gamma = 0.99$

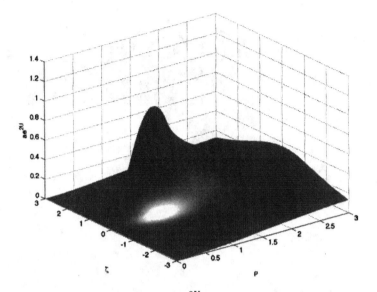

Fig. 7.3. Metric function ae^{2U} for $\varepsilon = 0.85$ and $\gamma = 0.99$

flatness of the spacetime. At the disk, the function has a jump which is zero at the rim of the disk since b is continuous there.

The metric function ae^{2U} (see Fig. 7.3) is equatorially symmetric and everywhere continuous. At the disk, the normal derivatives of a have a jump, in the remaining spacetime it is completely regular. On the axis and at infinity the function is identically zero.

The function e^{2k} in Fig. 7.4 has similar properties: it is equatorially symmetric and everywhere continuous, the normal derivatives have a jump at the disk. The function is identical to 1 on the axis ('elementary flatness') and at infinity (asymptotic flatness). The function is only significantly different from 1 in the vicinity of the disk. The metric function $e^{2(k-U)}$ is always positive even in the ergoregions which implies that the signature of the metric does not change.

7.2 Physical Properties of the Counter-rotating Dust Disk

In this section we discuss physical properties of the hyperelliptic solution for the example of the counter-rotating dust disks.

7.2.1 The Physical Parameters

We consider the metric as depending on the two physical parameters ε and γ. Mathematically more natural are the parameters λ and δ. These two sets

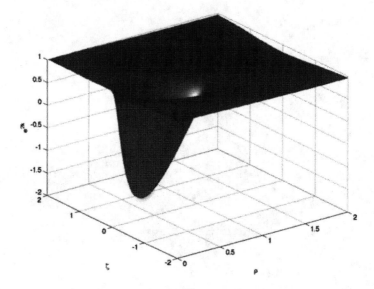

Fig. 7.4. Metric function e^{2k} for $\varepsilon = 0.85$ and $\gamma = 0.99$

can be converted through the following procedure. Formula (7.9) can be used to calculate the real part of the Ernst potential at the origin, e^{2U_0}, which is related to the redshift z_R of photons emitted from the center of the disk and detected at infinity, $z_R = e^{-U_0} - 1$,

$$e^{2U_0} = \frac{(1 + X^2)(\sqrt{1 + \lambda^2} - \lambda)}{X^2 - (\sqrt{1 + \lambda^2} - \lambda)^2} , \qquad (7.17)$$

where X is the purely imaginary quantity

$$X = \frac{\vartheta_3(u')\vartheta_4(0)}{\vartheta_1(u')\vartheta_2(0)} . \qquad (7.18)$$

The corresponding values of λ and δ follow from (5.26), (5.18) and (7.17). We get for $\varepsilon \neq 1$

$$\delta = \frac{1 - \gamma^2}{(1 - \varepsilon)^2} \frac{2}{\lambda} . \qquad (7.19)$$

With this value we enter equation (7.17) for e^{2U_0} and solve numerically for $\lambda(\varepsilon, \gamma)$. For $\delta = 0$ one finds that the first zero of e^{2U_0} is reached for $\lambda_c(0) = 4.62966\ldots$. The function has additional zeros for higher values of λ (see e.g. [118]). We are only interested in values $0 < \lambda < \lambda_c(\delta)$. For $\gamma < 1$ the quantity e^{2U_0} is a monotonous function in λ for $0 < \lambda < \infty$. Equation (7.19) then provides the corresponding value of $\delta(\varepsilon, \gamma)$.

For $\varepsilon = 1$ there are two cases: if $\gamma = 1$, then $\delta = 0$ and $\lambda = \lambda_c(0)$. For $\gamma \neq 1$, relation (7.19) implies that $\lambda_c(\delta)$ must be infinite. The corresponding value of δ follows with (5.26), (5.18) and (7.17) in the limit $\lambda \to \infty$ as the solution of the equation

$$\delta = \frac{4(1-\gamma^2)X^2}{1+X^2} . \tag{7.20}$$

Throughout this chapter we will consider the following limiting cases:
Newtonian limit: $\varepsilon = 0$ ($\lambda = 0$), i.e. small velocities $\Omega\varrho_0$ and small redshifts in the disk. For $\lambda \to 0$, the integral u' goes to zero. Thus the quantity X diverges since ϑ_1 is an odd function. Consequently one gets from (7.17) $U_0 = -\Omega^2$, the value for the Maclaurin disk (see [233] Theorem 5.1). There it was shown that in this limit e^{2U} tends to the Maclaurin disk solution, independently of γ. This solution can be written as

$$U(\varrho,\zeta) = -\frac{1}{4\pi i} \int_{-i}^{i} \frac{2\lambda(\tau^2+1)}{\sqrt{(\tau-\zeta)^2 + \varrho^2}} d\tau . \tag{7.21}$$

ultrarelativistic limit: $\varepsilon = 1$, i.e., diverging central redshift. For $\gamma = 1$ we have $\vartheta_4(u') = 0$ and thus $X = -i$ and $\mathcal{E}_0 = -i$, i.e. the value of the Ernst potential of the extreme Kerr metric at the horizon. For $\gamma \neq 1$, the ultrarelativistic limit is reached for $\lambda \to \infty$.
static limit: $\gamma = 0$ ($\delta = \delta_s(\lambda)$). In this limit, the branch points of Σ' collapse pairwise which leads to a diverging X and $e^{2U_0} = \sqrt{1+\lambda^2} - \lambda$. In [233] (Theorem 5.2) it was shown that this is the Morgan and Morgan solution [204] for constant Ω,

$$U(\varrho,\zeta) = -\frac{1}{4\pi i} \int_{-i}^{i} \frac{\ln G(\tau)}{\sqrt{(\tau-\zeta)^2 + \varrho^2}} d\tau \tag{7.22}$$

with

$$G = 1 - \frac{4}{\delta}(\tau^2+1) . \tag{7.23}$$

At the disk one has

$$e^{2U} = \sqrt{\frac{1}{4} - \frac{1}{\delta}} + \sqrt{\frac{1}{4} - \frac{1}{\delta} + \frac{\varrho^2}{\delta}} \tag{7.24}$$

with $\Omega^2 \delta = 1$.
one component: $\gamma = 1$ ($\delta = 0$), i.e., no counter-rotating matter in the disk. This is the disk which was studied numerically by Bardeen and Wagoner [205]. The analytic solution is due to Neugebauer and Meinel [109].

The parameter λ can be viewed as a 'relativity' parameter: for small values of λ, one is in the Newtonian regime, for larger values relativistic effects become more and more dominant up to the ultrarelativistic limit where the central redshift diverges. The values of λ itself, however, do not have an invariant meaning. Thus it seems better to use the central redshift z_R in $\varepsilon = z_R/(1+z_R)$ as a parameter as in [205],

$$\varepsilon = 1 - e^{U_0} , \qquad (7.25)$$

where e^{U_0} is taken from (7.18).

In the ultrarelativistic limit, the values of δ must be between 0 (the one-component case) and 4 (the static limit, where $\gamma = 0$ and $X^2 \to \infty$). We plot ε as a function of λ for $\gamma = 1$ and $\gamma = 0$ in Fig. 7.5. In the case $\gamma = 1$, the function goes to 1 at finite values of λ whereas for $\gamma \neq 1$ it goes monotonically to 1 as λ goes to infinity as in the static case $\gamma = 0$.

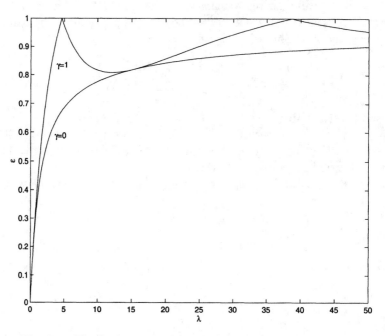

Fig. 7.5. The function ε in dependence of λ for $\gamma = 1$ and $\gamma = 0$

7.2.2 Mass and Angular Momentum

The ADM mass M and the angular momentum J of the spacetime (see e.g. [146]) can be obtained by expanding the axis potential (4.8) in the vicinity of infinity. The real part of the Ernst potential for $\varepsilon < 1$ reads $e^{2U} = 1 - 2M/\zeta + o(1/\zeta)$ and the imaginary part $b = 2J/\zeta^2 + o(1/\zeta^2)$. The ADM mass and the angular momentum can be calculated on the axis as was done in Proposition 4.5. For genus 2 we find

$$M = -D_{\infty^-} \ln \vartheta_4(u') - \frac{1}{4\pi i} \int_\Gamma \ln G \, d\omega'^{(1)}_{\infty^+} , \qquad (7.26)$$

for the ADM mass and

$$J = -\frac{\gamma}{\Omega}\left(D_{\infty^-}\ln\vartheta_4(u') + D_{\infty^-}\ln\vartheta_2(u') + \frac{1}{2\pi i}\int_\Gamma \ln G\, d\omega'^{(1)}_{\infty^+}\right) \quad (7.27)$$

for the angular momentum.

In the Newtonian limit this leads to

$$M = \frac{4\Omega^2}{3\pi}, \quad (7.28)$$

the value of the Maclaurin disk, and

$$J = \frac{8\gamma\Omega^3}{15\pi}. \quad (7.29)$$

In the ultrarelativistic limit of the one component disk, $\vartheta_4(u') = 0$, both the mass and the angular momentum diverge. In this limit the dimensionless quotient M^2/J remains bounded and goes to 1, the value of the extreme Kerr metric.

We plot the dimensionless quantity M^2/J in Fig. 7.6. As a function of ε it varies monotonically between the Newtonian value

$$\frac{M^2}{J} = \frac{10\Omega}{3\pi\gamma} \quad (7.30)$$

and the value in the ultrarelativistic case which is always bigger than 1 for $\gamma < 1$. For fixed ε it increases monotonically with γ.

7.2.3 Energy-momentum Tensor

The energy-momentum tensor of the disk is given by (5.14) which will be discussed as outlined in Sect. 5.2 by introducing the FIOs of [117]. If we introduce the four-velocities $\tilde{u}_\pm = N_\pm u_\pm$, the quantities $\sigma_\pm N_\pm^2$ are proper densities in the sense of [205]. The quantity σ which appears in the Einstein equations in Theorem 5.3 is related to $\tilde\sigma = \sigma_+ + \sigma_-$ via $\sigma = e^{k-U}\tilde\sigma$. Here σ is given by

$$\sigma = \frac{b_{,\varrho}}{8\pi\varrho\Omega^2(a - a_0)e^{2U}}. \quad (7.31)$$

It vanishes for $\varrho \to 1$ with infinite slope: in the non-static case it was shown in [233] that $b_{,\varrho}$ is always proportional to $\sqrt{1-\varrho^2}$ while in the static case one gets

$$\sigma = \frac{1}{4\pi^2\Omega\left(\frac{\delta}{4} - 1 + \varrho^2\right)}\arctan\sqrt{\frac{1-\varrho^2}{\frac{\delta}{4} - 1 + \varrho^2}}. \quad (7.32)$$

Since $b = b_0 + O(\varrho^2)$ in the vicinity of the origin for $\varepsilon \neq 1$, the density is regular in the whole disk for $\varepsilon < 1$ and $\gamma \neq 0$. This is however not true in the ultrarelativistic limit of the static disks which we will discuss in more detail in the following section.

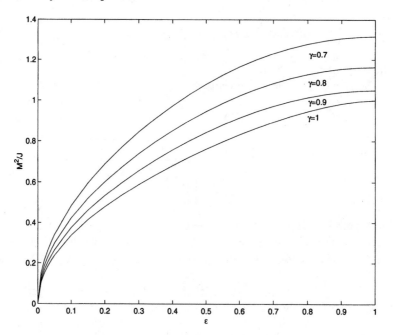

Fig. 7.6. The dimensionless quantity M^2/J in dependence of ε for several values of γ

The FIOs can interpret the matter in the disk as having a purely azimuthal pressure or as a disk of two counter-rotating dust streams if $p_p^*/\sigma_p^* < 1$. One can show numerically that p_p^*/σ_p^* is a monotonically decreasing non-negative function of γ which vanishes identically only for $\gamma = 1$. Thus, it is maximal in the static case as expected. There we have

$$1 - \frac{p_p^*}{\sigma_p^*} = 1 - \Omega^2 \varrho^2 e^{-4U} = e^{2(U_0 - U)} \geq 0 . \qquad (7.33)$$

The last equation follows from (5.24). The only case where $p_p^* = \sigma_p^*$ is the ultrarelativistic limit of the static disks. In this case the matter rotates with the velocity of light while in all other cases, the velocity $\sqrt{p_p^*/\sigma_p^*}$ is smaller than 1. Consequently the FIOs can interpret the matter in the disk as two streams of dust with proper energy density $\sigma_p^*/2$ which are counter-rotating with the same angular velocity $\Omega_c \doteq (N_2/N_1)\sqrt{p_p^*/\sigma_p^*}$. This is the interpretation we will refer to in our discussion.

Except for the static case $\gamma = 0$ the FIOs are not at rest with respect to the locally non-rotating frames which rotate with angular velocity

$$\omega_l = -\frac{g_{03}}{g_{33}} \qquad (7.34)$$

with respect to the inertial frame at infinity. Therefore, the quantities we will discuss in the following are the angular velocities ω_l, ω_ϕ, Ω_c, and the energy density $\sigma^* \doteq e^{(k-U)}\sigma_p^*$.

We discuss the angular velocities in units of Ω which has no invariant meaning but which provides a natural scale for the angular velocities in the disk. It is constant with respect to ϱ but depends on the parameters ε and γ. In the Newtonian limit it is small since $U_0 = -\Omega^2$. Thus independently of γ, the angular velocity Ω behaves as $\sqrt{\varepsilon}$ for $\varepsilon \approx 0$. The fact that the ultrarelativistic limit for the one-component disk is reached for a finite value of λ implies via (5.26) that Ω must vanish in this limit. This behavior will be discussed in more detail in Sect. 7.3.3. Thus, as ε varies between 0 and 1, for $\gamma = 1$, Ω starts near zero in the Newtonian regime, reaches a maximum smaller than 1 and then goes to zero. For $0 < \gamma < 1$, it reaches a maximum, too, but then it does not go to zero in the ultrarelativistic limit. In the static case ($\gamma = 0$) one has

$$\Omega(\varepsilon, 0) = \frac{1}{2}\sqrt{1-(1-\varepsilon)^4}\,, \qquad (7.35)$$

which grows monotonically from zero to $1/2$ in the ultrarelativistic limit. We plot Ω as function of ε for several values of γ in Fig. 7.7.

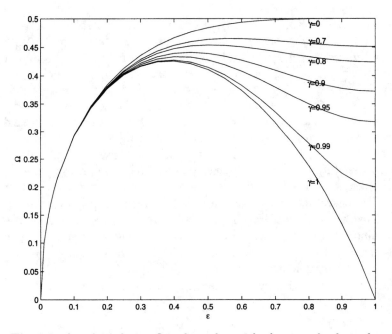

Fig. 7.7. Angular velocity Ω in dependence of ε for several values of γ

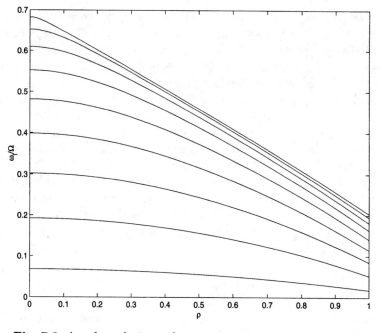

Fig. 7.8. Angular velocity ω_l for $\gamma = 0.7$ and $\varepsilon = 0.05, 0.15, ..., 0.85$

The angular velocity ω_l of the locally non-rotating observers is a measure for the frame dragging due to the rotating disk. We depict ω_l in dependence of ϱ at the disk for $\gamma = 0.7$ and several values of ε in Fig. 7.8. There is obviously no frame dragging in the Newtonian case, ω_l is of order Ω^3 for small Ω. The angular velocity ω_l increases monotonically with ε for fixed ϱ and γ. However the curves for $\varepsilon \geq 0.85$ are so close to the curve with $\varepsilon = 0.85$ that we omitted them in Fig. 7.8. Since the density (see below) is peeked at the center of the disk for $\varepsilon \to 1$, the frame dragging increases strongly near the center. In Fig. 7.9 we plot ω_l at the disk for $\varepsilon = 0.8$ for several values of γ. In the static case it is identical to zero. The frame dragging increases monotonically with γ for fixed ϱ and ε since more counter-rotating matter makes the spacetime more static. Since the central density decreases with γ for fixed ε, the frame dragging at the center is for $\gamma < 1$ closer to the one-component case than at the rim of the disk. The angular velocity ω_l is always smaller than Ω for $\gamma < 1$. In the ultrarelativistic limit for $\gamma = 1$ the ratio ω_l/Ω becomes identical to 1 in the disk.

In terms of the components of the energy-momentum tensor, the angular velocity ω_ϕ reads

$$\omega_\phi = \frac{1}{2S_3^0}\left(S_3^3 - S_0^0 - \sqrt{(S_3^3 - S_0^0)^2 + 4S_3^0 S_0^3}\right) . \tag{7.36}$$

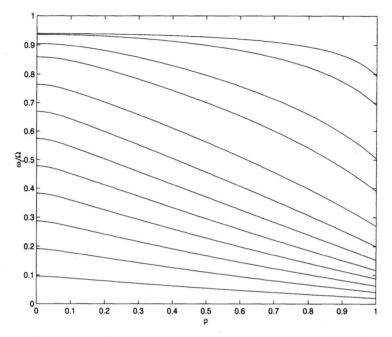

Fig. 7.9. Angular velocity ω_l for $\varepsilon = 0.8$ and $\gamma = 0.1, 0.2, ..., 0.9, 0.95, 0.99, 1$

For fixed ϱ and ε, the angular velocity ω_ϕ is monotonically increasing in γ from zero in the static case to Ω in the one-component limit. For $\varrho = 0$ it is identical to $\gamma\Omega$ which is also the value in the Newtonian limit. The ratio ω_ϕ/Ω is depicted in dependence of ϱ for $\gamma = 0.7$ for several values of ε in Fig. 7.10.

The angular velocity of the dust streams Ω_c with respect to the FIOs follows from

$$\Omega_c = \sqrt{\frac{\omega_\phi^2 - 2\omega_\phi\gamma\Omega + \Omega^2}{1 - 2\kappa\gamma\Omega + \Omega^2\kappa^2}} \ . \tag{7.37}$$

For fixed ϱ and ε the angular velocity Ω_c increases monotonically in γ from 0 in the one-component case to 1 in the static case. In the former case the observer follows the dust and can interpret the dust which is at rest in his coordinate system as 'two' non-rotating dust components. For $\varrho = 0$ the function Ω_c is identical to $\Omega\sqrt{1 - \gamma^2}$ which is also the value in the Newtonian limit. We plot Ω_c in dependence of ϱ for $\gamma = 0.7$ for several values of ε in Fig. 7.11.

The proper density σ_p^* for a FIO is given by

$$\sigma_p^* = \frac{\tilde{\sigma}}{1 - \kappa\omega_\phi} \frac{\varrho^2}{\kappa g_{03} + g_{33}} (1 - 2\kappa\gamma\Omega + \kappa^2\Omega^2) \ . \tag{7.38}$$

The density is finite except in the ultrarelativistic limit of the static disks. In the Newtonian limit, the density reads

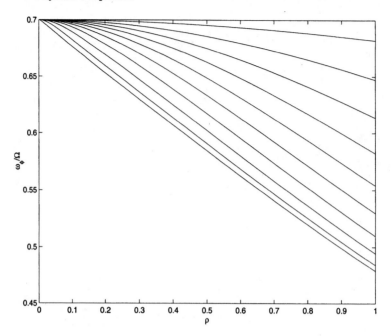

Fig. 7.10. Angular velocity ω_ϕ for $\gamma = 0.7$ and from top to bottom $\varepsilon = 0.05, 0.15, \ldots, 0.95$

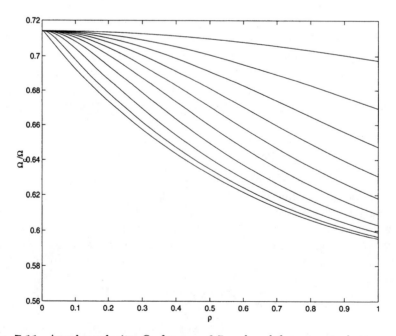

Fig. 7.11. Angular velocity Ω_c for $\gamma = 0.7$ and and from top to bottom $\varepsilon = 0.05, 0.15, \ldots, 0.95$

$$\sigma^* = \tilde{\sigma}(1 + \Omega^2((1-\gamma^2)\varrho^2 - 2)) = \frac{2\Omega^2}{\pi^2}\sqrt{1-\varrho^2}\,, \tag{7.39}$$

the value for the Maclaurin disk. The dependence of σ^* on ϱ is shown for $\gamma = 0.7$ for several values of ε in Fig. 7.12. With increasing ε, the central density grows and the matter is more and more concentrated at the center of the disk. For $\varepsilon = 0.8$ the density is plotted for several values of γ in Fig. 7.13. With increasing γ, the central density increases.

In [204] and [205] the observer dependent 'rest mass density' $\sigma_{0,\pm}$ of the dust streams was defined as $\sigma_{0,\pm} = \sigma^*/2U_{\pm}^0$ which leads to the total rest mass density σ_0 in the asymptotically fixed frame

$$\sigma_0 = \sigma^* \frac{N_1}{U^*}\,. \tag{7.40}$$

The total rest mass of the disk M_0 is then the integral

$$M_0 = 2\pi \int_0^1 \sigma_0 \varrho \mathrm{d}\varrho\,. \tag{7.41}$$

The binding energy of the disk is defined in [205] and [204] as the difference between the total rest mass and the ADM mass, $E_b = M_0 - M$. We plot E_b/M_0 as a function of ε for several values of γ in Fig. 7.14. In the Newtonian limit, the binding energy is independent of γ,

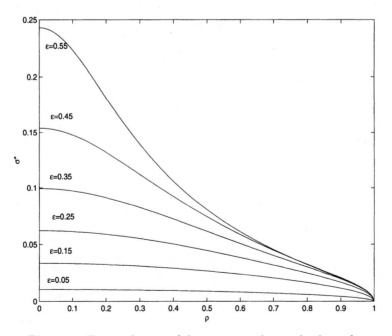

Fig. 7.12. Energy density σ^* for $\gamma = 0.7$ and several values of ε

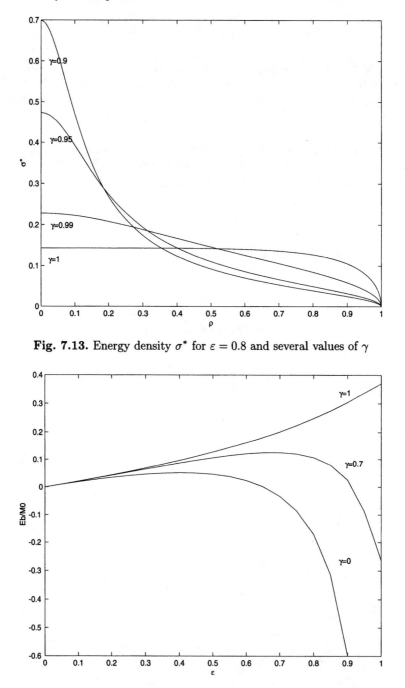

Fig. 7.13. Energy density σ^* for $\varepsilon = 0.8$ and several values of γ

Fig. 7.14. Binding energy of the disks in dependence of ε for several values of γ

$$E_b/M = \frac{1}{5}\Omega^2 \ . \tag{7.42}$$

In the case $\gamma = 1$, the binding energy increases monotonically up to a value of $E_b/M_0 \approx 0.37$ in the ultrarelativistic limit. For $\gamma < 1$ it reaches a maximum for a finite value of ε and can become even negative. In the static limit E_b/M_0 diverges to $-\infty$ in the ultrarelativistic limit since the rest mass of the disk goes to zero.

7.2.4 Ergospheres

In strongly relativistic situations it is possible that the asymptotically time-like Killing vector ∂_t becomes null or even spacelike. The vanishing of e^{2U} defines an ergosphere (although it does not have the topology of a sphere here) i.e. the boundary of a region of spacetime where there can be no static observer with respect to infinity.

The surface plot of the metric function e^{2U} in Fig. 7.1 shows the typical behavior of these functions: they are completely smooth in the exterior of the disk while the normal derivatives are discontinuous at the disk. The function does not assume a local extremum in the exterior of the disk and goes to 1 at infinity, $e^{2U} = 1 - 2M/|\xi| + \dots$. Since the ADM mass is always positive in the physical range of the parameters, the real part of the Ernst potential is always less than 1. At the disk, however, the function may have a global minimum.

In the Newtonian regime, the so-called gravito-magnetic effects such as ergospheres do not play a role. When the parameter ε increases from zero to one, the function e^{2U} may vanish at some points in the spacetime. Since it assumes its minimum value at the disk, this means that an ergosphere necessarily first appears at the disk when the minimum value becomes zero. For larger values of ε the minimum drops below zero in these cases so that the ergosphere grows for increasing values of ε. In the ultrarelativistic limit $\varepsilon = 1$ it reaches the axis.

To illustrate the dependence of ergospheres on the parameter ε for fixed γ, we plot them in Fig. 7.15 for $\gamma = 1$. The plot shows the (ϱ, ζ)-plane with the disk on the ϱ-axis between zero and one. The potential is regular in the equatorial plane in the exterior of the disk which implies that the equipotential surfaces hit the plane orthogonally there. At the disk, however, the normal derivatives have a jump which leads to a cusp of the equipotential contours at the disk. The ergosphere grows with ε and includes the whole spacetime in the ultrarelativistic limit which will be discussed in the next section.

Qualitatively, one would expect that counter-rotation makes a solution more static, i.e. that effects like ergospheres are suppressed. Thus in situations with the same central redshift but different γ, the ergoregion will always be smaller in the case of more counter-rotation if there is an ergoregion at all.

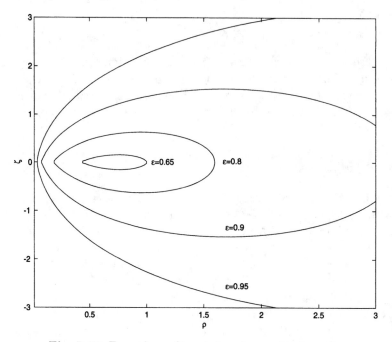

Fig. 7.15. Ergospheres for $\gamma = 1$ and several values of ε

In Fig. 7.16 we show the ergospheres for $\varepsilon = 0.95$ and several values of γ. It follows from (7.16) that the ergosphere goes through the rim of the disk if

$$\delta = 1 - \frac{2}{\lambda}\,. \tag{7.43}$$

This means that for disks with $\delta > 1$ possible ergoregions are confined to values of $\varrho < 1$. One finds numerically that smaller values of γ i.e. more counter-rotating matter imply that the ergoregion forms at bigger values of ε i.e., in stronger relativistic situations if it is to appear at all. The ergoregions are also formed closer to the axis. In the static case there is obviously no ergosphere. The function e^{2U} only vanishes in the ultrarelativistic limit at the center of the disk. There are no ergoregions for values of $\gamma < \gamma_c = 0.707\ldots$.

7.3 Ultrarelativistic Limit

In this section we discuss the limit in which the central redshift of the disks diverges.

7.3.1 Ultrarelativistic Limit of the Static Disks

The main features of the ultrarelativistic limit can already be found in [204]. The potential e^{2U} in the disk and its normal derivative there have the form

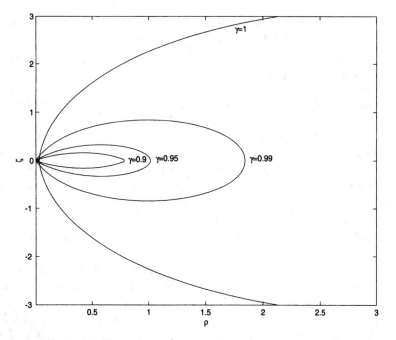

Fig. 7.16. Ergospheres for $\varepsilon = 0.95$ and several values of γ

$$e^{2U} = \frac{\varrho}{2} , \quad \left(e^{2U}\right)_{,\zeta} = \frac{1}{\pi} \arctan \sqrt{\frac{1 - \varrho^2}{\varrho^2}} , \qquad (7.44)$$

whereas the metric function k is of order ϱ^2 for small ϱ. The behavior of the metric functions can be obtained from (7.24) and (2.42). The angular velocity in the disk is $\Omega = 1/2$. The matter in the disk moves with the velocity of light since the four-velocity becomes null in the whole disk. The energy-density σ (7.32) diverges at the center as $1/\varrho^2$, the density $\sigma^* = -g_{00}\sigma$ diverges as $1/\varrho$. The ADM mass is however finite, $M = 1/(4\pi)$. Since the matter moves with the velocity of light, the rest mass of the disk must vanish. Thus the gravitational binding energy is negative.

The linear proper radius

$$\varrho_p \doteq \int_0^\varrho e^{k-U} d\varrho' \qquad (7.45)$$

is finite in the disk since the integrand behaves near the center (see [233], Corollary 4.1) as $1/\sqrt{\varrho}$ and is finite in the rest of the disk. The proper circumferential radius in the disk,

$$\varrho_c = \sqrt{g_{33}(\varrho)} = \sqrt{2\varrho} , \qquad (7.46)$$

is also finite. Thus the ultrarelativistic limit of the static disks with uniform rotation is a disk of finite radius with diverging central redshift and diverging

central density but finite mass. The matter in the disk consists of particles
with zero rest mass which move with the velocity of light.

7.3.2 Ultrarelativistic Limit for $0 < \gamma < 1$

The ultrarelativistic limit of stationary counter-rotating disks bears similar-
ities with the static case in the sense that the axis remains regular: the constants a_0 and C in (7.6) and (7.8) which are 0 and 1 respectively in the static
case remain finite here since they can only diverge if $\vartheta_4(u') = 0$ which cannot
happen for $\gamma \neq 1$. The integrals in the respective exponents of (7.6) and (7.8)
are always finite though $\ln G(\tau)$ has a term $\ln \tau$ in the limit $\lambda \to \infty$ as can
be easily seen. Thus the axis remains elementary flat in the case $\gamma < 1$ even
in the ultrarelativistic limit. Since $a_0 = -\gamma/\Omega$ is non-zero for $0 < \gamma < 1$, the
angular velocity Ω remains finite in the limit, too, as can be seen in Fig. 7.7.

In [233] (Corollary 4.1) it was shown that the potential e^{2U} is linear in ϱ
near the origin unless $\gamma = \gamma_c$ (which is just defined by this condition) where
it is quadratic in ϱ. For $\gamma > \gamma_c$ there are ergospheres in the spacetime, for
$\gamma < \gamma_c$ the potential e^{2U} is positive in the whole spacetime. We plot e^{2U}
at the disk for several values of γ in the ultrarelativistic limit in Fig. 7.17.
We note that the metric function ae^{2U} in the disk is also linear in ϱ in the
vicinity of the origin if e^{2U} is. For $\gamma \to 0$, the metric function e^{2U} in the disk
approaches $\varrho/2$. For $\gamma \to 1$ the limiting function is linear in ϱ in the whole

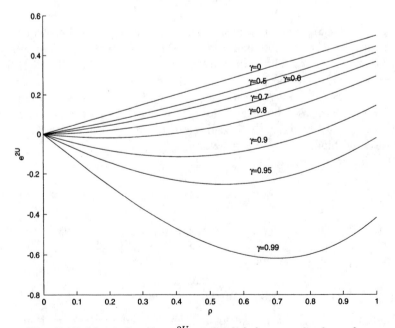

Fig. 7.17. Metric function e^{2U} at the disk for several values of γ

disk. One has to note that the limits $\gamma \to 1$ and $\varepsilon \to 1$ do not commute. The ultrarelativistic limit of the case $\gamma = 1$ is discussed in Sect. 7.3.3 below. The limit $\gamma \to 1$ of the ultrarelativistic solutions for $\gamma < 1$ is always obtained for $\lambda \to \infty$. If one goes with $\gamma \to 1$ ($\delta \to 0$) in this cases, the limiting function is one of the 'overextreme' solutions which are discussed in Sect. 7.3.4.

In contrast to the static case, the energy density σ^* is finite even in the ultrarelativistic limit. The proper linear radius (7.45) and the proper circumferential radius (7.46) are both finite in the disk. The velocity of the counter-rotating streams in the disk $\sqrt{p_p^*/\sigma_p^*}$ is less than 1, i.e. the velocity of light, in the limit $\varepsilon = 1$ for $0 < \gamma < 1$.

7.3.3 Ultrarelativistic Limit of the One-component Disks

The ultrarelativistic limit of the case $\gamma = 1$ is different from the previously discussed cases since it is reached for $\vartheta_4(u') = 0$. This implies with (7.6) and (7.8) that both constants a_0 and C diverge as $\varepsilon \to 1$. These constants do not have a direct physical importance. The fact that they diverge merely indicates that the axis cannot remain elementary flat in the ultrarelativistic limit. A consequence of the diverging constant a_0 is that the angular velocity Ω, which is the coordinate angular velocity in the disk as measured from infinity, vanishes. A diverging constant C implies that all linear proper distances (7.45) diverge. The function $e^{2(k-U)+2U_0}$ is however bounded.

The axis is in fact singular in the sense that the metric function e^{2U} vanishes there identically which can be seen from (7.9). The Ernst potential is identical to $-i$ on the axis for $\zeta > 0$. In the limit $\varepsilon \to 1$, the ergosphere becomes bigger and bigger. When it finally hits the axis for $\varepsilon = 1$, the whole axis and infinity form the ergosphere and the function e^{2U} is negative in the remainder of the spacetime. We plot the potential in Fig. 7.18. The fact that e^{2U} vanishes on the whole axis implies moreover that all multipole moments diverge. The dimensionless quotient M^2/J remains however finite and tends to 1, the value of the extreme Kerr metric (see Sect. 4.3).

The vanishing of $\Omega = \Omega\varrho_0$ in the limit $\varepsilon = 1$ indicates that either the angular velocity or the radius of the disk go to zero in this case. Bardeen and Wagoner [205] argued that the spacetime can be interpreted in the limit $\varepsilon \to 1$ and $\varrho_0 \to 0$ as the extreme Kerr metric in the exterior of the disk. In [118] it was shown that such a limit (diverging multipoles, singular axis,...) can occur in general hyperelliptic solutions and can always be interpreted as an extreme Kerr spacetime. For an algebraic treatment of the ultrarelativistic limit of the Bardeen–Wagoner disk see [234]. In the ultrarelativistic limit of the above disks for $\gamma = 1$, the spacetime becomes an extreme Kerr spacetime with $m = \frac{1}{2\Omega}$. The physical interpretation of this fact as already given in [205] is that the disks become more and more redshifted for increasing ε. Its radius shrinks and the disk finally vanishes behind the horizon of the extreme Kerr metric which forms in the ultrarelativistic limit.

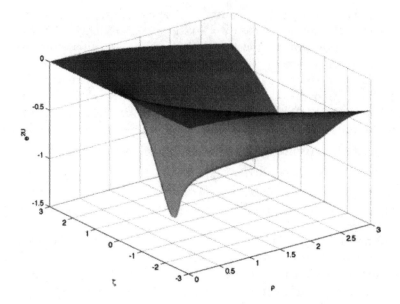

Fig. 7.18. Metric function e^{2U} in the ultrarelativistic limit of $\gamma = 1$

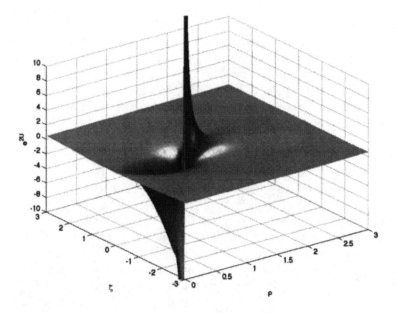

Fig. 7.19. Metric function e^{2U} in the over-extreme region of $\gamma = 1$

7.3.4 Over-extreme Region

Since the ultrarelativistic limit of the one-component disks is reached for a finite value λ_c of λ, the question arises what the solution (7.1) describes for $\lambda > \lambda_c$, the smallest value of λ where $\varepsilon = 1$. In Theorem 5.18 it was shown that the boundary conditions at the disk are still satisfied. Moreover the relations between the metric functions at the disk ensure that the functions are bounded at the disk (they have at most a jump discontinuity there). The proof for global regularity given in Theorem 5.18 does not hold in the 'over-extreme' region $\lambda > \lambda_c$. It indicates that a singularity in the equatorial plane is probable which in fact can be verified numerically. A typical plot is presented in Fig. 7.19. In the ultrarelativistic limit, the ergosphere stretches to infinity, in the over-extreme region with $\varepsilon < 1$ it is confined to a finite region of spacetime. The singularity in the equatorial plane is of the form $1/(\varrho - \varrho_s)$ at ϱ_s since the elliptic theta functions in the equatorial plane have zeros of first order. From formula (7.26) it can be seen that the singularity leads to a negative ADM mass for certain $\lambda > \lambda_c$. The spacetime is thus physically unacceptable. This is a striking example that it is not sufficient to solve a boundary value problem locally at the disk within the class of solutions [52], but that one has to find in addition the range of the physical parameters where the solution is globally regular outside the disk.

8 Open Problems

In the previous chapters we have discussed a subclass of Korotkin's hyperelliptic solutions to the Ernst equation with physically interesting properties. Physical and mathematical properties of the solutions have been studied analytically and numerically for in principle arbitrary genus of the solution. As an example we have presented the counter-rotating dust disk [130] which is given on a surface of genus 2, and which was obtained as the solution to a boundary value problem. What remains unclear is how to solve general boundary value problems with these Riemann surface techniques. In the case of dust disks the question is which rotation laws $\Omega(\varrho)$ are possible on a surface of given genus, and whether the solutions to these problems can be given in terms of quadratures. It is also unclear whether characteristic quantities of the Riemann surface can be directly related to physical quantities[1].

From an astrophysical point of view the most interesting models in this context are systems of black holes and surrounding thin disks of collisionless matter. They are discussed as models for accretions disks around black holes and for galaxies with supermassive black holes in the nucleus, see e.g. [235–237] for the observational evidence and [106, 238–240] for the theoretical background. Since a black hole is a genuinely relativistic object, a fully relativistic treatment of such situations is necessary. In cases where the mass of the disk is not negligible compared to the mass of the black hole, the self gravity of the disk plays a role. Disks around black holes must have an inner radius strictly larger than the photon radius, the radius where photons move on a sphere around the black hole, to exclude superluminal velocities of the particles.

In the context of the algebro-geometric solutions discussed in this book, the integrable non-linear equation is linearized on the Jacobian of a plane algebraic curve which implies that different solutions can be combined in a non-linear way (sometimes called 'non-linear superposition'). But since this is a non-linear operation, the combined solutions will have in general singularities as singular rings, Weyl struts on the axis and singular horizons. A well known example is the 'superposition' of two Kerr black holes which results

---

[1] In the case of the KdV equation this is only possible in the limit of small amplitudes, i.e. in the linear regime, where a genus g solution can be seen locally as a superposition of g solitary waves.

in a spacetime with a Weyl strut, a conical singularity on the axis in between the holes, or singular rings in the spacetime, see e.g. [35]. Such techniques are thus only helpful if the analytical properties of the solutions can be obtained in a general way as has been done in the preceding chapters.

In the Einstein–Maxwell case, similar techniques as in the stationary axisymmetric vacuum can be used since the equations are again completely integrable. However it turns out that theta-functional solutions in this case can no longer be constructed on the well understood hyperelliptic surfaces. Instead they are given on three-sheeted surfaces where it is not straight-forward to identify the needed analytical quantities. From an astrophysical point of view the importance of magnetic fields is known for the formation of neutron stars and for stars with a strong magnetic field, so-called magnetars. Charges typically compensate in astrophysical settings. Therefore electric fields are of less importance on larger scales. For a review of disk solutions to the stationary axisymmetric Einstein-Maxwell equations see [242].

In Sect. 8.1 we briefly review the algebraic approach of [217] which yields an integrated version of the Picard–Fuchs system in Chap. 5. This method could offer a more direct approach to the solution of boundary value problems in terms of theta functions. In Sect. 8.2 we discuss a subclass of Korotkin's solutions which are obtained on partially degenerate Riemann surfaces where precise statements on the analyticity of the solutions can be made. The solutions describe annular disks of infinite extension but finite mass and inner radius $\varrho_0 = 1$ in the equatorial plane. The black hole in the center of the annulus is characterized by a regular Killing horizon. The solutions are asymptotically flat, equatorially symmetric and regular outside the horizon and the disk. They contain a free function and a set of free parameters, the branch points of the Riemann surface. If the energy conditions are satisfied, the matter in the disk can be interpreted as in [117] as made up of one or more components of collisionless matter. Since the spacetimes are asymptotically flat, the matter in the asymptotic region behaves as in [117] as free particles which move on Keplerian orbits. The approach presented in [134] makes it possible to map the problem for an infinite disk around a black hole to the problem of a disk of finite radius which allows the use of the techniques of the previous chapters in this context. For the class of solutions presented here which can be seen as a combination of the Kerr solution with an infinite disk, we are thus able to establish regularity of the horizon and the exterior of the disk.

In Sect. 8.3 we study the Einstein–Maxwell equations in the presence of one Killing vector. We explore the group structure of the equations and give the Harrison transformation which generates electro-vacuum solutions from pure vacuum solutions. The solutions contain an additional real parameter related to the total charge. General properties of the transformed spacetimes as the asymptotics are discussed. In the stationary axisymmetric case, complete integrability of the equations is established. Solutions in terms of theta functions were obtained in [52] on three-sheeted Riemann surfaces.

8.1 Integrated version of the Picard–Fuchs system

In Chap. 5 we have used the Picard–Fuchs system of the hyperelliptic solutions to the Ernst equation together with an algebraic approach to the corresponding Ernst potentials to solve the boundary value problem for the counter-rotating dust disk. The disadvantage of this approach is the occurrence of a system of differential equations the order of which increases rapidly with the genus of the Riemann surface. Therefore it is in general difficult to solve boundary value problems with this approach.

This is why we have studied in [217] an integrated version of this system. The idea is to use the linear system (2.47) of the Ernst equation for the matrix Ψ and the explicit construction of this matrix Ψ in terms of theta functions in (3.48). If we consider the logarithmic derivative $\Psi_{,\xi}\Psi^{-1}$ in (2.47) it is equal to some matrix which is completely determined by the Ernst potential. The logarithmic derivative itself is given in terms of hyperelliptic functions for the class we are studying in this book. To establish integrated relations for the algebro-geometric solutions, the idea is to expand the matrix Ψ in the local parameter δ near ∞^+. To this end we consider the linear system (2.47) for the matrix $\tilde{\Psi}$ given by $\tilde{\Psi}(P) := \Psi(\infty^+)^{-1}\Psi(P)$,

$$\tilde{\Psi}_{,\xi} = \frac{1}{(\mathcal{E}+\bar{\mathcal{E}})^2}\begin{pmatrix} (\mathcal{E}\bar{\mathcal{E}})_{,\xi} & (\mathcal{E}-\bar{\mathcal{E}})_{,\xi} \\ \mathcal{E}^2\bar{\mathcal{E}}_{,\xi} - \bar{\mathcal{E}}^2\mathcal{E}_{,\xi} & -(\mathcal{E}\bar{\mathcal{E}})_{,\xi} \end{pmatrix}\left(\sqrt{\frac{K-\bar{\xi}}{K-\xi}}-1\right)\tilde{\Psi}, \qquad (8.1)$$

which is of the form $\tilde{\Psi}_{,\xi} = \mathcal{K}(\xi,\bar{\xi})\left(\sqrt{\frac{1-\delta\bar{\xi}}{1-\delta\xi}}-1\right)\tilde{\Psi}$, and similarly for $\tilde{\Psi}_{,\bar{\xi}}$. The advantage of this form of the linear system (2.47) is that the right-hand side vanishes for $\delta = 0$. If we expand $\tilde{\Psi}$ in δ, $\tilde{\Psi} = \hat{1} + \delta\Psi^1 + \delta^2\Psi^2 + \dots$, we obtain with constant matrices C_i

$$\Psi^1 = \int_{\Gamma_\xi} d\xi \mathcal{K}\frac{\xi-\bar{\xi}}{2} + C_1 \qquad (8.2)$$

and

$$\Psi^2 = \int_{\Gamma_\xi} d\xi \mathcal{K}\Psi^1 + \frac{1}{8}\int_{\Gamma_\xi} d\xi \mathcal{K}(3\xi+\bar{\xi})(\xi-\bar{\xi}) + C_2 \qquad (8.3)$$

and so on. The integrals $\int_{\Gamma_\xi} d\xi$ denote an integral along a contour in the ξ-plane from some point ξ_0 to ξ. If one is interested in boundary value problems, it is convenient to choose Γ_ξ to be the contour where the boundary values are prescribed. In [217] we have proven the following proposition in order δ:

Proposition 8.1. *The following relations are valid for $g > 1$:*

$$\Psi_{21}^1 + \Psi_{12}^1 \mathcal{E}\bar{\mathcal{E}} = \frac{\mathcal{E}\bar{\mathcal{E}}}{\mathcal{E} + \bar{\mathcal{E}}} \sum_{i=1}^{g} (K_i - \bar{K}_i) \tag{8.4}$$

and

$$\Psi_{21}^1 - \Psi_{12}^1 \mathcal{E}\bar{\mathcal{E}} = -\frac{\mathcal{E} - \bar{\mathcal{E}}}{4} \left(\sum_{i=1}^{g} (E_i + \bar{E}_i) - \xi - \bar{\xi} \right) - \frac{\mathcal{E} + \bar{\mathcal{E}}}{2} x_{g-2} . \tag{8.5}$$

The formulas hold also for $g = 1$ if one sets formally $x_{-1} = 0$. For a proof using the techniques of Chap. 5 see [217].

Remark 8.2. The 12-component of Ψ_1 is with (3.10) related to the metric function a, i.e. $\Psi_{12}^1 = -\mathrm{i}(a - a_0)/2$. The relation for the function Ψ_{21}^1 is new. Both functions can be considered as an integral of the Picard–Fuchs system (3.59) and (3.60), which can be seen by differentiating (8.5) with respect to ξ.

For the genus 2 solutions with equatorial symmetry and $E_1^2 = \alpha + \mathrm{i}\beta$ which we have discussed in detail in Chap. 5, we get in the equatorial plane with (8.4) and (8.5) in addition to (5.52) and

$$Z = \frac{\mathrm{i}x_0}{(1 - x^2)} - I_1 , \tag{8.6}$$

a relation between x, Z and the quantity $Y := -\mathrm{i}(\mathcal{E} + \bar{\mathcal{E}})\Psi_{21}^1/(\mathcal{E}\bar{\mathcal{E}})$ of the form

$$\left(\frac{1}{2} YZ - \frac{\alpha + \varrho^2/2}{1 - x^2} + \frac{\varrho^2}{2} \right)^2 (Y - Z)^2$$

$$- \left(\frac{1}{2} YZ - \frac{\alpha + \varrho^2/2}{1 - x^2} - \frac{\varrho^2}{2} \right)^2 x^2 (Y + Z)^2$$

$$= \frac{\alpha^2 + \beta^2}{1 - x^2} (Y - Z)^2 + \frac{\varrho^2}{4} (1 - x^2)(Y^2 - Z^2)^2 . \tag{8.7}$$

Equations (5.52) and (8.6) were used to determine x_0, R_i and I_i, $i = 1, 2$ in dependence of Z and x. The problem is that Y is not directly related to the boundary data as is Z. If one chooses an integration path Γ_ξ in (8.2) along the axis from infinity to the origin and then in the equatorial plane to the point $(\varrho, 0)$, the integration along the axis leads to a constant matrix Ψ^1 since the integrand in (8.2) vanishes there. The integration in the disk gives integrals over the boundary relations. Thus relation (8.7) only contains contributions from the boundary data as well as two integration constants and the constants α and β. It has to be checked whether these constants can be chosen in a way that (8.7) holds for some $\Omega(\varrho)$ in Theorem 5.3.

8.2 Black-hole Disk Systems

In [243] we have shown that annular disks with a finite radius lead to a difficult boundary value problem at the disk even in the Newtonian case. The reason is that one has to prescribe the normal derivative of the gravitational potential (vacuum and thus vanishing normal derivatives of the potential) in the exterior and the interior of the annular disk. At the disk one typically prescribes the tangential derivative (the angular velocity in the disk) which leads to a so-called three-part boundary value problem. If the disk stretches to infinity, the related two-part boundary value problem can be however solved. Therefore we consider in this section only infinite annular disks in asymptotically flat settings.

8.2.1 Newtonian Case

It is instructive to consider first a Newtonian analogue for a system consisting of a black hole and a disk, i.e. a point mass with a surrounding annular disk. As in Sect. 5.1 we describe the gravitational potential of disks in the equatorial plane $\zeta = 0$ between radii 0 and 1 in the form

$$U(\varrho, \zeta) = -\frac{1}{4\pi i} \int_\Gamma \frac{\mathcal{A}(\tau) d\tau}{\sqrt{(\tau - \zeta)^2 + \varrho^2}}, \tag{8.8}$$

where Γ is the part of the imaginary axis between $-i$ and i on the first sheet of the Riemann surface $\mathcal{L}_0$; $\mathcal{A}$ is a Hölder–continuous function on Γ independent of the physical coordinates. It has to vanish at the end points of Γ to avoid a ring singularity at the rim of the disk. Relation (8.8) can be used to construct an annular disk of infinite extension with finite inner radius with the help of a Kelvin transformation as in [243, 244], a reflection at the inner circle in the ϱ, ζ plane. It is a property of the Laplace operator that a function $\hat{U} = U(\varrho/r, \zeta/r)/r$ with $r^2 = \varrho^2 + \zeta^2$ is a solution to the Laplace equation if $\Delta U(\varrho, \zeta) = 0$. Using the linearity of the Laplace equation, we get that

$$U(\xi, \bar{\xi}) = -\frac{m}{r} - \frac{1}{4\pi i r} \int_\Gamma \frac{\mathcal{A}(\tau)}{\sqrt{(\tau - 1/\xi)(\tau - 1/\bar{\xi})}} d\tau \tag{8.9}$$

is a solution to the Laplace equation which describes a point mass surrounded by a disk of infinite outer radius and of inner radius 1. It can be easily seen that the solution U is equatorially symmetric if $\mathcal{A}$ is an even function. We note for later use that the integral in expression (8.9) can be written in the form

$$U(\varrho, \zeta) = -\frac{1}{4\pi i} \int_{\Gamma_\infty} \frac{\ln G(\tau) d\tau}{\sqrt{(\tau - \zeta)^2 + \varrho^2}} \tag{8.10}$$

after a transformation $\tau \to 1/\tau$ where $\Gamma_\infty = (-i\infty, -i] \cup [i, i\infty)$ and $\mathcal{A}(\tau) = \ln G(1/\tau)/\tau$ (the integral has to be understood as a standard contour integral in the complex plane after appropriate choice of the sign). This establishes the relation to the form of the solutions discussed in the previous chapters.

The total mass M of the system (8.9) is given by $M = m + \mathcal{A}(0)/4$. Thus the contribution of the disk to the mass is due to the value of $\mathcal{A}$ at the origin. We assume that the matter in the disk consists of pressureless matter rotating with angular velocity Ω. Thus the centrifugal force is the only force to stabilize the disk against gravitational collapse and has to compensate the gravitational attraction, $U_\varrho = \Omega^2(\varrho)\varrho$. In this Newtonian setting the velocity $\Omega\varrho$ of particles with radius ϱ in the disk is obviously related to the point mass. By limiting the value of m one can thus impose upper limits on the velocities of the particles. Asymptotically the angular velocity reads $\Omega^2 = M/\varrho^3$, the Keplerian value for test particles.

8.2.2 Relativistic Case

In Sect. 4.3 we have shown that the Kerr solution can be also obtained on a degenerate hyperelliptic Riemann surface of genus 2. In fact the Kerr solution can be obtained on a partially degenerate Riemann surface of higher genus. Consider families of hyperelliptic surfaces $\mathcal{L}$ of genus $g+2$ defined by the algebraic relation (A.82). The corresponding Ernst potential is given by (4.19). In [133] it was shown that the Kerr solution can be obtained from (4.19) for even g in the absence of a disk ($u_i = I = 0$) in the limit $E_{g+j} \to \bar{E}_{g+j} = K_j$, $j = 1, 2$ with $K_1 = -K_2 = m\cos\varphi$. The additional branch points are subject to the condition $E_i = -\bar{E}_{g+1-i}$, $i = 1, \ldots, g/2$, and the characteristic reads (we use the cut-system of Fig. A.4)

$$\begin{bmatrix} 0 \ldots 0 & \frac{1}{2} & \frac{1}{2} \\ \frac{1}{2} \cdots \frac{1}{2} & -\frac{1}{4} + \frac{\ln A}{2\pi i} & \frac{1}{4} - \frac{\ln A}{2\pi i} \end{bmatrix}, \tag{8.11}$$

where with A real

$$e^{-i\varphi} = \frac{iA+1}{iA-1} \prod_{i=1}^{g} \sqrt{\frac{K_1 - F_i}{K_1 - E_i}}. \tag{8.12}$$

Here the Kerr solution is obtained by partially degenerating the surface $\mathcal{L}$ which leads to an Ernst potential given on a surface $\tilde{\mathcal{L}}$ of genus g where the cuts $[E_{g+j}, \bar{E}_{g+j}]$, $j = 1, 2$ are removed from $\mathcal{L}$. The theta function on $\mathcal{L}$ is in this limit proportional to

$$\mathcal{F}(x) = \theta_{pq}\left(\tilde{x} + \int_{K_2^-}^{K_1^+} d\tilde{\omega}\right) + e^{x_{g+1}+x_{g+2}}\theta_{pq}\left(\tilde{x} + \int_{K_1^+}^{K_2^-} d\tilde{\omega}\right) \tag{8.13}$$

$$-ie^{-P}\left(Ae^{x_{g+1}}\theta_{pq}\left(\tilde{x} + \int_{K_1^+}^{K_2^+} d\tilde{\omega}\right) - A^{-1}e^{x_{g+2}}\theta_{pq}\left(\tilde{x} + \int_{K_2^+}^{K_1^+} d\tilde{\omega}\right)\right),$$

where $P = \mathbf{B}_{(g+1)(g+2)}/2$, i.e. it can be expressed completely in terms of quantities defined on $\tilde{\mathcal{L}}$, the corresponding theta function θ and the differentials $d\tilde{\omega}$. This makes it possible to combine the Kerr solution and the disks of infinite extension in a non-linear way by considering the above limit of the Ernst potential (4.19). The important point is that we can establish the regularity of the horizon and the range of the free parameters where the exterior of the disk is regular in a general way. These features which are summarized in the following theorem establish the physical relevance of this class of solutions.

Theorem 8.3. *The potential*

$$\mathcal{E} = \frac{\mathcal{F}(u + \omega(\infty^+))}{\mathcal{F}(u + \omega(\infty^-))} e^{\tilde{I} + \omega_{g+1}(\infty^+) + \omega_{g+2}(\infty^+)} \, , \quad \tilde{I} = \frac{1}{2\pi i} \int_{\Gamma_\infty} \ln G \mathrm{d}\tilde{\omega}_{\infty^+ \infty^-} \, ,$$

(8.14)

where $\mathrm{d}\omega_{g+j} = \mathrm{d}\omega_{K_j^- K_j^+}$, $j = 1, 2$, *is an equatorially symmetric solution to the Ernst equation which has a finite jump at the disk between 1 and infinity in the equatorial plane and a regular Killing horizon on the axis between* $-m \cos \varphi$ *and* $m \cos \varphi$. *The solution is regular in the exterior of the horizon and the disk if* $\mathcal{F}(u + \omega(\infty^-)) \neq 0$.

We briefly sketch the proof which follows from Theorem 4.10 and [133].

Proof. The fact that (8.14) solves the Ernst equation is a direct consequence of working out the limit of (4.19) as in [133]. The behavior at the disk was demonstrated in Chap. 4, here the discontinuity is located in the equatorial plane between 1 and infinity because of the Kelvin transformation. The potential on the axis can be calculated as in Sect. 4.1.4 by a further degeneration of the Riemann surface $\tilde{\mathcal{L}}$. The corresponding formulas also hold on the horizon where it is a consequence of [133] that the metric is identical to the Kerr metric which establishes the regularity of the horizon. All metric functions being given explicitly (see Sect. 3.6), the same is true on the axis where one finds that the function e^{2U} is zero at the endpoints of the horizon where the ergosphere touches the horizon and is negative in between. The functions a and k are zero on the axis in the exterior of the horizon and have non-vanishing constant values with respect to ζ on the horizon which characterizes a regular horizon (see e.g. Carter [241]). In the rest of the spacetime the Ernst potential is regular unless there is a zero in the denominator of (8.14) as was shown in Theorem 4.10.

□

Since the disk extends to infinity, the spacetime corresponding to (8.14) is in general not asymptotically flat. Choosing $\mathcal{A}(0)$ to be finite, one assures that $\mathcal{E}$ is finite at infinity. If $\mathcal{F}'(u')$ does not vanish on the regular part of the axis where it is a constant, the Ernst potential tends to 1 and one can define in a standard way multipole moments on the axis, $\mathcal{E} = 1 - 2M/\zeta - 2iJ/\zeta^2 + \dots$. For a general choice of the function G and the additional parameters, the mass M will not be real which implies a NUT parameter. For a vanishing NUT parameter, the spacetime is asymptotically flat in the standard sense. This procedure implies that the rotation state of the hole and the disk must be synchronized. In other words the black hole and the disk cannot rotate in an arbitrary way in an asymptotically flat setting.

8.2.3 The Case $g = 0$

For illustration we will consider the simplest case $g = 0$ in more detail even though it will turn out that the corresponding disks always consist of exotic matter. Since the Ernst potential is defined on a surface of genus zero it can be expressed in terms of elementary functions. Writing the Ernst potential in the form $\mathcal{E} = (\mathcal{G} - 1)/(\mathcal{G} + 1)e^I$, formula (8.14) takes the form

$$
\mathcal{G} = \frac{1 + e^{u_1 + u_2}}{1 - e^{u_1 + u_2} - i(Ae^{u_1} + e^{u_2}/A)} \frac{r_1 - r_2}{E_1 - E_2}
$$
$$
+ \frac{i(Ae^{u_1} - e^{u_2}/A)}{1 - e^{u_1 + u_2} - i(Ae^{u_1} + e^{u_2}/A)} \frac{r_1 + r_2}{E_1 - E_2} . \tag{8.15}
$$

We put $A = \cot \frac{\varphi}{2}$ to obtain for $u = I = 0$ the Kerr solution in the form (1.8). The static limit $\varphi = 0$ of these solutions describes the superposition of a Schwarzschild black hole and an annular disk as in [243, 244]. Since this is a solution to the Laplace equation in the exterior of the horizon and the disk, the solution is always regular there and asymptotically flat. This implies that the solutions (4.32) are also regular in this sense for small φ. It depends however on the choice of the function $\mathcal{A}$ whether the extreme limit $\varphi \to \frac{\pi}{2}$ can be reached without generating naked singularities as in the case of the over-extreme Kerr solutions. In non-static situations the spacetime will have a NUT parameter unless

$$
\int_{-i}^{i} \frac{\mathcal{A}(\tau)d\tau}{\tau^2 m^2 \cos^2 \varphi - 1} = 0 . \tag{8.16}
$$

This is obviously only possible if $\mathcal{A}$ changes sign on the path of integration. Since the density in the Newtonian limit is proportional to $\int_\varrho^1 dt \mathcal{A}(it)/\sqrt{t^2 - 1}$ (see Chap. 7), it must in this case also change sign. Numerically one finds that in the relativistic case, too, there must be regions with negative density in the disk which implies that the case $g = 0$ does not lead to asymptotically flat spacetimes with non-exotic matter in the disk.

The ADM mass is given by $M = m + \mathcal{A}(0)/4$, the angular momentum by $J = m^2 \sin \varphi (1 + \mathcal{A}(0)/(2m))$. The values for the invariant surface $\mathcal{A}_{BH}$ of the horizon and the constant value $1/\Omega_{BH}$ of the metric function a at the horizon are unchanged with respect to the pure Kerr case. The constant Ω_{BH} can be interpreted as the angular velocity of the black hole (see e.g. [238]).

We have plotted in Fig. 8.1 the real part of the Ernst potential for the simplest possible choice of $\mathcal{A}$, a polynomial in τ^2 of second order, in the case where the contribution of the disk and the hole to the mass are equal. It can be seen that the function is equatorially symmetric. It is negative or zero on

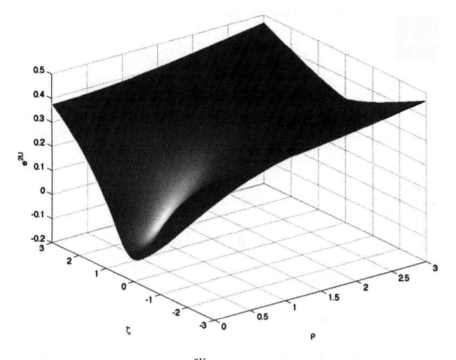

Fig. 8.1. Metric function e^{2U} in the case $g = 0$ for $m = 1$, $\varphi = \pi/2$

the horizon located on the axis and continuous at the disk between 1 and infinity in the equatorial plane. The normal derivatives of the function are discontinuous at the disk which leads to a cusp.

In contrast to the genus 0 case, the condition of a vanishing NUT parameter can in principle be satisfied for higher genus by choosing the parameter A appropriately in dependence of the other parameters whereas the density can be positive in the whole disk. For a given function $\mathcal{A}$, the parameter m has to be chosen in a way that the energy conditions in the disk are satisfied. The analytically known expressions in terms of theta constants for the horizon surface $\mathcal{A}_{BH}$ and the angular velocity Ω_{BH} are in general different from the pure Kerr case.

Thus it appears possible that physically acceptable disks can be found for $g \geq 2$. Since it is straight-forward to extend the formalism of [113] and the numerical treatment of theta functions in [131] to the case discussed here, it should be possible to identify physically interesting black hole disk systems in this class. Whether the algebraic difficulties in solving boundary value problems due to a prescribed matter distribution in the disk can be handled analytically is an open question.

8.3 Einstein–Maxwell Equations

In the Einstein–Maxwell case, the Maxwell equations have in principle the same form as in the absence of gravitation, only the partial derivatives have to be replaced by covariant derivatives since the spacetime is no longer flat (see e.g. [35]),

$$F_{\mu\nu}{}^{;\nu} = 0 , \quad {}^*F_{\mu\nu}{}^{;\nu} = 0 . \tag{8.17}$$

The tracefree energy-momentum tensor of the electromagnetic field is given by

$$T_{\mu\nu} = F_{\mu\alpha}F_\nu{}^\alpha - \frac{1}{4}g_{\mu\nu}F_{\alpha\beta}F^{\alpha\beta} . \tag{8.18}$$

Since the Einstein equations have the form

$$R_{\mu\nu} - \frac{1}{2}g_{\mu\nu}R = T_{\mu\nu} , \tag{8.19}$$

we get

$$R_{\mu\nu} = F_{\mu\lambda}F_\nu^\lambda - \frac{1}{4}g_{\mu\nu}F_{\kappa\lambda}F^{\kappa\lambda} . \tag{8.20}$$

They can be derived from the action

$$S = \frac{1}{2}\int dx^4 \sqrt{-g}\left(R - \frac{1}{2}F_{\mu\nu}F^{\mu\nu}\right) , \quad g = \det(g_{\mu\nu}) . \tag{8.21}$$

Equations (8.17) and (8.20) form the Einstein–Maxwell equations. Since the Maxwell fields only enter the Einstein equations via the energy-momentum tensor, the well known $U(1)$ symmetry of the Maxwell equations carries over to the Einstein–Maxwell case.

In general one would expect that the above $U(1)$ invariance of the Einstein–Maxwell equations is the only symmetry of the equations even in the presence of Killing symmetries in the spacetime. However, it turns out that a much bigger symmetry group exists already for a single Killing vector. It is convenient to use again the projection formalism of [121, 211] as in Sect. 2.1. The metric is written in the form (2.2) for the stationary case.

The vector potential is decomposed as the metric into pieces parallel and orthogonal to the Killing vector, $A_\mu = (A, \mathcal{A}_m + k_m A)$. The Lagrangian of (8.21) can then be written in the form

$$L^{(3)} = \frac{1}{2}\sqrt{h}\left(\mathcal{R} - \frac{1}{2f^2}h^{ab}f_{,a}f_{,b} + \frac{f^2}{4}\mathcal{K}_{ab}\mathcal{K}^{ab} + \frac{1}{f}h^{ab}A_{,a}A_{,b}\right.$$
$$\left. - \frac{f}{2}(\mathcal{F}_{ab} + A\mathcal{K}_{ab})(\mathcal{F}^{ab} + A\mathcal{K}^{ab})\right) , \tag{8.22}$$

where $L^{(3)}$ is a three-dimensional Lagrangian density, where $\mathcal{F}_{ab} = \mathcal{A}_{a,b} - \mathcal{A}_{b,a}$, and where $\mathcal{K}_{ab} = k_{a,b} - k_{b,a}$. All indices are raised and lowered with h_{ab}.

The first part of the Maxwell equations (8.17) can be written in the form

$$\frac{1}{\sqrt{-g}}(\sqrt{-g}F^{\mu\nu})_{,\nu} = 0 \qquad (8.23)$$

which implies $(\sqrt{h}F^{ab}/f)_{,b} = 0$. With this relation or by varying (8.22) with respect to $\mathcal{A}_a$, we obtain

$$(\sqrt{h}f(\mathcal{F}^{ab} + A\mathcal{K}^{ab}))_{,b} = 0 . \qquad (8.24)$$

We can define the potential B as in magnetostatics via

$$B_{,c} = -\frac{1}{2}\varepsilon_{cab}\sqrt{h}f(\mathcal{F}^{ab} + A\mathcal{K}^{ab}) . \qquad (8.25)$$

Again A and B can be combined to the complex electromagnetic potential $\Phi = A + iB$.

Similarly we get by varying (8.22) with respect to k_a

$$\left(\sqrt{h}\left(\frac{f^2}{2}\mathcal{K}^{ab} - Af(\mathcal{F}^{ab} + A\mathcal{K}^{ab})\right)\right)_{,b} = 0 . \qquad (8.26)$$

This can be dualized as in the vacuum case by introducing the twist potential b via

$$b_{,c} = \varepsilon_{cab}\sqrt{h}\frac{f^2}{2}\mathcal{K}^{ab} + BA_{,c} - AB_{,c} . \qquad (8.27)$$

The potentials f and b can again be combined to the complex Ernst potential,

$$\mathcal{E} = f - \Phi\bar{\Phi} + ib . \qquad (8.28)$$

The scalars b and B replace the vectors k_a and $\mathcal{A}_a$. The corresponding three-dimensional Lagrangian reads with $w_a = b_{,a} - 2BA_{,a} + 2AB_{,a}$

$$L^{(3)} = \frac{\sqrt{h}}{2}\left(\mathcal{R} - h^{ab}\left(\frac{1}{2f^2}(f_{,a}f_{,b} + w_aw_b) - \frac{1}{f}(A_{,a}A_{,b} + B_{,a}B_{,b})\right)\right) . \qquad (8.29)$$

The line element

$$ds^2 = \frac{1}{2f^2}((df)^2 + (db + 2BdA - 2AdB)^2) - \frac{1}{f}((dA)^2 + (dB)^2)) \qquad (8.30)$$

describes the invariant metric of the Riemannian symmetric space $S = SU(2,1)/S[U(1,1) \times U(1)]$ in some coordinates. The stationary Einstein–Maxwell equations can thus be interpreted as three-dimensional gravity coupled to some matter model. The 'matter' is an $SU(2,1)/S[U(1,1) \times U(1)]$ nonlinear sigma model [246] which can again be related to harmonic maps, see [247]. Following [248] we parametrize the space S by trigonal 3×3 matrices V,

$$V = \begin{pmatrix} \sqrt{f} & 0 & 0 \\ i\sqrt{2}\Phi & 1 & 0 \\ (b+i|\Phi|^2)/\sqrt{f} & (\sqrt{2}\bar{\Phi})/\sqrt{f} & 1/\sqrt{f} \end{pmatrix}. \tag{8.31}$$

The matrix V satisfies

$$V^\dagger \eta V = \eta, \quad \eta = \begin{pmatrix} 0 & 0 & i \\ 0 & 1 & 0 \\ -i & 0 & 0 \end{pmatrix}, \tag{8.32}$$

i.e. it is unitary with respect to the metric η of $SU(2,1)$. The action of $\mathcal{G} \in SU(2,1)$ on V is

$$V \to H(V,\mathcal{G})V\mathcal{G}^{-1}, \quad H(V,\mathcal{G}) \in S[U(1,1) \times U(1)], \tag{8.33}$$

where H restores the triangular gauge of V. To obtain a gauge invariant parametrization, one introduces

$$\chi := \Xi V^\dagger \Xi V, \quad \Xi = \mathrm{diag}(1,-1,1), \tag{8.34}$$

on which the action of $\mathcal{G} \in SU(2,1)$ is given by

$$\chi \to \Xi(\mathcal{G}^{-1})^\dagger \Xi \chi \mathcal{G}^{-1}. \tag{8.35}$$

We have

$$\chi = \begin{pmatrix} f - 2|\Phi|^2 + (b^2 + |\Phi|^4)/f & \sqrt{2}\bar{\Phi}(b - i|\Phi|^2 + if)/f & (b - i|\Phi|^2)/f \\ -\sqrt{2}\Phi(b + i|\Phi|^2 - if)/f & 1 - 2|\Phi|^2/f & -\sqrt{2}\Phi/f \\ (b + i|\Phi|^2)/f & \sqrt{2}\bar{\Phi}/f & 1/f \end{pmatrix}. \tag{8.36}$$

The $SU(2,1)$ symmetry can be used to generate solutions by the action of an element $\mathcal{G}$. We list the infinitesimal transformations and their consequences:

$$\begin{pmatrix} 0 & 0 & 0 \\ \theta_1 & 0 & 0 \\ \theta_2 & \theta_3 & 0 \end{pmatrix}$$

lead to gauge transformations which add physically irrelevant constants to $\Im\mathcal{E}$ and $\Im\Phi$,

$$\begin{pmatrix} 0 & 0 & \theta \\ 0 & 0 & 0 \\ 0 & 0 & 0 \end{pmatrix}$$

is an Ehlers transformation [121] which changes $f \to b$, i.e. which generates stationary from static solutions (if the ADM mass of the spacetime is non-zero, the transformed solution will have a NUT parameter)

$$\begin{pmatrix} i\theta & 0 & 0 \\ 0 & -2i\theta & 0 \\ 0 & 0 & i\theta \end{pmatrix}$$

is an electromagnetic duality transformation, i.e. the $U(1)$ symmetry of the sourcefree Maxwell equations,

$$\begin{pmatrix} \theta & 0 & 0 \\ 0 & 0 & 0 \\ 0 & 0 & -\theta \end{pmatrix}$$

is a scale transformation, $f, b, \Phi \rightarrow e^\theta f, e^\theta b, e^{\theta/2}\Phi$, and

$$\begin{pmatrix} 0 & -i\theta & 0 \\ 0 & 0 & \bar{\theta} \\ 0 & 0 & 0 \end{pmatrix}$$

is a Harrison transformation [136] which changes $f \rightarrow \Phi$, i.e. generates solutions with electromagnetic fields from pure vacuum solutions.

8.3.1 Harrison Transformations

Harrison transformations offer the possibility to generate solutions with charge from pure vacuum solution. This leads to solutions to the Einstein–Maxwell equations containing one additional constant parameter which is related to the charge. For physical reasons we are interested in solutions which are equatorially symmetric and asymptotically flat, i.e. $f \rightarrow 1$, $\Phi \rightarrow 0$ and $b \rightarrow 0$ for $|\xi| \rightarrow \infty$.

We assume that the pure vacuum solutions which we want to submit to a Harrison transformation satisfy these conditions. To ensure that the transformed solutions have the same asymptotic behavior, one has to use a scale transformation ($f \rightarrow 1$) together with a transformation which changes Φ and b by some constant ($\Phi, b \rightarrow 0$). We are interested in transformations which preserve the equatorial symmetry, i.e. $f(-\zeta) = f(\zeta)$, $b(-\zeta) = -b(\zeta)$ and $\Phi(-\zeta) = \bar{\Phi}(\zeta)$. By exponentiating the matrices of the $SU(2,1)$ transformations, we thus consider a transformation of the form

$$\mathcal{G} = \frac{1}{1 - q^2} \begin{pmatrix} 1 & i\sqrt{2}q & -iq^2 \\ -i\sqrt{2}q & 1 + q^2 & -\sqrt{2}q \\ iq^2 & -\sqrt{2}q & 1 \end{pmatrix} . \tag{8.37}$$

If we transform an Ernst potential of a pure vacuum solution ($\Phi = 0$), we end up with

$$f' = \frac{(1 - q^2)^2 f}{(1 - q^2 f)^2 + q^4 b^2} , \tag{8.38}$$

$$b' = \frac{(1 - q^4) b}{(1 - q^2 f)^2 + q^4 b^2} , \tag{8.39}$$

$$\Phi' = -q \frac{(1 - f)(1 - q^2 f) + q^2 b^2 + ib(1 - q^2)}{(1 - q^2 f)^2 + q^4 b^2} . \tag{8.40}$$

The real parameter q has to be in the region $0 < |q| < 1$, for $q > 1$ the transformed spacetime would have a negative mass if the original mass was positive. The value $q = 0$ corresponds to the untransformed solution. The above formulas imply that the functions f', b' and Φ' are analytic where the original functions are analytic.

A well-known example is the Harrison transformation of the Schwarzschild solution which leads to the Reissner–Nordström solution. In the Ernst picture the Schwarzschild solution reads in cylindrical Weyl coordinates with $r_\pm = \sqrt{(\zeta \pm m)^2 + \varrho^2}$

$$f = \frac{r_+ + r_- - 2m}{r_+ + r_- + 2m}, \quad b = 0, \tag{8.41}$$

where the horizon ($f = 0$) is located on the axis between $-m$ and m. For the transformed solution we get with (8.40)

$$f' = \frac{(r_+ + r_-)^2 - 4m^2}{(r_+ + r_- + 2m')^2}, \quad \Phi' = \frac{2Q}{r_+ + r_- + 2m'},$$

$$m' = m\frac{1 + q^2}{1 - q^2}, \quad Q = -\frac{2mq}{1 - q^2}, \tag{8.42}$$

and $b' = 0$ which is the Reissner–Nordström solution. This is a static space-time with mass m' and charge Q subject to the relation $m'^2 - Q^2 = m^2$. Both m' and Q diverge for $q \to 1$. The extreme Reissner–Nordström solution with $m' = Q$ is only possible in the limit $m \to 0, |q| \to 1$. The horizon of the solution is again located on the axis between $\pm m$ which illustrates that the horizon degenerates in the extreme case.

8.3.2 Asymptotic Behavior of the Harrison Transformed Solutions

We assume that the asymptotic behavior of the original solution, which can be read off on the axis, is of the form $f = 1 - 2M/|\zeta|$, $b = -2J/\zeta^2$ and $\Phi = Q/|\zeta| - iJ_M/\zeta^2$ plus terms of lower order in $1/|\zeta|$ where M is the ADM mass, J the angular momentum, Q the electric charge and J_M the magnetic moment. The same will hold for the Harrison transformed potentials. We find (see [137])

$$M' = M\frac{1 + q^2}{1 - q^2} - \frac{2q}{1 - q^2}Q, \quad J' = J\frac{1 + q^2}{1 - q^2} - \frac{2q}{1 - q^2}J_M, \tag{8.43}$$

and

$$Q' = Q\frac{1 + q^2}{1 - q^2} - \frac{2q}{1 - q^2}M, \quad J'_M = J_M\frac{1 + q^2}{1 - q^2} - \frac{2q}{1 - q^2}J. \tag{8.44}$$

It is interesting to note that the quantities $M^2 - Q^2$ and $J^2 - J_M^2$ are invariants of the transformation. They are related to the Casimir operator of the $SU(2,1)$-group. If the original solution was uncharged, the extreme relation $M' = \pm Q'$ is only possible in the limit $M \to 0$.

A consequence of the relations (8.44) is the presence of a non-vanishing charge if the ADM mass of the original solution is non-zero whereas the charge is. Since charges normally compensate in astrophysical settings, the astrophysical relevance of Harrison transformed solutions is limited.

A further invariant is the combination $J_M M - JQ$ which is of importance in relation to the gyromagnetic ratio

$$g_M = \frac{2M J_M}{JQ} . \tag{8.45}$$

Relation (8.43) implies that g'_M is equal to 2 if $Q = J_M = 0$ and $q \neq 0$. Thus all solutions which can be generated via a Harrison transformation from solutions with vanishing electromagnetic fields as the Kerr–Newman family from Kerr (and this includes most of the known exact solutions) have a gyromagnetic ratio of 2. Due to the invariance of $J_M M - JQ$ under Harrison transformations, a gyromagnetic ratio of 2 is not changed under the transformation.

Whether this property is an indication of a deep relation between relativistic quantum mechanics and general relativity as claimed in [249, 250] is an open question. Here it is just related to an invariant of the Harrison transformation, a subgroup of $SU(2,1)$. Numerical calculations of charged neutron stars [251] indicate, however, that values well below 2 are to be expected in astrophysically realistic situations.

8.3.3 The Stationary Axisymmetric Case

In the astrophysically important stationary axisymmetric case, the symmetry group of the equations increases again, this time an infinite-dimensional group as shown by Kinnersley [135]. This means that the equations are completely integrable as in the vacuum case. In the Weyl–Lewis–Papapetrou metric (2.40) the Einstein–Maxwell equations reduce to the electromagnetic Ernst equations [245]

$$f \Delta \mathcal{E} = (\nabla \mathcal{E} + 2\bar{\Phi}\nabla\Phi)\nabla\mathcal{E} ,$$
$$f \Delta \Phi = (\nabla \mathcal{E} + 2\bar{\Phi}\nabla\Phi)\nabla\Phi , \tag{8.46}$$

where Δ and ∇ are the standard differential operators in cylindrical coordinates, and where the potentials $\mathcal{E}$ and Φ are independent of ϕ. The duality relations (8.25) and (8.27) read

$$(\Im\Phi)_{,\xi} = \frac{i}{\varrho} f(A_{\phi,\xi} - aA_{t,\xi}) , \tag{8.47}$$

$$a_{,\xi} = \frac{\varrho}{f^2} \left(i(\Im\mathcal{E})_{,\xi} + \Phi\bar{\Phi}_{,\xi} - \bar{\Phi}\Phi_{,\xi} \right) , \tag{8.48}$$

which implies that a and A_3 follow from $\mathcal{E}$ and Φ. We choose a gauge where $A_1 = A_2 = 0$. The equations for $\mathcal{R}_{ab}$ of (8.29) are equivalent to

$$k_{,\xi} = \frac{\xi - \bar{\xi}}{f} \left(\frac{1}{4f} (\mathcal{E}_{,\xi} + 2\bar{\Phi}\Phi_{,\xi})(\bar{\mathcal{E}}_{,\xi} + 2\Phi\bar{\Phi}_{,\xi}) - \Phi_{,\xi}\bar{\Phi}_{,\xi} \right) . \tag{8.49}$$

Thus the complete metric and the electromagnetic potential can be obtained from given potentials $\mathcal{E}$ and Φ via quadratures.

Since the electromagnetic Ernst equations are completely integrable, there is an associated linear differential system for a 3×3 matrix-valued function Ψ which reads in the form [252]

$$\Psi_{,\xi}\Psi^{-1} = \begin{pmatrix} \mathcal{D}_1 & 0 & \mathcal{M}_1 \\ 0 & \mathcal{C}_1 & 0 \\ -\mathcal{N}_1 & 0 & \frac{1}{2}(\mathcal{C}_1 + \mathcal{D}_1) \end{pmatrix} + \frac{K - \bar{\xi}}{\mu_0} \begin{pmatrix} 0 & \mathcal{D}_1 & 0 \\ \mathcal{C}_1 & 0 & -\mathcal{M}_1 \\ 0 & -\mathcal{N}_1 & 0 \end{pmatrix} ,$$

$$\Psi_{,\bar{\xi}}\Psi^{-1} = \begin{pmatrix} \mathcal{D}_2 & 0 & \mathcal{M}_2 \\ 0 & \mathcal{C}_2 & 0 \\ -\mathcal{N}_2 & 0 & \frac{1}{2}(\mathcal{C}_2 + \mathcal{D}_2) \end{pmatrix} + \frac{K - \xi}{\mu_0} \begin{pmatrix} 0 & \mathcal{D}_2 & 0 \\ \mathcal{C}_2 & 0 & -\mathcal{M}_2 \\ 0 & -\mathcal{N}_2 & 0 \end{pmatrix} , \tag{8.50}$$

where Ψ depends on the spectral parameter K which varies on the Riemann surface $\mathcal{L}_0$ of genus zero given by the relation $\mu_0^2(K) = (K - \bar{\xi})(K - \xi)$.

The expressions for $\mathcal{C}_i$, $\mathcal{D}_i$ and $\mathcal{M}_i$ ($i = 1, 2$) follow from the condition

$$\Psi(\infty^+, \xi, \bar{\xi}) = \begin{pmatrix} \bar{\mathcal{E}} + 2\Phi\bar{\Phi} & 1 & \sqrt{2}i\Phi \\ \mathcal{E} & -1 & -\sqrt{2}i\Phi \\ -2i\bar{\Phi}e^U & 0 & \sqrt{2}e^U \end{pmatrix} . \tag{8.51}$$

As in the vacuum case in Chap. 2, the existence of the linear system can be used to generate solutions. Since the matrix Ψ in (8.50) is now a 3×3-matrix, the same holds for the monodromy matrix L of (3.52). The characteristic equation

$$Q(\hat{\mu}, K) = \det(L - \hat{\mu}\hat{1}) = 0 \tag{8.52}$$

is thus cubic in $\hat{\mu}$ which can be always brought into normal form by a redefinition of $\hat{\mu}$:

$$\hat{\mu}^3 + \mathcal{P}(K)\hat{\mu} + \mathcal{Q}(K) = 0 . \tag{8.53}$$

The functions $\mathcal{P}$ and $\mathcal{Q}$ are analytic in K. Equation (8.53) defines a three-sheeted Riemann surface which will in general have infinite genus. For polynomial $\mathcal{P}$ and $\mathcal{Q}$, the surface will be compact and will have finite genus. On a given surface the solutions to the Ernst equations can be given in terms of the corresponding theta functions which was first done in [52].

The theory of these surfaces is not as well understood as the theory of hyperelliptic surfaces which occur in the pure vacuum case, for instance it is not straight forward to construct the holomorphic differentials for a Riemann surface where only the branching is known. Therefore we have taken an intermediate step in [137] where we have considered a Harrison transformation of the counter-rotating dust disk [130]. The found charged disks can be interpreted for a range of the physical parameters as charged dust moving on electro-geodesics, i.e. on solutions to the geodesic equation in the presence of a Lorentz force. However as shown in Sect. 8.3.2 above, these solutions will always have a non-vanishing charge. Astrophysically interesting solutions with strong magnetic field and vanishing total charge can thus be only expected on three-sheeted surfaces or suitable degenerations. To identify disk solutions with strong magnetic fields, but vanishing total charge will be the subject of further research. In this context the Riemann–Hilbert techniques of Korotkin [253] for multi-sheeted Riemann surfaces will be important, see also [254, 255] for the related τ-function.

A Riemann Surfaces and Theta Functions

In this appendix we collect some basic facts on theta functions on Riemann surfaces. The idea is to give a comprehensive presentation of the mathematics on theta functions and the notation used in this book without providing lengthy proofs. For more detailed accounts of the subject, the reader is referred to [128], [138], [139], [187], and [256] to [270], for topics related to the Ernst equation see also [271].

A.1 Riemann Surfaces and Algebraic Curves

A *Riemann surface* Σ is a connected complex one-dimensional manifold. Riemann surfaces can be associated to multi-valued functions. Let f be an analytic function on $\mathbb{C}^2$ with arguments w and z. The equation

$$f(w, z) = 0 \tag{A.1}$$

defines a one-dimensional complex submanifold of $\mathbb{C}^2$. Let the complex gradient $\mathrm{grad}_\mathbb{C} f$ be given by

$$\mathrm{grad}_\mathbb{C} f := \left(\frac{\partial f}{\partial w}, \frac{\partial f}{\partial z} \right) \tag{A.2}$$

and call (w_0, z_0) with $f(w_0, z_0) = 0$ regular iff

$$\mathrm{grad}_\mathbb{C} f|_{(w_0, z_0)} \neq 0 . \tag{A.3}$$

It can be shown that the corresponding Riemann surface admits a compactification if equation (A.1) describes a plane algebraic curve, i.e. if f is of the form

$$f(w, z) = \sum_{i=0}^{k} w^i \, a_i(z) , \tag{A.4}$$

with a_i $(i = 0, \ldots, k)$ being polynomials in z. Then $w = w(z)$ is a multiple-valued algebraic function. It can be shown that any compact Riemann surface can be represented as an algebraic curve. A complex structure is introduced on the algebraic curve in the following way: in the neighborhoods of the points where $\partial f/\partial w \neq 0$, the variable z is taken as a local parameter, in the neighborhoods of the points $\partial f/\partial z \neq 0$, the parameter w is taken as a local parameter.

In this volume we shall be mainly concerned with functions of the form

$$f(w, z) = w^2 - P_n(z) , \qquad (A.5)$$

with P_n being a polynomial of degree n in z. The corresponding multiple-valued function is denoted by $w = \sqrt{P_n(z)}$ and the Riemann surface $\mathcal{L}$ is called *hyperelliptic* (for $n = 3, 4$ the surface is called elliptic). The hyperelliptic surface is everywhere regular if and only if $P_n(z)$ has no multiple roots. For $w \neq 0$ one chooses z as a local parameter, for $z \sim z_i$, where $P_n(z_i) = 0$, the variable $\tau_{z_i} = \sqrt{z - z_i}$ can be used as a local parameter.

For Σ being hyperelliptic the *hyperelliptic involution* σ is defined by

$$\sigma : \Sigma \ni P = (z, w) \to \sigma(P) \equiv P^\sigma = (z, -w) \in \Sigma , \qquad (A.6)$$

i.e., σ interchanges the two sheets of the Riemann surface.

Any compact Riemann surface Σ_g of genus g is topologically equivalent to a sphere with g handles. Any compact Riemann surface has finite genus. The first homology group of Σ_g is denoted by $H_1(\Sigma_g, \mathbb{Z})$. A standard basis of generators of $H_1(\Sigma_g, \mathbb{Z})$ consists of g pairs of cycles $(a_1, b_1), \ldots, (a_g, b_g)$ where a pair (a_i, b_i) encircles the i-th handle (or surrounds the ith hole) so that a_i intersects b_i, see Fig. A.1

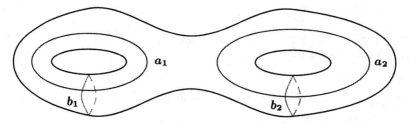

Fig. A.1. A homology basis for a Riemann surface of genus two

The way cycles intersect is described by intersection numbers. In Fig. A.2 we show for two cycles γ_1 and γ_2 when the intersection number $\gamma_1 \cdot \gamma_2$ is $+1$ or -1.

We choose oriented closed curves $a_1, \ldots, a_g, b_1, \ldots, b_g$ such that their intersection numbers are

$$a_i \cdot a_j = b_i \cdot b_j = 0 ,$$
$$a_i \cdot b_j = -b_i \cdot a_j = \delta_{ij} . \qquad (A.7)$$

A basis with the above intersection numbers is called *canonical basis*. The choice of such a basis is not unique: Any other basis $(\tilde{a}, \tilde{b})$ of $H_1(\Sigma_g, \mathbb{Z})$

Fig. A.2. The orientation dependence of the intersection numbers

(where a and b denote the g-dimensional vector $a = (a_1, \ldots, a_g)^T$ and $b = (b_1, \ldots, b_g)^T$) is given by the transformation

$$\begin{pmatrix} \tilde{a} \\ \tilde{b} \end{pmatrix} = A \begin{pmatrix} a \\ b \end{pmatrix}, \quad A \in SL(2g, \mathbb{Z}). \tag{A.8}$$

From the requirement that the new basis is also canonical we find that the matrix A is symplectic, $A \in Sp(g, \mathbb{Z})$,

$$J = AJA^T, \quad J = \begin{pmatrix} 0 & -I \\ I & 0 \end{pmatrix}. \tag{A.9}$$

A canonical basis is also referred to as a cut-system. If one cuts the Riemann surface starting from one point along the canonical cycles, the resulting surface is simply connected, a $4g$-gon called the fundamental polygon. For the surface of Fig. A.1 one gets the fundamental polygon shown in Fig. A.3.

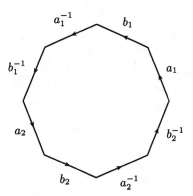

Fig. A.3. The fundamental polygon $\tilde{\Sigma}_g$ of a Riemann surface Σ_g of $g = 2$

A.2 Differentiation and Integration on Riemann Surfaces

A differential (or one form) $d\Omega = a dx + b dy = \alpha dz + \beta d\bar{z}$, where z is a local complex coordinate in the neighborhood of some point P on the Riemann surface Σ_g of genus g is called an *Abelian differential* if we have

$$d\Omega = f(z) \, dz \, , \tag{A.10}$$

with f being a meromorphic function in the vicinity of P. Since

$$d(d\Omega) = \frac{\partial f}{\partial z} dz \wedge dz + \frac{\partial f}{\partial \bar{z}} d\bar{z} \wedge dz = 0 \, , \tag{A.11}$$

Abelian differentials are always closed. We may classify Abelian differentials as follows:

– Abelian differentials of the first kind (or holomorphic differentials): f is a holomorphic function in any local chart.
– Abelian differentials of the second kind: have a single pole of order higher than 1 with vanishing residue.
– differentials of the third kind: have two first-order poles with residues ± 1.

Furthermore, it turns out that each Abelian differential can be decomposed into Abelian differentials of the above mentioned kinds. We denote a differential of the second kind with pole p of order $n + 1$ by $d\Omega_P^{(n)}$ and a differential of the third kind with poles P (residue $+1$) and Q (residue -1) by $d\Omega_{PQ}$. We have for the differential $d\Omega_{PQ}$

$$d\Omega_{PQ} = \left(\frac{1}{\tau_P} + O(1) \right) d\tau_P \, , \quad \text{near } P \, ,$$

$$d\Omega_{PQ} = \left(-\frac{1}{\tau_Q} + O(1) \right) d\tau_Q \, , \quad \text{near } Q \, , \tag{A.12}$$

where τ_P and τ_Q are local parameters at P and Q with $\tau_P(P) = 0$ and $\tau_Q(Q) = 0$. For the differential of the second kind we have

$$d\Omega_P^{(n)} = \left(\frac{1}{\tau_P^{n+1}} + O(1) \right) d\tau_P \, , \quad \text{near } P \, , \tag{A.13}$$

where τ_P is a local parameter at P with $\tau_P(P) = 0$. A differential of the second kind can be obtained from a differential of the third kind by differentiating

$$d\Omega_P^{(n)} = \frac{1}{n!} \partial_P^n d\Omega_{PQ} \, . \tag{A.14}$$

An *Abelian integral* on a Riemann surface is an integral of an Abelian differential.

It can be shown that the vector space $H^1(\Sigma_g)$ of holomorphic differentials on Σ_g is g-dimensional. For example if Σ_g is a hyperelliptic Riemann surface of the form $\mu^2 = \prod_{i=1}^{2g+2} (\lambda - \lambda_i)$, we may define a basis in $H^1(\Sigma_g)$ by

$$d\nu_k = \frac{\lambda^{k-1} d\lambda}{\mu} , \tag{A.15}$$

for $i = 1, \ldots, g$. The hyperelliptic differentials of the third kind have the form

$$d\omega_{PQ}(R) = \left(\frac{\mu + \mu_P}{\lambda - \tau_P} - \frac{\mu + \mu_Q}{\lambda - \tau_Q}\right) \frac{d\lambda}{2\mu} , \quad \text{if } \mu_P \neq 0, \mu_Q \neq 0 ,$$

$$d\omega_{PQ}(R) = \left(\frac{\mu + \mu_P}{\lambda - \tau_P} - \frac{1}{\lambda - \tau_Q}\right) \frac{d\lambda}{2\mu} , \quad \text{if } \mu_P \neq 0, \mu_Q = 0 ,$$

$$d\omega_{PQ}(R) = \left(\frac{1}{\lambda - \tau_P} - \frac{1}{\lambda - \tau_Q}\right) \frac{d\lambda}{2\mu} , \quad \text{if } \mu_P = 0, \mu_Q = 0 , \tag{A.16}$$

where the argument of $d\omega_{PQ}$ is the point $R = (\lambda, \mu) \in \Sigma_g$.

Definition A.1. *The periods along the cycles $a_1, \ldots, b_g$ of a closed differential $d\Omega$ are defined by*

$$A_i \doteq \oint_{a_i} d\Omega ,$$

$$B_i \doteq \oint_{b_i} d\Omega , \tag{A.17}$$

for $i = 1, \ldots, g$.

The periods are independent of the representatives of the cycles with the given homology classes since the differentials are closed. Let $d\Omega$ respectively $d\Omega'$ be closed differentials and denote the corresponding periods by A_i and B_i respectively A_i' and B_i' ($i = 1, \ldots, g$). Let $P_0 \in \Sigma_g$ be fixed and define a function f on $\tilde{\Sigma}_g$ (the fundamental polygon) by

$$f(P) = \int_{P_0}^{P} d\Omega , \tag{A.18}$$

$\forall P \in \Sigma_g$. With these settings one obtains Riemann's bilinear identities,

Theorem A.2. *The following relation holds:*

$$\int_{\Sigma_g} d\Omega \wedge d\Omega' = \oint_{\partial\tilde{\Sigma}_g} f d\Omega' = \sum_{i=1}^{g} (A_i B_i' - A_i' B_i) , \tag{A.19}$$

where $\partial\tilde{\Sigma}_g$ is the boundary of the $4g$-gon $\tilde{\Sigma}_g$, oriented in positive direction.

The first identity in (A.19) follows from Stokes' theorem, the last identity follows from an evaluation of the integrand at the boundary of the fundamental domain (for details see [186]).

The Riemann bilinear relations of Theorem A.2 imply useful relations for the A- and B-periods of Abelian differentials. Applying them to two holomorphic differentials $d\Omega$ and $d\Omega'$ one finds

$$\sum_{j=1}^{g}\left(A_j B_j' - B_j A_j'\right) = 0 . \tag{A.20}$$

and

$$\Im\left(\sum_{j=1}^{g} A_j \overline{B_j}\right) \leq 0 . \tag{A.21}$$

Thus one has

Corollary A.3. *(i) An Abelian differential of the first kind where all A-*

(ii) or all B-periods vanish is identically zero.
(iii) An Abelian differential of the first kind with only real periods is identically zero.

Remark A.4. With the above relations one may show that for an integral of the third kind the poles and integration limits can be interchanged

$$\int_P^Q d\Omega_{RS} = \int_S^R d\Omega_{QP} , \tag{A.22}$$

$P, Q, R, S \in \Sigma_g$.

We normalize the Abelian differentials in the following way:

– The holomorphic differentials are normalized by the condition

$$\oint_{a_i} d\omega_j = 2\pi i \delta_{ij} . \tag{A.23}$$

The so normalized basis of the holomorphic differentials is called canonical.
– Abelian differentials of the second and third kind are only determined up to a linear combination of holomorphic differentials. This ambiguity will be fixed by demanding that all a-periods of the normalized differentials of the second and third kind vanish.

The Riemann bilinear relations (A.19) imply that the matrix $\mathbf{B}_{ij}$,

$$\mathbf{B}_{ij} \doteq \oint_{b_i} d\boldsymbol{\omega}_j \,, \tag{A.24}$$

of b-periods of the normalized holomorphic differentials is a Riemann matrix, i.e. it is symmetric $\mathbf{B}_{ij} = \mathbf{B}_{ji}$ and has a negative definite real part. We note that the Riemann matrix transforms under a change of the homology basis from (a, b) to $(\tilde{a}, \tilde{b})$ according to

$$\begin{pmatrix} \tilde{a} \\ \tilde{b} \end{pmatrix} = \begin{pmatrix} A & B \\ C & D \end{pmatrix} \begin{pmatrix} a \\ b \end{pmatrix}, \quad \begin{pmatrix} A & B \\ C & D \end{pmatrix} \in Sp(g, \mathbb{Z}) \tag{A.25}$$

as

$$\mathbf{B} = 2\pi i (\mathbf{DB} + 2\pi i \mathbf{C})(\mathbf{BB} + 2\pi i \mathbf{A})^{-1} \,. \tag{A.26}$$

For the b-periods of the normalized differentials of the third kind, the bilinear relations imply

$$\oint_{b_k} d\Omega_{PQ} = \int_Q^P d\boldsymbol{\omega}_k \,. \tag{A.27}$$

The corresponding relations for the b-periods of the normalized differentials of the second kind follow from (A.14).

In physical applications so-called real surfaces (surfaces with an anti-holomorphic involution) play an important role. In the hyperelliptic case the surface Σ_g of genus g is given by the relation

$$\mu^2 = \prod_{i=1}^{g+1} (K - E_i)(K - F_i) \,, \tag{A.28}$$

where $E_i, F_i \in \mathbb{R}$ or $E_i = \bar{F}_i$ for $i = 1, \ldots, g + 1$. On this surface the anti-holomorphic involution has the form

$$\tau : \Sigma \ni P = (K, \mu) \to \tau(P) \equiv \bar{P} = (\bar{K}, \bar{\mu}) \in \Sigma \,, \tag{A.29}$$

i.e.. it acts as a complex conjugation on each sheet of $\mathcal{L}$. In the context of the Ernst equation we are interested in surfaces with $E_{g+1} = \bar{F}_{g+1} = \xi$. Since in the cut-system of Fig. A.4 the closed curves a_k remain in one sheet, they are not invariant under the hyperelliptic involution σ whereas the curves b_k are (as a point set). The hyperelliptic involution acts on the Abelian differentials of the first kind, as one easily finds from (A.15) and (A.23), as multiplication by -1. Similarly, the anti-involution τ acts on $H_1(\Sigma_g)$ as follows:

$$\tau(a_i) = -a_i \,,$$
$$\tau(b_i) = b_i - \sum_{k \neq i} a_k \,, \tag{A.30}$$

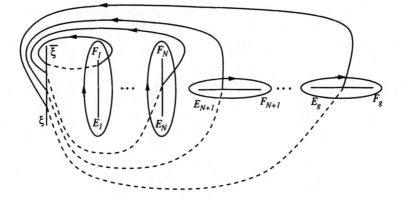

Fig. A.4. Homology basis for real hyperelliptic Riemann surfaces

for curves a_i surrounding two complex conjugated branch points and

$$\tau(a_i) = -a_i \,,$$
$$\tau(b_i) = b_i - \sum_{k=1} a_k \,, \tag{A.31}$$

for curves a_i surrounding two real ones. Since the non-normalized differentials $\mathrm{d}\nu_i$ (A.15) transform according to $\tau^* \mathrm{d}\nu_i = \mathrm{d}\bar\nu_i$ we find with (A.23) and (A.30) for the normalized Abelian differentials of the first kind

$$\tau^*(\mathrm{d}\omega_i) = \mathrm{d}\bar\omega_i \,. \tag{A.32}$$

Therefore, the components of the Riemann matrix **B** transform under τ in the following way

$$\bar{\mathbf{B}}_{ij} = \oint_{b_i} \mathrm{d}\bar\omega_j = \oint_{b_i} \tau^*(\mathrm{d}\omega)_j = \oint_{\tau(b_i)} \mathrm{d}\omega_j = \oint_{b_i} \mathrm{d}\omega_j - \sum_{l \neq i} \oint_{a_l} \mathrm{d}\omega_k$$
$$= \mathbf{B}_{ij} - 2\pi\mathrm{i} \sum_{l \neq i} \delta_{jl} \,, \tag{A.33}$$

for ovals a_i surrounding two complex conjugated branch points, and

$$\bar{\mathbf{B}}_{ij} = \oint_{b_i} \mathrm{d}\bar\omega_j = \oint_{b_i} \tau^*(\mathrm{d}\omega)_j = \oint_{\tau(b_i)} \mathrm{d}\omega_j = \oint_{b_i} \mathrm{d}\omega_j - \sum_{l=1}^{g} \oint_{a_l} \mathrm{d}\omega_k$$
$$= \mathbf{B}_{ij} - 2\pi\mathrm{i} \,, \tag{A.34}$$

for ovals a_i surrounding two real branch points. Notice that the periods of the holomorphic differentials on a hyperelliptic surface can be expressed in terms of differences of the Abel map between branch points. At the end of this section we introduce the notion of the Jacobian of a Riemann surface and of the Abel map. Let Λ be the lattice

$$\Lambda = \{2\pi i \mathbf{N} + \mathbf{B}\mathbf{M}, \quad \mathbf{N}, \mathbf{M} \in \mathbb{Z}^g\} \tag{A.35}$$

generated by the periods of Σ_g. This defines an equivalence relation in $\mathbb{C}^g$: two points of $\mathbb{C}^g$ are equivalent if they differ by an element of Λ.

Definition A.5. *The complex torus*

$$Jac(\Sigma_g) = \mathbb{C}^g/\Lambda \tag{A.36}$$

is called the Jacobi variety or the Jacobian of Σ_g.

Definition A.6. *The map*

$$\omega : \Sigma_g \rightarrow Jac(\Sigma_g), \quad \omega(P) = \int_{P_0}^{P} d\omega, \tag{A.37}$$

where $d\omega$ is the canonical basis of the holomorphic differentials and where $P_0 \in \Sigma_g$, is called the Abel map.

Let $\{e^{(i)}\}$, $i = 1, \ldots, g$ be a basis in $\mathbb{C}^g$ with $(e^{(i)})_j = \delta_{ij}$ and define

$$(f^{(i)})_j \doteq \mathbf{B}_{ij}. \tag{A.38}$$

Then the vectors $2\pi i e^{(1)}, \ldots, e^{(g)}, f^{(1)}, \ldots, f^{(g)}$ are lineary independent since the real part of the Riemann matrix is negative definite. A point $e \in Jac(\Sigma_g)$ is uniquely determined by the $2g$ real numbers p_i, q_j $(i, j = 1, \ldots, g)$,

$$e = 2\pi i \sum_{k=1}^{g} q_k e^{(k)} + \sum_{k=1}^{g} p_k f^{(k)}. \tag{A.39}$$

The numbers p_i, q_j form the characteristic $[e]$ of the point e,

$$[e] \doteq \begin{bmatrix} p \\ q \end{bmatrix}. \tag{A.40}$$

A.3 Divisors and the Theorems of Abel and Jacobi

A *divisor* on a Riemann surface Σ_g is a formal sum of points,

$$\mathfrak{A} = \sum_{i=1}^{N} n_i P_i \tag{A.41}$$

with $P_i \in \Sigma_g$ and $n_i \in \mathbb{Z}$. The sum

$$\deg \mathfrak{A} = \sum_{i=1}^{N} n_i \qquad \qquad \text{(A.42)}$$

is called the *degree* of $\mathfrak{A}$. The set of all divisors with the obviously defined group action

$$n_1 P + n_2 P = (n_1 + n_2)P , \quad -\mathfrak{A} = \sum_{i=1}^{N} (-n_i)P_i \qquad \text{(A.43)}$$

forms an Abelian group $\mathrm{Div}(\Sigma_g)$. A divisor with all $n_i \geq 0$ is called *positive* (or *integral* or *effective*). This notion allows a partial ordering in $\mathrm{Div}(\Sigma_g)$,

$$\mathfrak{A} \leq \mathfrak{A}' \iff \mathfrak{A}' - \mathfrak{A} \geq 0 . \qquad \qquad \text{(A.44)}$$

Let f be a meromorphic function on Σ_g with zeros $P_1, \ldots, P_M$ of multiplicities $p_1, \ldots, p_M > 0$ and poles $Q_1, \ldots, Q_N$ of multiplicities $q_1, \ldots, q_N > 0$. Then the divisor

$$\mathfrak{A} = p_1 P_1 + \ldots + p_M P_M - q_1 Q_1 - \ldots - q_N Q_N = (f) \qquad \text{(A.45)}$$

is called the divisor of f and denoted by (f). A divisor is called *principal* if there exists a function f with $(f) = \mathfrak{A}$. For two meromorphic functions f and g one has obviously $(fg) = (f) + (g)$. Two divisors $\mathfrak{A}$ and $\mathfrak{A}'$ are called equivalent if the divisor $\mathfrak{A} - \mathfrak{A}'$ is principal which is denoted by $\mathfrak{A} \equiv \mathfrak{A}'$. The corresponding equivalence class is called the *divisor class*.

Similarly one can define the divisor of an Abelian differential $d\Omega = f(z)dz$. Since the quotient of two Abelian differentials is a meromorphic function, any two divisors of Abelian differentials are lineary equivalent. The corresponding class is called *canonical*.

The Abel map is defined for divisors of the form (A.41) in a natural way,

$$\omega(\mathfrak{A}) = \sum_{i=1}^{N} n_i \int_{P_0}^{P_i} d\omega . \qquad \qquad \text{(A.46)}$$

If the divisor is of degree zero, $\omega(\mathfrak{A})$ is obviously independent of P_0. This leads to the important

Theorem A.7 (Abel's theorem). *The divisor $\mathfrak{D} \in Div(\Sigma_g)$ is principal if and only if:*
1. $\deg \mathfrak{D} = 0$,
2. $\omega(\mathfrak{D}) = 0$ on $Jac(\Sigma_g)$.

Since the divisor $\mathfrak{D}$ is principal, it defines a class of meromorphic functions. Let F be such a function. Then we have the useful

Corollary A.8. *Let the conditions of Abel's theorem hold and let* $\mathfrak{D} = \mathfrak{A} - \mathfrak{B}$ *where* $\mathfrak{A}$ *and* $\mathfrak{B}$ *are positive divisors. Then the following identity holds*

$$\int_{\mathfrak{B}}^{\mathfrak{A}} d\omega_{PQ} = \ln \frac{F(P)}{F(Q)} . \tag{A.47}$$

Let $\mathfrak{D}_\infty$ be a positive divisor on Σ_g. A natural question is to describe the vector space of meromorphic functions with poles at $\mathfrak{D}_\infty$ only. For a general divisor $\mathfrak{D}$ on Σ_g one can define the vector space

$$L(\mathfrak{D}) = \{f \text{ meromorphic functions on } \Sigma_g | (f) \geq \mathfrak{D} \text{ or } f \equiv 0\} . \tag{A.48}$$

If we split $\mathfrak{D}$ in positive $(\mathfrak{D}_0)$ and negative $(\mathfrak{D}_\infty)$ parts

$$\mathfrak{D}_0 = \sum n_i P_i , \quad \mathfrak{D}_\infty = \sum m_k Q_k , \tag{A.49}$$

where $n_i, m_k > 0$, then the space $L(\mathfrak{D})$ (we call its dimension $r(\mathfrak{D})$) consists of the meromorphic functions with zeros of order at least n_i at P_i and with poles of order at most m_k at Q_k. Similarly one can define the corresponding vector space of the differentials

$$H(\mathfrak{D}) = \{\Omega \text{ Abelian differential on } \Sigma_g | (\Omega) \geq \mathfrak{D} \text{ or } \Omega \equiv 0\} , \tag{A.50}$$

the dimension of which is denoted by $i(\mathfrak{D})$. This dimension is called the index of speciality.

Definition A.9. *A positive divisor* $\mathfrak{D}$ *of degree* $\deg\mathfrak{D} = g$ *is called special if there exists a holomorphic differential* $d\omega$ *with*

$$(d\omega) \geq \mathfrak{D} . \tag{A.51}$$

It can be shown that the condition (A.51) is equivalent to the existence of a non-constant function f with $(f) \geq -\mathfrak{D}$. Since the space of holomorphic differentials is g-dimensional, equations (A.51) form a linear system of g equations for g variables. This shows that most of the positive divisors are non-special. In the hyperelliptic case equations (A.51) imply that a divisor is special if and only if it contains two or more points lying on different sheets but having the same projection into the complex plane.

With these notions we can state the important

Theorem A.10 (Jacobi's inversion theorem). *Let* $\mathfrak{D}_g$ *be the set of positive divisors of degree* g *on* Σ_g. *The Abel map on this set* $\omega : \mathfrak{D}_g \mapsto \text{Jac}(\Sigma_g)$ *is surjective, i.e. for any* $\boldsymbol{x} \in \text{Jac}(\Sigma_g)$ *there exists a positive divisor of degree* g *with* $P_1 + \ldots + P_g \in \mathfrak{D}_g$ *(the* P_i *are not necessarily different) satisfying*

$$\sum_{i=1}^{g} \int_{P_0}^{P_i} d\omega = \boldsymbol{x} . \tag{A.52}$$

A.4 Theta Functions of Riemann Surfaces

Theta functions are a convenient tool to work with meromorphic functions
on Riemann surfaces. We define them as an infinite series.

Definition A.11. *Let* **B** *be a* $g \times g$ *Riemann matrix. The theta function with
characteristics* $[p, q]$ *is defined as*

$$\Theta_{pq}(x, \mathbf{B}) = \sum_{N \in \mathbb{Z}^g} \exp \left\{ \frac{1}{2} \langle \mathbf{B}(N + p), N + p \rangle + \langle x + 2\pi i q, N + p \rangle \right\},$$
(A.53)

with $x \in \mathbb{C}^g$ *and* $p, q \in \mathbb{C}^g$, *where* $\langle \cdot, \cdot \rangle$ *denotes the Euclidean scalar product*
$\langle N, x \rangle = \sum_{i=1}^{g} N_i x_i$.

The properties of the Riemann matrix ensure that the series converges absolutely and that the theta function is an entire function on $\mathbb{C}^g$. A characteristic is called *singular* if the corresponding theta function vanishes identically. Of special importance are half-integer characteristics with $2p, 2q \in \mathbb{Z}^g$. A half-integer characteristic is called *even* if $4\langle p, q \rangle = 0$ mod 2 and *odd* otherwise. Theta functions with odd (even) characteristic are odd (even) functions of the argument x. The theta function with characteristics is related to the Riemann theta function, the theta function with zero characteristics $\Theta \doteq \Theta_{00}$, via

$$\Theta_{pq}(x, \mathbf{B}) = \Theta(x + \mathbf{B}p + 2\pi i q) \exp \left\{ \frac{1}{2} \langle \mathbf{B}p, p \rangle + \langle p, x + 2\pi i q \rangle \right\}. \quad (A.54)$$

The theta function has the periodicity properties

$$\Theta_{pq}(x + 2\pi i e_j) = e^{2\pi i p_j} \Theta_{pq}(x), \quad \Theta_{pq}(x + \mathbf{B}e_j) = e^{-2\pi i q_j - z_j - \frac{1}{2} B_{jj}} \Theta_{pq}(x),$$
(A.55)

where e_j is the g-dimensional vector consisting of zeros except for a 1 in
jth position, and satisfies the heat equation (under the assumption that all
entries of the matrix **B** are independent)

$$2\partial_{\mathbf{B}_{\alpha\beta}} \Theta_{pq}(z, \mathbf{B}) = \partial_{z_\alpha} \partial_{z_\beta} \Theta_{pq}(z, \mathbf{B}). \quad (A.56)$$

The above definitions are possible for Riemann matrices **B** which are not
associated to a Riemann surface. In the following we will only consider theta
functions on Riemann surfaces. Then we have

Proposition A.12. *The theta function* $\Theta(\omega(P) - e)$ *either vanishes identically on* Σ_g *or has exactly g zeros (counting multiplicities).*

Proposition A.13. *Let* $\mathcal{K}$ *be the Riemann vector defined by*

$$K = \frac{2\pi i + \mathbf{B}_{jj}}{2} - \frac{1}{2}\sum_{l\neq j}\oint_{a_l}\left(d\omega_l(P)\int_{P_0}^{P}d\omega_j\right) , \tag{A.57}$$

where P_0 is the base point of the Abel map, and let $\mathfrak{D} = P_1 + \ldots + P_g$ be a non-special divisor. Then the theta function $\Theta(\omega(P) - \omega(\mathfrak{D}) - K)$ vanishes if and only if $P \in \mathfrak{D}$.

For general Riemann surfaces it is difficult to find an explicit form for the Riemann vector. On hyperelliptic surfaces it is related to an odd characteristic which can be expressed in terms of the Abel map of some divisor of branch points. For example for the cut-system in Fig. A.4 one has

$$K = \sum_{i=1}^{g}\omega(F_i) . \tag{A.58}$$

In general there is no simple relation between the complex conjugate of a theta function $\overline{\Theta}(x)$ and the theta function of the complex conjugated argument $\Theta(\bar{x})$. However for real hyperelliptic surfaces such relations exist. For the cut-system of Fig. A.4 the relations (A.33) for the matrix of b-periods imply for the Riemann theta function,

$$\overline{\Theta}(x) = \Theta\left(\bar{x} + i\pi\Delta\right) , \tag{A.59}$$

where $\Delta_i = 1$ if E_i and F_i are real and $\Delta_i = 0$ otherwise.

Abelian integrals can be expressed in terms of theta functions. Since for the applications in Chap. 3 the integrals of the third kind are particularly interesting, we want to show how a sum of g such integrals can be expressed via theta functions, see [187]. Let f be a meromorphic function on $\tilde{\Sigma}_g$ of the form

$$f(P) = \frac{\Theta(\omega(P) - \omega(\mathfrak{D}) - K)}{\Theta(\omega(P) - \omega(\mathfrak{E}) - K)} , \tag{A.60}$$

where $\mathfrak{D} = P_1 + \cdots + P_g$ and $\mathfrak{E} = \tilde{P}_1 + \cdots + \tilde{P}_g$, $P_i, \tilde{P}_j \in \Sigma_g$ are two non-special divisors of degree g and K is the Riemann vector. This function has g zeros respectively poles at the points of the divisor $\mathfrak{D}$ respectively $\mathfrak{E}$. Furthermore, $f(P)$ has at the cut a_i the jump 1 and at b_i the jump $\exp\{2(\omega_i(\mathfrak{D}) - \omega_i(\mathfrak{E}))\}$. On the other hand, the function

$$\hat{f}(P) = \exp\left\{\sum_i\int_{P_0}^{P}d\omega_{\tilde{P}_i P_i}\right\} , \tag{A.61}$$

where $P_0 \in \Sigma_g$ is fixed, has the same properties as f: we have $\hat{f}(P_i) = 0$ and $\hat{f}(\tilde{P}_j) = \infty$. The jumps at the cuts a_j respectively b_j are the same as for the function f. Therefore, both functions coincide, up to a P-independent factor. By taking $P = P_0$ one gets

$$\sum_i \int_{P_0}^{P} d\omega_{\tilde{P}_i P_i} = \sum_i \int_{\tilde{P}_i}^{P_i} d\omega_{P_0 P} \tag{A.62}$$

$$= \ln \frac{\Theta(\omega(P) - \omega(\mathfrak{D}) - \mathcal{K})\Theta(\omega(P_0) - \omega(\mathfrak{E}) - \mathcal{K})}{\Theta(\omega(P) - \omega(\mathfrak{E}) - \mathcal{K})\Theta(\omega(P_0) - \omega(\mathfrak{D}) - \mathcal{K})} .$$

Here we have used that for an Abelian integral of the third kind the poles and limits of integration may be interchanged.

A.4.1 Elliptic Surfaces

A special case are surfaces of genus 1 which are called *elliptic*, see e.g. [199]. Elliptic curves can always be brought into the standard form

$$\mu^2 = (1 - \lambda^2)(1 - k^2 \lambda^2) , \tag{A.63}$$

by using a Möbius transformation. The periods of the holomorphic differentials can be expressed in terms of *complete elliptic integrals*

$$K(k) = \int_{-1}^{1} \frac{d\lambda}{\mu} , \quad \tilde{K}(k) = K(\sqrt{1 - k^2}). \tag{A.64}$$

The Jacobi elliptic theta functions ϑ_i where $i = 1, \dots, 4$ have the characteristics $\frac{1}{2}\begin{bmatrix} 1 \\ 1 \end{bmatrix}$, $\frac{1}{2}\begin{bmatrix} 1 \\ 0 \end{bmatrix}$, $\frac{1}{2}\begin{bmatrix} 0 \\ 0 \end{bmatrix}$ and $\frac{1}{2}\begin{bmatrix} 0 \\ 1 \end{bmatrix}$ respectively.

A.5 The Trisecant Identity for Theta Functions on Riemann Surfaces

Theta functions are subject to a number of addition theorems. A typical example is the ternary addition theorem which can be cast into the form, see [1, 270]:

Theorem A.14. *Let* $[m_i] = [m_i^1, m_i^2]$ $(i = 1, \dots, 4)$ *be arbitrary real 2g-dimensional vectors. Then*

$$\Theta[m_1](\boldsymbol{u} + \boldsymbol{v})\Theta[m_2](\boldsymbol{u} - \boldsymbol{v})\Theta[m_3](0)\Theta[m_4](0) \tag{A.65}$$

$$= \frac{1}{2^g} \sum_{2a \in (Z_2)^{2g}} \exp(-4\pi i \langle m_1^1, a^2 \rangle)$$

$$\times \Theta[n_1 + a](\boldsymbol{u})\Theta[n_2 + a](\boldsymbol{u})\Theta[n_3 + a](\boldsymbol{v})\Theta[n_4 + a](\boldsymbol{v}) ,$$

where $a = (a^1, a^2)$, *and* $(m_1, \dots, m_4) = (n_1, \dots, n_4)T$ *with*

$$T = \frac{1}{2} \begin{pmatrix} 1 & 1 & 1 & 1 \\ 1 & -1 & -1 & -1 \\ 1 & -1 & 1 & -1 \\ 1 & -1 & -1 & 1 \end{pmatrix} . \tag{A.66}$$

Each 1 in T denotes the $g \times g$ identity matrix.

The above addition theorem holds for general theta functions. A very useful identity due to Fay [128] (see also [139, 272]), however, holds only for theta functions on Riemann surfaces.

To generalize the cross ratio function (3.1) to the Riemann surface Σ_g one needs as a building block an object on Σ_g which has exactly one zero as the difference of two points $z - z_0$ in $\mathbb{CP}^1$. There is no function with this property on a Riemann surface, but the so called prime form which is the $(-\frac{1}{2}, -\frac{1}{2})$-differential on $\Sigma_g \times \Sigma_g$ defined by

$$E(P, Q) = \frac{\Theta_\star(\int_Q^P)}{h_\Delta(P)h_\Delta(Q)} ,$$

where $h_\Delta^2(P) = \sum_{\alpha=1}^{g} \frac{\partial \Theta_\star}{\partial z_\alpha}(0) d\omega_\alpha(P)$, and where $\star \equiv [p^\star q^\star]$ is an odd non-singular half-integer characteristic (note that the prime form is independent of the choice of the characteristic $\star$). $\int_Q^P$ denotes the line integral from Q to P of the vector $d\omega(\tau)$. With this notation we can define the cross ratio function

$$\lambda_{1234} = \frac{E(P_1, P_2)E(P_3, P_4)}{E(P_1, P_4)E(P_3, P_2)} , \tag{A.67}$$

which is a function on Σ_g that vanishes for $P_1 = P_2$ and $P_3 = P_4$ and has poles for $P_1 = P_4$ and $P_2 = P_3$. The generalized identity (3.2) is Fay's trisecant identity

Theorem A.15. *Let $P_1, P_2, P_3, P_4 \in \Sigma_g$ be four arbitrary points and let $p, q \in \mathbb{C}^g$ be two characteristic vectors. Then the following identity holds:*

$$\lambda_{3124} \Theta_{pq} \left(z + \int_{P_2}^{P_3} \right) \Theta_{pq} \left(z + \int_{P_1}^{P_4} \right) + \lambda_{3214} \Theta_{pq} \left(z + \int_{P_1}^{P_3} \right) \Theta_{pq} \left(z + \int_{P_2}^{P_4} \right)$$

$$= \Theta_{pq}(z) \Theta_{pq} \left(z + \int_{P_2}^{P_3} + \int_{P_1}^{P_4} \right) , \tag{A.68}$$

where all integration contours are chosen not to intersect the canonical basic cycles; this requirement completely fixes all terms of the identity (A.68).

In the sequel we will use degenerate versions of Fay's identity which lead to identities for derivatives of theta functions. Let τ be a local coordinate near P. Then we can write the Abel map $\int_P^X$ for $X \sim P$ as a series in τ,

$$\int_P^X = U\tau + \frac{1}{2}V\tau^2 + \frac{1}{6}W\tau^3 + \dots . \tag{A.69}$$

Let us denote by D_P the operator for the directional derivative along the basis of holomorphic differentials, acting on theta functions, and similarly D_P' and D_P'' the directional derivatives along V and W,

$$D_P\Theta_{pq}(z) = \langle \nabla\Theta_{pq}(z), U \rangle ,$$
$$D_P'\Theta_{pq}(z) = \langle \nabla\Theta_{pq}(z), V \rangle ,$$
$$D_P''\Theta_{pq}(z) = \langle \nabla\Theta_{pq}(z), W \rangle . \tag{A.70}$$

Since the theta function (A.53) depends only on the sum of the vectors z and q, the action of the operator D_P on a theta function with characteristics can be written alternatively as

$$D_P\Theta_{pq}(z) = \sum_\alpha \partial_{q_\alpha}\{\Theta_{pq}(z)\}\frac{d\omega_\alpha(P)}{d\tau_P} . \tag{A.71}$$

This form of D_P can be easily extended to any object depending on a vector q.

Differentiating (A.68) with respect to the argument P_4 and taking the limit $P_4 \to P_2$ one obtains

Corollary A.16. *The following identity for first derivatives of theta functions on Riemann surfaces holds:*

$$D_{P_2} \ln \frac{\Theta_{pq}(z + \int_{P_1}^{P_3})}{\Theta_{pq}(z)}$$

$$= c_1(P_1, P_2, P_3) + c_2(P_1, P_2, P_3)\frac{\Theta_{pq}(z + \int_{P_1}^{P_2})\Theta_{pq}(z + \int_{P_2}^{P_3})}{\Theta_{pq}(z)\Theta_{pq}(z + \int_{P_1}^{P_3})} , \tag{A.72}$$

where the functions of three variables c_1 and c_2 are given by

$$c_1(P_1, P_2, P_3) = \frac{d\omega_{P_1 P_3}(P_2)}{d\tau_{P_2}} , \tag{A.73}$$

and

$$c_2(P_1, P_2, P_3) = \frac{E(P_1, P_3)}{E(P_1, P_2)E(P_2, P_3)d\tau_{P_2}} . \tag{A.74}$$

The derivative of (A.72) with respect to argument P_3 gives in the limit $P_3 \to P_1$

Corollary A.17. *The following identity for second derivatives of theta functions on Riemann surfaces holds:*

$$D_{P_1}D_{P_2}\ln\Theta_{pq}(z) = d_1(P_1,P_2) + d_2(P_1,P_2)\frac{\Theta_{pq}(z+\int_{P_2}^{P_1})\Theta_{pq}(z+\int_{P_1}^{P_2})}{\Theta_{pq}^2(z)} ,$$

$$(A.75)$$

where the functions of two variables d_1 and d_2 are given by

$$d_1(P_1,P_2) = -\frac{W(P_1,P_2)}{\mathrm{d}\tau_{P_1}\mathrm{d}\tau_{P_2}} , \qquad (A.76)$$

$$d_2(P_1,P_2) = \frac{1}{E^2(P_1,P_2)\mathrm{d}\tau_{P_1}\mathrm{d}\tau_{P_2}} ; \qquad (A.77)$$

$W(P_1,P_2) = \mathrm{d}_{P_1}\mathrm{d}_{P_2}\ln E(P_1,P_2)$ is the canonical meromorphic bidifferential.

In the limit $P_2 \to P_1 = P$ we obtain for an expansion of the terms in (A.75) in the difference of the local parameters near P_1 and P_2

Corollary A.18. *The following identity holds:*

$$D_P^4\ln\Theta(z) + 6(D_P^2\ln\Theta(z))^2 + 3D_P'D_P'\ln\Theta(z) - 2D_PD_P''\ln\Theta(z)$$
$$-24e_1(P)D_P^2\ln\Theta(z) + 12(10e_2(P) - 3e_1^2(P)) = 0 . \qquad (A.78)$$

Here the functions $e_1(P)$, $e_2(P)$ *turn up in the Taylor expansion of the normalized differential* $\mathrm{d}\Omega_P^{(1)}$ *of the second kind with a pole of second order at* P *(τ is the local parameter in the vicinity of P with $\tau(P) = 0$),*

$$\mathrm{d}\Omega_P^{(1)} = \left(-\frac{1}{\tau^2} + 2e_1(P) - (6e_1^2(P) - 12e_2(P))\tau^2 + \ldots\right)\mathrm{d}\tau . \qquad (A.79)$$

A.6 Rauch's Formulas and Root Functions

So far we have only studied functions on a given Riemann surface with fixed branch points. In the context of the Ernst equation it is however necessary to study certain functions on a whole family of Riemann surfaces in dependence on the branch points. Rauch's variational formulas [179] describe the dependence of the basic normalized holomorphic differentials $\mathrm{d}\omega_\alpha$ and the matrix of b-periods $\mathbf{B}_{\alpha\beta}$ on the moduli of the Riemann surface. The moduli space of hyperelliptic curves can be parameterized by the positions of the branch points. Let $\tau_{\lambda_m}(P) = \sqrt{\lambda - \lambda_m}$ be a local parameter in the neighborhood of P. Then Rauch's formulas read:

$$\partial_{\lambda_m}(\mathrm{d}\omega_\alpha(P)) = \frac{1}{2}\frac{W(P,\lambda_m)}{\mathrm{d}\tau_{\lambda_m}}\frac{\mathrm{d}\omega_\alpha(\lambda_m)}{\mathrm{d}\tau_{\lambda_m}} , \qquad (A.80)$$

$$\partial_{\lambda_m}\mathbf{B}_{\alpha\beta} = \frac{1}{2}\frac{\mathrm{d}\omega_\alpha(\lambda_m)}{\mathrm{d}\tau_{\lambda_m}}\frac{\mathrm{d}\omega_\beta(\lambda_m)}{\mathrm{d}\tau_{\lambda_m}} . \qquad (A.81)$$

In the case of hyperelliptic Riemann surfaces, quotients of theta functions with the same argument but different half integer characteristics are equivalent to so-called *root functions*, see Chap. 1 of [1], [267]. If we write the surface $\mathcal{L}$ in the form

$$\mu^2 = \prod_{m=1}^{2g+2} (\lambda - \lambda_m) , \tag{A.82}$$

the following identity for root functions holds [1, 267] for any point $P = (K, \mu(K)) \in \mathcal{L}$:

$$\frac{E(P, \lambda_m)\sqrt{d\tau_{\lambda_m}}}{E(P, \lambda_n)\sqrt{d\tau_{\lambda_n}}} = C\sqrt{\frac{K - \lambda_m}{K - \lambda_n}} , \tag{A.83}$$

where C is a constant with respect to $\lambda(P)$.

We will also need these functions in a form free of prime forms: Let Q_i, $i = 1, \ldots, 2g+2$, be the branch points of a hyperelliptic Riemann surface Σ_g of genus g and $A_j = \omega(Q_j)$ with $\omega(Q_1) = 0$. Furthermore let $\{i_1, \ldots, i_g\}$ and $\{j_1, \ldots, j_g\}$ be two sets of numbers in $\{1, 2, \ldots, 2g + 2\}$. Then the following equality holds for an arbitrary point $P \in \mathcal{L}$,

$$\frac{\Theta \left[\mathcal{K} + \sum_{k=1}^{g} A_{i_k} \right] (\omega(P))}{\Theta \left[\mathcal{K} + \sum_{k=1}^{g} A_{j_k} \right] (\omega(P))} = C_1 \sqrt{\frac{(K - E_{i_1}) \ldots (K - E_{i_g})}{(K - E_{j_1}) \ldots (K - E_{j_g})}} , \tag{A.84}$$

where C_1 is a constant independent of K. Let $\mathfrak{X} = X_1 + \ldots + X_g$ with $X_j = (K_j, \mu(K_j))$ be a divisor of degree g on $\mathcal{L}$. Then the following identity is satisfied,

$$\frac{\Theta \left[\mathcal{K} + A_i \right] (\omega(\mathfrak{X}))}{\Theta \left[\mathcal{K} + A_j \right] (\omega(\mathfrak{X}))} = C_2 \prod_{k=1}^{g} \sqrt{\frac{(K_k - Q_i)}{(K_k - Q_j)}} , \tag{A.85}$$

where C_2 is a function independent of the K_k.

B Ernst Equation and Twistor Theory

In this appendix we establish the relation between twistor theory and the solutions to the Ernst equation discussed in this volume. We basically follow the approach of Mason and Woodhouse [32]. In Sect. B.1 we review some basic facts on the quaternionic Hopf bundle. This will be used in Sect. B.2 to perform a symmetry reduction of the Penrose–Ward transform which leads to the linear system of the Yang equation discussed in Sect. 2.4. In Sect. B.3 we construct explicitly the holomorphic bundles over the reduced twistor space R_V for the algebro-geometric solutions of the stationary and axisymmetric solutions to the Einstein equations according to Woodhouse and Mason [88]. We equip the space R_V associated to a region V of the (ϱ, ζ)-plane with a standard cover consisting of four charts. It turns out that for solutions to the Yang equation which yield a symmetric and real Yang matrix and which have a regular behavior on the symmetry axis, one patching matrix (and a couple of integers) characterizes the bundle over R_V completely. Then we pass to the construction of the holomorphic bundles associated to the solutions to the Ernst equation discussed in Chap. 4. If the solution has a regular axis, we can read off the corresponding patching matrix directly from the Ernst potential at the symmetry axis.

B.1 The Quaternionic Hopf Bundle and the Twistor Transform

It is well known that methods of complex analysis are also suited for the description of real analysis of two variables. This is due to the fact that one may identify $\mathbb{R}^2 \simeq \mathbb{C}$ (i. e. one may introduce in $\mathbb{R}^2$ a complex structure) uniquely, if an orientation and a metric in $\mathbb{R}^2$ are given. If one wants to make a similar identification for $\mathbb{R}^4$ one would naively identify $\mathbb{R}^4 \simeq \mathbb{C}^2$. Unfortunately, for $\mathbb{R}^4$ this identification is not possible, because in this case there is no natural complex structure in $\mathbb{C}^2$ induced by orientation and metric in $\mathbb{R}^4$. It turns out that the correct four-dimensional analogue of $\mathbb{C}$ is the three-dimensional complex projective space $\mathbb{CP}^3$, the so called *Penrose twistor space*. Using this complex manifold it turns out that one may describe (Euclidean) (anti-)self dual Yang–Mills fields by algebraic constructions. To be more precise, we first introduce the basic notions of quaternion geometry. We define three formal symbols $\mathbf{i}$, $\mathbf{j}$ and $\mathbf{k}$ which fulfil the following requirements

$$\mathbf{i}^2 = \mathbf{j}^2 = \mathbf{k}^2 = -1 \ ,$$
$$\mathbf{ij} = -\mathbf{ji} = \mathbf{k}, \ \ \mathbf{jk} = -\mathbf{kj} = \mathbf{i}, \ \ \mathbf{ki} = -\mathbf{ik} = \mathbf{j} \ . \qquad (B.1)$$

A general quaternion $q \in \mathbb{H}$ is then given as a linear combination

$$q = x_0 + x_1\mathbf{i} + x_2\mathbf{j} + x_3\mathbf{k} \ , \qquad (B.2)$$

with $x_0, \ldots, x_3 \in \mathbb{R}$. The *conjugated* quaternion $\bar{q}$ is defined by

$$\bar{q} = x_0 - x_1\mathbf{i} - x_2\mathbf{j} - x_3\mathbf{k} \ . \qquad (B.3)$$

The conjugation is an anti-involution, i.e. $\overline{(q\tilde{q})} = \bar{\tilde{q}}\bar{q}$. With (B.1) one finds

$$q\bar{q} = \bar{q}q = \sum_{\mu=0}^{3} x_\mu^2 \ . \qquad (B.4)$$

We denote the above expression, which is zero only for $q = 0$, by $|q|^2$. Each q with $|q| \neq 0$ has a unique inverse q^{-1}, which is given by

$$q^{-1} = \bar{q}/|q|^2 \ . \qquad (B.5)$$

The unit quaternions, i.e. all quaternions with $|q| = 1$, form a multiplicative group which is geometrically the sphere S^3

$$\sum_{\mu=0}^{3} x_\mu^2 = 1 \ . \qquad (B.6)$$

If we identify $\mathbf{i}$ with the usual complex number we may regard the complex numbers $\mathbb{C}$ as contained in $\mathbb{H}$ (by taking $x_2 = x_3 = 0$). Furthermore, each q has a unique representation

$$q = z_1 + z_2\mathbf{j} \ , \qquad (B.7)$$

with $z_1 = x_0 + x_1\mathbf{i}$ and $z_2 = x_2 + x_3\mathbf{i}$. Therefore, we may identify $\mathbb{H} \simeq \mathbb{C}^2$.

In analogy to real and complex projective spaces, one may define *quaternionic projective spaces* $\mathbb{HP}^n$. An element of $\mathbb{HP}^n$ is an equivalence class of lines in $\mathbb{H}^{n+1}$, i.e. we call $q = (q_0, \ldots, q_n) \in \mathbb{H}^{n+1}$ equivalent to $\tilde{q} = (\tilde{q}_0, \ldots, \tilde{q}_n) \in \mathbb{H}^{n+1}$, $q \sim \tilde{q}$, iff $q = \lambda \tilde{q}$ for some $\lambda \in \mathbb{H}^* = \mathbb{H} \setminus \{0\}$.

For the further analysis it is useful to introduce two important principal fibre bundles, the *Hopf bundle* and the *quaternionic Hopf bundle*. The Hopf bundle is the simplest example of a family of U(1)-bundles $S^{2n+1} \to \mathbb{CP}^n$, which can be defined as follows. From the equivalence class of $(n+1)$-vectors $z = (z_0, \ldots, z_n)$ in $\mathbb{C}^{n+1} \simeq \mathbb{R}^{2n+2}$ we may always choose a representative whose tip lies on the unit sphere in S^{2n+1}, i.e. it satisfies

$$|z_0|^2 + \cdots + |z_n|^2 = 1 \ , \tag{B.8}$$

by multiplying z by the scalar $\lambda = \left(\sum_{\mu=0}^{n} |z_\mu|\right)^{-1/2}$. Of course, the resulting vector is only unique up to multiplication by scalars of the form $e^{i\phi}$. In other words, we find that $\mathbb{CP}^n$ can be obtained from S^{2n+1} by identifying all points $e^{i\phi}z$ on S^{2n+1} with z. Therefore, we have a principal fibre bundle $S^{2n+1}(\mathbb{CP}^n, U(1))$ with base space $\mathbb{CP}^3$ and structure group $U(1)$. Now, for $n = 1$ (using the identifications $\mathbb{CP}^1 \simeq S^2$ and $U(1) \simeq S^1$) the corresponding bundle is called the Hopf bundle $S^3(S^2, S^1)$, i.e. S^3 is a principal fibre bundle with base space S^2 and structure group S^1.

Analogously, one may define the quaternionic Hopf bundle as follows. We first endow $\mathbb{R}^{4n+4}$ with the structure of the $(n+1)$-dimensional quaternion space $\mathbb{H}^{n+1}$ with coordinates $q_0, \ldots, q_n$. In these coordinates the unit sphere S^{4n+3} in $\mathbb{R}^{4n+4}$ is given by

$$\sum_{\mu=0}^{n} |q_\mu|^2 = 1 \ . \tag{B.9}$$

We have a natural (left) action of unit quaternions on S^{4n+3} as follows

$$SU(2) \times S^{4n+3} \ni (q, (q_0, \ldots, q_n)) \to (qq_0, \ldots, qq_n) \in S^{4n+3} \ , \tag{B.10}$$

because with $\sum_{\mu=0}^{n} |q_\mu|^2 = 1$ we have $\sum_{\mu=0}^{n} |qq_\mu|^2 = 1$. The resulting orbit space $S^{4n+3}/SU(2)$ is just the quaternion projective space $\mathbb{HP}^n$. Thus we have

Proposition B.1. $S^{4n+3}(\mathbb{HP}^n, SU(2))$ *is a principal fibre bundle with standard fibre SU(2).*

Remark B.2. For $n = 1$ we have the identification $\mathbb{HP}^1 \simeq S^4$. With $SU(2) \simeq S^3$ we find the quaternionic Hopf bundle $S^7(S^4, S^3)$, i.e. S^7 is a principal fibre bundle with base space S^4 and structure group S^3.

With these notions at hand we can now introduce the *twistor bundle*, a fibre bundle which is associated to the quaternionic Hopf bundle. We construct this fibre bundle with standard fibre $SU(2)/U(1) \simeq S^2$ as follows. We remark that $SU(2)$ acts naturally on the homogeneous space $SU(2)/U(1)$ by adjoining to $[g] \in SU(2)/U(1)$ and $g' \in SU(2)$ the equivalence class $[g'g] \in SU(2)/U(1)$. Now we construct the bundle

$$E = S^7 \times_{SU(2)} SU(2)/U(1) = S^7/U(1) = \mathbb{CP}^3 \ . \tag{B.11}$$

Here the second equality reflects a standard proposition in the theory of associated bundles [273], and the third equality is nothing but the above construction of $U(1)$-bundles $S^{2n+1} \to \mathbb{CP}^n$ for $n = 3$. The above construction motivates the following

Definition B.3. *The* twistor bundle $\mathbb{CP}^3(S^4, S^2)$ *is a bundle with base space* S^4 *and standard fibre* S^2 *associated to the quaternionic Hopf bundle* $S^7(S^4, S^3)$.

The twistor bundle is of interest for the construction of solutions to the Yang–Mills equations for the following reason. Suppose we are given a Yang–Mills field over the four-dimensional Euclidean space $\mathbb{E}_4$, i.e. we are given a principal fibre bundle $P(\mathbb{E}^4, G)$ with structure group (gauge group G) and bundle space P. If one compactifies $\mathbb{E}_4$ by adding the point "∞", one gets a bundle with compact base space S^4 (this is just the four-dimensional analogue to the one point compactification in complex analysis). This is due to a fundamental result of Uhlenbeck [29], which states that all Euclidean finite-action Yang–Mills fields over $\mathbb{E}_4$ are smoothly extendible to Yang–Mills fields over S^4. Thus, we are led to the investigation of principal fibre bundles over S^4. The projection $\pi : \mathbb{CP}^3 \to S^4$ in the twistor bundle can be applied to the principal bundle $P(S^4, G)$ and gives an induced bundle over $\mathbb{CP}^3$, the pullback bundle $\pi^* P(\mathbb{CP}^3, G)$. A fundamental result states then, see e.g. [274],

Proposition B.4. *There is a natural one-to-one correspondence between*

(i) *anti-self-dual $U(n)$-gauge potentials over S^4 (up to gauge equivalence) and*

(ii) *holomorphic vector bundles E of rank n over $\mathbb{CP}^3$ with a positive definite real form (up to isomorphism).*

B.2 Symmetry Reductions
of the Penrose–Ward Transform

B.2.1 The Reduced Twistor Space

An interesting feature of the Penrose–Ward transform (see for instance [30]) is that it allows for symmetry reductions. It turns out that the Yang equation can be obtained as such a reduction. In fact, in [32] most of the known integrable non-linear equations have been solved by a symmetry reduction of the above Penrose–Ward transform. For the stationary, axisymmetric Einstein equations this procedure has been worked out in [88], see also [163] and [164]. A detailed description of the procedure for the Kerr solution can be found in [165]. The important point for us is that in the solution process linear systems for the Yang equation are generated and get a geometric meaning.

Let us recall this symmetry reduction procedure. To start with we consider the action of the Abelian isometry group

$$G = S^1 \times \mathbb{R} ,$$ (B.12)

on the four-dimensional Euclidean space $\mathbb{E}^4$, defined by

$$(\phi_0, t_0) \cdot (t, \phi, \varrho, \zeta) := (t + t_0, \phi + \phi_0, \varrho, \zeta) ,$$ (B.13)

for all $\phi_0 \in [0, 2\pi)$, $t_0 \in \mathbb{R}$. This action extends to a conformal action on the compactification S^4 of $\mathbb{E}^4$ and we may lift this conformal action to an action on $\mathbb{CP}^3$. In terms of local coordinates $(t, \phi, \varrho, \zeta, \delta, \bar{\delta})$ on $\mathbb{CP}^3$, adapted to the Hopf bundle structure, which are related to holomorphic coordinates (ξ^1, ξ^2, ξ^3) by

$$\xi^1 = \delta \ , \quad \xi^2 = \zeta + it - \delta\varrho e^{-i\phi} \ , \quad \xi^3 = \varrho e^{i\phi} + \delta(\zeta - it) \ , \qquad \text{(B.14)}$$

we find for the generators Φ and T of this action

$$\Phi = \frac{\partial}{\partial\phi} + i\delta\frac{\partial}{\partial\delta} - i\bar{\delta}\frac{\partial}{\partial\bar{\delta}} \ ,$$

$$T = \frac{\partial}{\partial t} \ . \qquad \text{(B.15)}$$

In holomorphic coordinates (ξ^1, ξ^2, ξ^3) on $\mathbb{CP}^3$ we also find $\Phi = 2\Re(Y_\Phi)$ and $T = 2\Re(Y_T)$, with

$$Y_\Phi = i\xi^3\frac{\partial}{\partial\xi^3} + i\xi^1\frac{\partial}{\partial\xi^1} \ ,$$

$$Y_T = i\frac{\partial}{\partial\xi^2} - i\xi^1\frac{\partial}{\partial\xi^3} \ . \qquad \text{(B.16)}$$

The holomorphic vector fields Y_Φ and Y_T generate the action of the complex-ification $G^{\mathbb{C}} = \mathbb{C}^* \times \mathbb{C}$ of G on $\mathbb{CP}^3$. Let us consider the orbits of these Killing fields. We have

Proposition B.5. *The orbits of the Killing fields are parametrized by $w \in \mathbb{C}$ and given by subsets of*

$$Q_w = \left\{ (\xi^1, \xi^2, \xi^3) \in \mathbb{CP}^3 \left| \frac{\xi^3}{\xi^1} + \xi^2 = \delta^{-1}\varrho e^{i\phi} + 2\zeta - \delta\varrho e^{-i\phi} = 2w \right. \right\} \ . \qquad \text{(B.17)}$$

Proof. Let

$$Y := \alpha Y_\Phi + \beta Y_T \qquad \text{(B.18)}$$

with $\alpha, \beta \in \mathbb{C}$, denote an arbitrary linear combination of Y_Φ and Y_T. In the holomorphic coordinates (ξ^1, ξ^2, ξ^3) we have

$$Y = i\alpha\xi^1\frac{\partial}{\partial\xi^1} + i\beta\frac{\partial}{\partial\xi^2} + i\left(\alpha\xi^3 - \beta\xi^1\right)\frac{\partial}{\partial\xi^3} \ . \qquad \text{(B.19)}$$

For the integral curve (with curve parameter t) of Y we have to solve the following system of equations (a dot denotes the derivative with respect to t)

$$\dot{\xi}^1 = \alpha i\xi^1 \ ,$$
$$\dot{\xi}^2 = \beta i \ ,$$
$$\dot{\xi}^3 = i\left(\alpha\xi^3 - \beta\xi^1\right) \ , \qquad \text{(B.20)}$$

from which we immediately obtain for $\xi^1(t)$ and $\xi^2(t)$

$$\xi^1(t) = e^{i\alpha t + C_1} ,$$
$$\xi^2(t) = i\beta t + C_2 , \tag{B.21}$$

with arbitrary complex constants C_1 and C_2. Thus, we obtain

$$\dot\xi^3 - i\alpha\xi^3 = -i\beta e^{i\alpha t + C_1} , \tag{B.22}$$

with general solution

$$\xi^3(t) = e^{i\alpha t}\left(C_3 - i\beta t e^{C_1}\right) , \tag{B.23}$$

$(C_3 \in \mathbb{C})$. Inserting (B.21) yields

$$\frac{\xi^3(t)}{\xi^1(t)} + \xi^2(t) = C_3 e^{-C_1} + C_2 =: 2w . \tag{B.24}$$

$\square$

It is well known, that if G acts freely and properly on a manifold M then the space of orbits is a Hausdorff manifold. Since Y_Φ and Y_T are the lifts of the G-action on $\mathbb{E}^4$ which is obviously not free, the situation is a bit more involved. First, we have for the zeros of the Killing vector fields

Proposition B.6. *For the Killing vectors of the action of $G^{\mathbb{C}}$ on $\mathbb{CP}^3$ we have*

(i)

$$(Y_\Phi(p) = 0) \Longleftrightarrow p \in \begin{cases} L_0 = \{\varrho = \delta = 0\} \\ L_1 = \{\varrho = 0, \delta = \infty\} \end{cases} , \tag{B.25}$$

(ii)

$$(Y_T(p) = 0) \Longleftrightarrow p \in I = \pi^{-1}(\infty) . \tag{B.26}$$

Proof. ad (i) From (B.16) we find $Y_\Phi = 0$ for $\xi^1 = \xi^3 = 0$, i. e., for $r = \delta = 0$. In order to prove that Y_Φ vanishes on L_1 we consider its local expression in the chart $(\mathbb{CP}^3 \setminus \{Z^1 = 0\})$ with coordinates

$$\tilde\xi^1 = \frac{Z^0}{Z^1} = \frac{1}{\delta} ,$$

$$\tilde\xi^2 = \frac{Z^2}{Z^1} = \frac{1}{\delta}(\zeta + it) - \varrho e^{-i\phi} ,$$

$$\tilde\xi^3 = \frac{Z^3}{Z^1} = (\zeta - it) + \frac{1}{\delta}\varrho e^{i\phi} . \tag{B.27}$$

Thus, we find for Y_Φ in this local chart

$$Y_\Phi = -i\tilde\xi^1\frac{\partial}{\partial\tilde\xi^1} - i\tilde\xi^2\frac{\partial}{\partial\tilde\xi^2} , \tag{B.28}$$

which vanishes for $\tilde{\xi}^1 = \tilde{\xi}^2 = 0$, i.e., for $\varrho = 0$ and $\delta = \infty$. The proof of ii)
follows similarly.

$\square$

With this proposition we find that the different orbits of $G^{\mathbb{C}}$ on $\mathbb{CP}^3$ are
characterized by the intersections of Q_w with the singularities L_0, L_1, and I
of Y_Φ and Y_T.

In the following we will often consider regions $D \subset \mathbb{CP}^3$ which are related
to subsets of the (ϱ, ζ)-plane. Let $D \subset \mathbb{CP}^3$ be open. Then a holomorphic
function f on D, invariant under the action of G has also to be invariant
under the action of its complexification $G^{\mathbb{C}}$, i.e., we have

$$Y_\Phi(f) = Y_T(f) = 0 \ . \tag{B.29}$$

Of course, the orbits of Y_Φ and Y_T may intersect D in disconnected sets.
Then we find with (B.29) that f is locally constant on each component of the
intersection of D with the orbits of $G^{\mathbb{C}}$. But, on different intersections it can
take different values, i.e., f is not the pull-back of a function on the space
of orbits of $G^{\mathbb{C}}$ on $\mathbb{CP}^3$. But, since Y_Φ and Y_T generate an Abelian isometry
group, they commute

$$[Y_\Phi, Y_T] = 0 \ ,$$

and span, by Frobenius' theorem, an integrable distribution of TD which is
non-singular at all points $p \in D \setminus \{L_0 \cup L_1 \cup I\}$. In other words, we have a
codimension 1 foliation $\mathcal{F}$ of D, except at the intersections of D with L_0, L_1,
and I, generated by Y_Φ and Y_T.

Let $H \subset \mathbb{C}$ be the upper half of the (ϱ, ζ)-plane, i.e.,

$$H = \{w = \zeta + i\varrho \in \mathbb{C} | \varrho > 0\} \tag{B.30}$$

and $V \subset H$ be connected and simply connected and define $\bar{V}$ by $\bar{V} =$
$\{\bar{w} | w \in V\}$. We may consider $w = \zeta + i\varrho$ as a stereographic coordinate on a
Riemann sphere $\mathbb{CP}^1$. We define

$$\mathrm{pr} : \mathbb{E}^4 \setminus \{\varrho = 0\} \longrightarrow H \tag{B.31}$$

by

$$\mathrm{pr}((t, \phi, \varrho, \zeta)) := \zeta + i\varrho \ . \tag{B.32}$$

We denote $\mathbb{E}_V$ an open subset of $\mathbb{E}^4$ with

$$\mathrm{pr}(\mathbb{E}_V) = V \ . \tag{B.33}$$

Remark B.7. Often we shall be interested in a simply connected set $\mathbb{E}_V$, i.e.,
in general we have $\mathbb{E}_V \neq \mathrm{pr}^{-1}(V)$.

Let $V \subset H$ be fixed and choose $\mathbb{E}_V \subset \mathbb{E}^4$ to be an open subset of the
Euclidean four-space (see above) and, correspondingly, of S^4. We define

$$\mathbb{P}_V := \{\xi \in \mathbb{CP}^3 | \pi(\xi) \in \mathbb{E}_V\} \ , \tag{B.34}$$

where $\pi : \mathbb{CP}^3 \to S^4$ is the projection in the twistor bundle $\mathbb{CP}^3 \left(S^4, S^2\right)$.

Definition B.8. *Let* $\mathbb{P}_V \subset \mathbb{CP}^3$ *be given as above. The set of connected components of* $Q_w \cap \mathbb{P}_V$ *(w ∈ ℂ) is the reduced twistor space* R_V *associated with* V.

Let us now show that the reduced twistor space R_V can be represented as a compact but in general non-Hausdorff Riemann surface. We consider the surjection

$$\Gamma : \mathbb{CP}^3 \setminus (I \cup L_0 \cup L_1) \longrightarrow \mathbb{CP}^1 , \qquad (B.35)$$

defined by

$$\Gamma\left((\xi^1, \xi^2, \xi^3)\right) := \frac{1}{2}\left(\frac{\xi^3}{\xi^1} + \xi^2\right) = \frac{1}{2}\delta^{-1}\varrho e^{i\phi} + \zeta - \frac{1}{2}\delta\varrho e^{-i\phi} , \qquad (B.36)$$

or, with (B.17),

$$\Gamma\left((\xi^1, \xi^2, \xi^3)\right) = w . \qquad (B.37)$$

Let w be fixed and $x \in \mathbb{E}^4$ have the coordinates $x = (t, \phi, \varrho, \zeta)$. Let $\pi^{-1}(x) \simeq \mathbb{CP}^1$ be the fibre over x with stereographic coordinate δ. This fibre intersects the quadric Q_w labelled by w in points $\delta_\pm$ with

$$\frac{1}{2}\delta_\pm^2 \varrho e^{-i\phi} + (w - \zeta)\delta_\pm - \frac{1}{2}re^{i\phi} = 0 , \qquad (B.38)$$

see Fig. B.1.

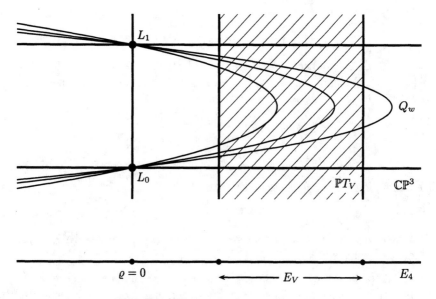

Fig. B.1. The twistor space above E_V

We have

Lemma B.9. *Let w be fixed. The intersection $Q_w \cap \mathbb{P}_V$ is connected if there exists a path $\gamma : [0,1] \to Q_w$ with $\gamma(0) = \delta_+$ and $\gamma(1) = \delta_-$ such that $\pi(\gamma(t)) \in \mathbb{E}_V$ for $t \in [0,1]$ and has two connected components otherwise.*

Proof. The proof follows directly from the definition of a connected set.

□

This lemma yields another possibility to characterize points of the reduced twistor space R_V. For $w = \zeta \pm i\varrho$ the discriminant $\Delta = (w - \zeta)^2 + \varrho^2$ of (B.38) vanishes such that $\delta_+ = \delta_-$. Thus we have

Lemma B.10. *If $w \in V \cup \bar{V}$ then $Q_w \cap \mathbb{P}_V$ has one connected component and two two connected components otherwise.*

A direct consequence of this Lemma is that R_V does not possess the Hausdorff property.

Lemma B.11. *For V being open the reduced twistor space R_V associated to V is not Hausdorff.*

Proof. Let $x \in \partial V$ and U_x a neighborhood of x. Then $x \notin V$ and according to Lemma B.10 there are two points x_1 and x_2 in R_V associated to x. Schematically we have

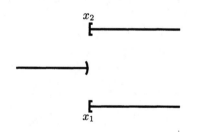

Fig. B.2. The non-Hausdorff property of R_V

Obviously, each neighborhood U_{x_1} of x_1 contains elements of the neighborhood U_{x_2} of x_2, i.e., $U_{x_1} \cap U_{x_2} \neq \emptyset$, and the Hausdorff property fails.

□

Now we want to show that R_V is a compact Riemann surface. More precisely, we have

Proposition B.12. *Let $V \subset H$ be simply connected. Then R_V is a compact Riemann surface, i.e., there is a covering map $\Gamma : R_V \to \mathbb{CP}^1$.*

Proof. We have just shown that $\Gamma^{-1}(w)$ has either one or two points depending on w. Thus, it remains to show that $\Gamma^{-1}\left(\mathbb{CP}^1 \setminus (V \cup \bar{V})\right)$ is a connected double cover of the complex projective plane $\mathbb{CP}^1$. Let $x \in \mathbb{E}^4$ be fixed. Let

$w(t)$ be a closed path which encircles V such that $w(0) = w(1) = w_0$. From (B.38) it follows that Δ changes its sign, i. e., $\delta_\pm$ changes continuously to $\delta_\mp$. Thus, a closed curve in $\mathbb{CP}^1$ lifts to a path in $\Gamma^{-1}\left(\mathbb{CP}^1 \setminus \left(V \cup \bar{V}\right)\right)$ which joins the two points in the preimage $\Gamma^{-1}(w_0)$.

$\square$

Thus, following [88], we can construct R_V associated to V in the following way:

(i)   Choose points $c \in V$ and $\bar{c} \in \bar{V}$ and make a cut C from c to $\bar{c}$.
(ii)   Take two copies S_0 and S_1 of the extended w-plane with cuts C_0 and C_1.
(iii) Identify points in V_0 with the corresponding points in V_1 and points in $\bar{V}_0$ with corresponding points in $\bar{V}_1$.
(iv)   Identify points on the $+$-branch of the cut C_0 in S_0 with points on the $-$-branch of the cut C_1 in S_1.

B.2.2 Holomorphic Vector Bundles over the Reduced Twistor Space

This subsection deals with the symmetry reduced Penrose–Ward transform. We will find that holomorphic vector bundles over R_V for some region V of spacetime will correspond to a solution of the Yang equation on V. Let us first recall the Penrose–Ward transform for our model. Let $V \subset H$ and $J = J(\varrho, \zeta)$ a solution to the Yang equation (2.85) over V. Suppose, we have

$$H, \hat{H} : E_V \longrightarrow \mathrm{Gl}(2; \mathbb{C}) . \tag{B.39}$$

such that

$$J = H \cdot \hat{H} . \tag{B.40}$$

Using the complex coordinates $p \equiv \tilde{w} = \varrho e^{i\phi}$ and $q \equiv \bar{z} = \zeta + it$ we define by

$$\Upsilon := H^{-1}\left(\partial_{\bar{p}} H \mathrm{d}\bar{p} + \partial_{\bar{q}} H \mathrm{d}\bar{q}\right) + \hat{H}^{-1}\left(\partial_p \hat{H} \mathrm{d}p + \partial_q \hat{H} \mathrm{d}q\right) \tag{B.41}$$

a $gl(2; C)$-valued connection $D = \mathrm{d} + \Upsilon$ in the trivial vector bundle $(B_V, E_V, \mathbb{C}^2)$ with total space

$$B_V = E_V \times \mathbb{C}^2 . \tag{B.42}$$

In Sect. 2.4 it has been shown

Lemma B.13. *The connection* $\mathrm{D} = \mathrm{d} + \Upsilon$ *in* $(B_V, E_V, \mathbb{C}^2)$ *is anti-self-dual if and only if J is a solution to the Yang equation (2.111).*

Let us now define by means of Υ a $gl(2; \mathbb{C})$-valued one-form on $\mathbb{P}_V$. We take

$$\Psi := \pi^* \Upsilon^{(0,1)} , \tag{B.43}$$

where π is the projection in the twistor bundle $\pi : \mathbb{CP}^3 \to S^4$ and where $\pi^* \Upsilon^{(0,1)}$ denotes the $(0,1)$-component of the one form $\pi^* \Upsilon$ on $\mathbb{P}_V$ in the decomposition $(\xi, \bar{\xi})$. With $\lambda := \delta e^{-i\phi}$ we have

Proposition B.14. *The connection one form Ψ is given by*

$$\Psi = \frac{1}{1+\lambda^2}\left(H^{-1}\left[(\partial_\varrho H - \lambda\partial_\zeta H)\,\bar{\partial}_\varrho + (\partial_\zeta H + \lambda\partial_\varrho H)\,\bar{\partial}_\zeta\right]\right.$$
$$\left.+\hat{H}^{-1}\left[\left(\lambda^2\partial_\varrho\hat{H} + \lambda\partial_\zeta\hat{H}\right)\bar{\partial}_\varrho + \left(\lambda^2\partial_\zeta\hat{H} - \lambda\partial_\varrho\hat{H}\right)\bar{\partial}_\zeta\right]\right) . \quad \text{(B.44)}$$

Proof. We will prove the proposition for $H = J$. Then we have

$$\pi^*\Upsilon = \frac{e^{i\phi}}{2}H^{-1}\frac{\partial H}{\partial\varrho}\left(e^{-i\phi}d\varrho - i\varrho e^{-i\phi}d\phi\right) + \frac{1}{2}H^{-1}\frac{\partial H}{\partial\zeta}\left(d\zeta - idt\right) , \quad \text{(B.45)}$$

such that we find

$$\Psi = \frac{e^{i\phi}}{2}H^{-1}\frac{\partial H}{\partial\varrho}\left(e^{-i\phi}\bar{\partial}_\varrho + \varrho\bar{\partial}\left(e^{-i\phi}\right)\right) + \frac{1}{2}H^{-1}\frac{\partial H}{\partial\zeta}\left(\bar{\partial}_\zeta - i\bar{\partial}t\right) . \quad \text{(B.46)}$$

From (B.14) we know

$$\xi^2 = \zeta + it - \delta\varrho e^{-i\phi} , \qquad \xi^3 = \varrho e^{i\phi} + \delta(\zeta - it) , \quad \text{(B.47)}$$

and using the fact that ξ^1 and ξ^2 are holomorphic coordinates, i. e., $\bar{\partial}\xi^1 = \bar{\partial}\xi^2 = 0$, we obtain

$$0 = \bar{\partial}\zeta + i\bar{\partial}t - \delta\left(\bar{\partial}\varrho\right)e^{-i\phi} - \delta\varrho e^{-2i\phi}\bar{\partial}\left(e^{i\phi}\right) ,$$
$$0 = e^{i\phi}\bar{\partial}\varrho + \varrho\bar{\partial}\left(e^{i\phi}\right) + \delta\bar{\partial}\zeta - i\delta\bar{\partial}t .$$

Therefore, we find

$$\bar{\partial}\left(e^{i\phi}\right) = \frac{\left(\lambda^2 - 1\right)\bar{\partial}\varrho - 2\lambda\bar{\partial}\zeta}{e^{-i\phi}\varrho\left(\lambda^2 + 1\right)} ,$$
$$i\bar{\partial}t = \frac{\left(\lambda^2 - 1\right)\bar{\partial}\zeta + 2\lambda\bar{\partial}\varrho}{\left(\lambda^2 + 1\right)} ,$$

and inserting into (B.46) yields (B.44).

$\square$

Remark B.15.
(i) Since $\bar{\partial}\varrho = \bar{\partial}\zeta = 0$ for $x \in \mathbb{E}_V$ fixed, we have

$$\Psi|_X = 0 \quad \text{(B.48)}$$

along all fibres $X = \pi^{-1}(x)$.
(ii) We have

$$L_T\Psi = L_\Phi\Psi = 0 . \quad \text{(B.49)}$$

In fact, since H and λ are independent of t, we have

$$T\left(\frac{1}{1+\lambda^2}H^{-1}\left(\frac{\partial H}{\partial\varrho} - \lambda\frac{\partial H}{\partial\zeta}\right)\right) = 0 . \quad \text{(B.50)}$$

Furthermore, $L_T\bar{\partial}\varrho = L_T\bar{\partial}\zeta = 0$. The proof of the $L_\Phi\Psi = 0$ follows similarly.

(iii) Let us consider the quadric $\lim\limits_{w\to\infty} Q_w$ in $\mathbb{P}_V$. Since

$$w = \zeta + \frac{1}{2}\varrho\left(\lambda^{-1} - \lambda\right) \ , \tag{B.51}$$

this means $\lambda, \delta \to 0$ or $\lambda, \delta \to \infty$. We denote

$$\begin{aligned}\Lambda_0 &:= \mathbb{P}_V \cap \{\lambda = 0\} \ ,\\ \Lambda_1 &:= \mathbb{P}_V \cap \{\lambda = \infty\} \ ,\end{aligned} \tag{B.52}$$

i.e., Λ_0 and Λ_1 are the points in R_V above ∞ in $\mathbb{CP}^1$. Then we find from (B.44)

$$\begin{aligned}\Psi(\Lambda_0) &= H^{-1}\left(\frac{\partial H}{\partial\varrho}\bar\partial\varrho + \frac{\partial H}{\partial\zeta}\bar\partial\zeta\right) = H^{-1}\bar\partial H \ ,\\ \Psi(\Lambda_1) &= \lim_{\lambda\to\infty}\frac{1}{1+\lambda^2}\left(A + \lambda B + \lambda^2\left(\hat H^{-1}\partial_\varrho\hat H\right)\bar\partial\varrho + \lambda^2\left(\hat H^{-1}\partial_\zeta\hat H\right)\bar\partial\zeta\right)\\ &= \left(\hat H^{-1}\partial_\zeta\hat H\right)\bar\partial\zeta + \left(\hat H^{-1}\partial_\varrho\hat H\right)\bar\partial\varrho\\ &= \hat H^{-1}\bar\partial\hat H \ .\end{aligned} \tag{B.53}$$

The curvature form of Υ is anti-self-dual and we want to investigate the consequences for Ψ. To do this, we use the following theorem, due to Atiyah, see [274].

Proposition B.16. *A two-form ω on $U \subset S^4$ is anti-self-dual if and only if $\pi^*\omega$ is a $(1,1)$-form over $\tilde U \subset \mathbb{CP}^3$.*

Remark B.17. This proposition is closely related to a lemma which states that a two-form ω over $U \subset \mathbb{R}^4$ is anti-self-dual if and only if ω is of type $(1,1)$ for all compatible complex structures on $\mathbb{R}^4$, see [274].

The above proposition implies the following result

Lemma B.18. *A necessary condition for the anti-self-duality of the curvature two-form $F = \mathrm{d}\Upsilon + \Upsilon \wedge \Upsilon$ is*

$$\bar\partial\Psi + \Psi \wedge \Psi = 0 \ . \tag{B.54}$$

Proof. Let us decompose

$$\pi^*\Upsilon = \Psi + \hat\Psi =: \tilde\Psi \ , \tag{B.55}$$

with $\hat\Psi$ being a $(1,0)$-form. Then we find for the pull-back π^*F of the curvature form F

$$\begin{aligned}\pi^*F &= \pi^*\left(\mathrm{d}\Upsilon\right) + \pi^*\left(\Upsilon \wedge \Upsilon\right) = \mathrm{d}\left(\pi^*\Upsilon\right) + \pi^*\Upsilon \wedge \pi^*\Upsilon\\ &= \left(\partial + \bar\partial\right)\left(\Psi + \hat\Psi\right) + \left(\Psi + \hat\Psi\right) \wedge \left(\Psi + \hat\Psi\right)\\ &= \partial\Psi + \bar\partial\Psi + \partial\hat\Psi + \bar\partial\hat\Psi + \Psi \wedge \Psi + \Psi \wedge \hat\Psi + \hat\Psi \wedge \Psi + \hat\Psi \wedge \hat\Psi \ .\end{aligned}$$

The $(2,0)$-part of the pull-back of F does not depend on Ψ and, according to Proposition B.16, the $(0,2)$-part of the pull-back of F has to vanish, i.e.,

$$\bar{\partial}\Psi + \Psi \wedge \Psi = 0 \ . \tag{B.56}$$

$\square$

In the following we will make use of the decomposition of the connection $\nabla = d + \tilde{\Psi}$ in the trivial C^∞-bundle $B = \mathbb{P}_V \times \mathbb{C}^2$ into a $(1,0)$-part and a $(0,1)$-part

$$\nabla = \nabla^{(1,0)} + \nabla^{(0,1)} \ . \tag{B.57}$$

Then we have the following proposition, for the proof of which we refer to [275], see also [276].

Proposition B.19. *The $(0,1)$-form $\Psi \in T^* (\mathbb{P}_V) \otimes gl(2; \mathbb{C})$ with $\bar{\partial}\Psi + \Psi \wedge \Psi = 0$ defines a holomorphic structure on B, if*

$$\nabla^{(0,1)} b = \bar{\partial} b + \Psi b = 0 \ , \tag{B.58}$$

for $b \in \Gamma (\mathbb{P}_V, B)$.

Remark B.20. The proof makes essential use of the following proposition

Proposition B.21. *Let (B, M, π) be a vector bundle over a complex manifold M and let ∇ be a connection in B with curvature being of type $(1,1)$. Then there exists a holomorphic structure on (B, M, π).*

Corollary B.22. *According to (B.48), the restriction $\Psi|_X$ of Ψ to the fibres $X = \pi^{-1}(x)$ vanishes for $x \in \mathbb{E}^4$. Thus (B.58) reduces along the fibres X to*

$$\bar{\partial} b = 0 \ . \tag{B.59}$$

From this it follows easily that the restriction $B|_X$ is holomorphically trivial.

Thus, Proposition B.19 gives the condition for locally holomorphic sections in the vector bundle B, such that $B|_X = X \times \mathbb{C}^2$ is trivial not only as a smooth bundle but also as holomorphic bundle. However, for a bundle over R_V, it is also necessary that the sections are invariant under the G-action, i.e., the sections should fulfil

$$\bar{\partial} b + \Psi b = 0 \ , \quad \Phi(b) = 0 = T(b) \ . \tag{B.60}$$

The existence of such sections is implied by (B.49) and (B.54). We will prove

Lemma B.23. *The sections in (B.60) depend only on ζ, ϱ and λ.*

Proof. That b is independent of t is a direct consequence of the definition of T as

$$T = \frac{\partial}{\partial t} \ . \tag{B.61}$$

Let us now show that b depends on ϕ and δ only via λ. We have

$$\frac{\partial}{\partial\lambda} = \frac{e^{i\phi}}{2}\frac{\partial}{\partial\delta} + \frac{ie^{i\phi}}{2\delta}\frac{\partial}{\partial\phi} \ , \tag{B.62}$$

and

$$\frac{\partial b}{\partial\phi} + i\delta\frac{\partial b}{\partial\delta} - i\bar{\delta}\frac{\partial b}{\partial\bar{\delta}} = 0 \ , \tag{B.63}$$

because $\Phi(b) = 0$. The derivative with respect to $\bar{\delta}$ vanishes due to the holomorphicity condition of Proposition B.19 and we have

$$\frac{\partial b}{\partial\phi} = -i\delta e^{-i\phi}\frac{\partial b}{\partial\lambda} \ , \qquad \frac{\partial b}{\partial\delta} = e^{-i\phi}\frac{\partial b}{\partial\lambda} \ . \tag{B.64}$$

$\square$

Since the leaves of the foliation of $\mathbb{P}_V$ are simply connected, the value of b on a leaf Λ is completely determined by its value in one point of Λ via

$$Y_\phi(b) - \left(\bar{Y}_\phi\Psi\right)b = 0 = Y_t(b) - \left(\bar{Y}_t\Psi\right)b \ , \tag{B.65}$$

and the complex conjugate equations which are a direct consequence of (B.60) and the definitions of Y_ϕ and Y_T. Let now E_Λ be the two-dimensional complex space of solutions to these equations.

Definition B.24. *The vector bundle $E \to R_V$ is generated in each point $\Lambda \in R_V$ by the two-dimensional fibre E_Λ.*

By taking the locally holomorphic invariant sections of B, i. e., solutions to (B.60), as locally holomorphic sections of E, the holomorphic structure of E is induced by that of B. The matrices H and $\hat{H}$ distinguish bases in the fibres over Λ_0 and Λ_1 in the following way. We define

$$f^0{}_i = \left(H^{-1} \circ \pi\big|_{\Lambda_0}\right)e_i \ ,$$
$$f^1{}_i = \left(\hat{H}^{-1} \circ \pi\big|_{\Lambda_1}\right)e_i \ , \tag{B.66}$$

$(i = 1, 2)$, with

$$e_1 = \begin{pmatrix} 1 \\ 0 \end{pmatrix} \ ,$$
$$e_2 = \begin{pmatrix} 0 \\ 1 \end{pmatrix} \ . \tag{B.67}$$

Now we have

Proposition B.25. *The sets $\{f^0{}_1, f^0{}_2\}$ and $\{f^1{}_1, f^1{}_2\}$ form fibre bases over Λ_0 and Λ_1.*

Proof. Since H and $\hat{H}$ have non-vanishing determinants, $f^0{}_1$ and $f^0{}_2$ respectively $f^1{}_1$ and $f^1{}_2$ are linearly independent. Then we have for the i-th column of H^{-1}

$$\bar{\partial}\left(H^{-1}\right)_i + \Psi\left(H^{-1}\right)_i = \bar{\partial}\left(H^{-1}\right)_i + \left(H^{-1}\bar{\partial}H\right)\left(H^{-1}\right)_i \qquad (B.68)$$
$$= \left(-H^{-1}\bar{\partial}(H)H^{-1}\right)_i + \left(H^{-1}\bar{\partial}H\right)\left(H^{-1}\right)_i = 0 \ .$$

The proof for $\{f^1{}_1, f^1{}_2\}$ follows similarly.

$\square$

Remark B.26. A different choice of H and $\hat{H}$ generates a bundle isomorphic to E.

To summarize, we have shown, how to associate to a solution to the Yang equation on some region V of spacetime a holomorphic vector bundle over the reduced twistor space R_V associated to V. In fact, the converse is also true. We have

Theorem B.27. *There is a 1-1-correspondence between:*

(i) $Gl(2;\mathbb{C})$-valued solutions J to the Yang equation on $V \subset H$, uniquely determined up to transformations $J \mapsto AJB^{-1}$ with $A, B \in Gl(2;\mathbb{C})$ and
(ii) holomorphic vector bundles $(E, R_V, Gl(2;\mathbb{C}))$ such that $\Pi^*E|_X$ is holomorphically trivial for any fibre $X = \pi^{-1}(x) \subset \mathbb{P}_V$ $(x \in \mathbb{E}_V)$.

The fixing of frames f^0 respectively f^1 over Λ_0 respectively Λ_1 in E determines J completely.

Remark B.28. The transformations $J \mapsto AJB^{-1}$ correspond to a left multiplication of H by A and of $\hat{H}$ by B. Since A and B are constant matrices the one-form Υ remains constant.

Proof. It remains to be shown how a solution to the Yang equation with the required properties can be constructed by means of a holomorphic vector bundle over R_V. Let $(E, R_V, Gl(2;\mathbb{C}))$ be a vector bundle fulfilling the above requirements. Due to the triviality in any fibre we may construct for B a globally smooth frame (b_1, b_2) such that b_i is a holomorphic section of $B|_X$ for all X. Taking the derivative of b_i with respect to the coordinates of $\mathbb{P}_V$ we define globally a $gl(2;\mathbb{C})$-valued $(0,1)$-form Ψ by

$$\bar{\partial}b_i + \Psi b_i = 0 \ , \qquad (B.69)$$

which in turn yields a holomorphic structure on B. Writing

$$\Psi = (\pi^*\Upsilon)^{(0,1)} \ , \qquad (B.70)$$

we obtain, according to Proposition B.16, a one-form Υ with anti-self-dual curvature. Obviously, the choice of b_i is not unique: with b_i we may also choose $a_i{}^j\, b_j$ with $(a_i{}^j) \in \mathrm{Gl}(2;\mathbb{C})$ depending smoothly on ϱ, ζ, ϕ and t. This transformation induces a gauge transformation of $\varUpsilon$. In order to determine J completely, we define frames over $\varLambda_0$ respectively $\varLambda_1$ by the following formulae

$$s^0 = \varPi^* f_0 \ , \quad s^0 = \varPi^* f_0 \ . \tag{B.71}$$

Here

$$\varPi : \mathbb{P}_V / \mathcal{F} \longrightarrow R_V \tag{B.72}$$

is the projection of $\mathbb{P}_V$ onto R_V which associates to each $p \in \mathbb{P}_V$ the corresponding leaf. We define

$$b|_{\varLambda_0} := s^0\, H \ , \tag{B.73}$$

and obtain

$$b|_{\varLambda_1} := s^1\, \hat{H} \ , \tag{B.74}$$

where $\varUpsilon$ is determined by (B.41). Then we have $\bar{\partial} b + \varPsi b = 0$ on $\varLambda_0$ and $\varLambda_1$ and obtain J by

$$J := H \cdot \hat{H}^{-1} \ . \tag{B.75}$$

Therefore, the chosen gauge fixes the splitting of J into H and $\bar{H}$ and the fixing of f^0 and f^1 determines the entries of J. In particular, for $b|_{\varLambda_1} = s^1$ we get

$$b|_{\varLambda_0} = s^0 \cdot \tilde{J} \ , \tag{B.76}$$

with $\tilde{J} : \varLambda_0 \to \mathrm{Gl}(2;\mathbb{C})$ given by E, f^0 and f^1. Then $\tilde{J}$ is just the solution of the Yang equation one is looking for. Furthermore, we have $\varPsi|_{\varLambda_0} = \tilde{J}^{-1}\bar{\partial}\tilde{J}$ and it follows with (B.75) that $H = \tilde{J} = J$. A change of the frames s_0 respectively s_1 on $\varLambda_0$ respectively $\varLambda_1$ can, according to the definition of these frames, only lead to a transformation $J \mapsto AJB^{-1}$ with constant matrices A and B. For the same reason we have $J = J(\varrho, \zeta)$.

$$\square$$

Of course, a solution to the Yang equation has to be real and symmetric in order to describe a stationary and axisymmetric solution to the Einstein equations. From Sect. 2.4 we know that with J being a solution to the Yang equation also J^{-1}, $\bar{J}$ and J^T are solutions ($J = \bar{J}$). Therefore, we may construct holomorphic vector bundles over R_V corresponding to these solutions and relate them to E. To do this we have to define some natural involutions on R_V:

(i) Let $i : R_V \to R_V$ be the holomorphic mapping, which interchanges the two spheres S_0 and S_1 in R_V, i.e.,

$$i(w_\pm) := w_\mp \ . \tag{B.77}$$

(ii) Let $j : R_V \to R_V$ denote the anti-holomorphic mapping

$$j(w_\pm) := \bar{w}_\mp \ . \tag{B.78}$$

Then we have the following

Proposition B.29. *Let J be a solution to the Yang equation corresponding to the triple (E, f^0, f^1) over the reduced twistor space R_V by Theorem B.27. Then*

(i) the inverse solution J^{-1} is generated by

$$i^*(E, f^0, f^1) := \left(i^*(E), i^*(f^0), i^*(f^1) \right) , \qquad (B.79)$$

(ii) the complex conjugate solution $\bar{J}$ is generated by the triple

$$\bar{j}^*(E, f^0, f^1) := \left(j^*(\bar{E}), j^*(\bar{f^0}), j^*(\bar{f^1}) \right) , \qquad (B.80)$$

(iii) and the solution $(J^{-1})^T$ is generated by the triple of dual objects

$$(E, f^0, f^1)^* := \left(E^*, (f^0)^*, (f^1)^* \right) . \qquad (B.81)$$

Proof. We will prove (i), the proof of (ii) and (iii) follows similarly. Let $J = H \cdot \hat{H}^{-1}$. Then $J^{-1} = \hat{H} \cdot H^{-1}$. Therefore, J^{-1} generates by (B.44) a $(0,1)$-form Ψ' in which H and $\hat{H}$ have changed their role. But, from the form of Ψ is obvious that $\Psi(\lambda) = \Psi'(-\lambda^{-1})$ and, therefore, for the fibre it follows $E'(\lambda) = E(-\lambda^{-1})$. With $i(w_\pm) = w_\mp$ it follows $i(\lambda) = -\lambda^{-1}$ and we find $E' = i^*(E)$.

□

Corollary B.30. *We have: J is symmetric iff*

$$(E, f^0, f^1)^* = i^*(E, f^0, f^1) , \qquad (B.82)$$

and J is real iff

$$(E, f^0, f^1)^* = \bar{j}^*(E, f^0, f^1) . \qquad (B.83)$$

Proof. The first assertion follows from the fact that $J = J^T$ is equivalent to $J^{-1} = (J^{-1})^T$. The second is just the definition.

□

B.3 Transition Matrices for the Holomorphic Vector Bundles

B.3.1 The Covering of the Reduced Twistor Space

It is well known, that the two-sphere S^2 can be covered completely by two local charts. As it has been shown in Sect. B.2.1, the reduced twistor space R_V, is a two-sheeted covering of the Riemann sphere. This suggests that R_V may be covered by four charts $R_0, \ldots, R_3$. These can be viewed as covers of S^2 induced by the compactification of $\mathbb{E}^2$ with the complex coordinate w. Let us now, following [88], but see also [166] for a detailed description, construct a standard cover of R_V. To begin with, we cover the Riemann sphere by two charts U and U' in the following way. Let U denote an open subset of the complex w-plane such that

(i) $u \in \implies \bar{u} \in U$,
(ii) $V \cup \bar{V} \subset U$,
(iii) $U \setminus V, U \setminus \bar{V}$, and U itself are connected and simply connected.

Furthermore, let U' be a neighborhood of $w = \infty$ such that

(i) $u' \in U \implies \bar{u}' \in U$ and
(ii) the intersection $A = U \cup U'$ is an annular region, see Fig. B.3.

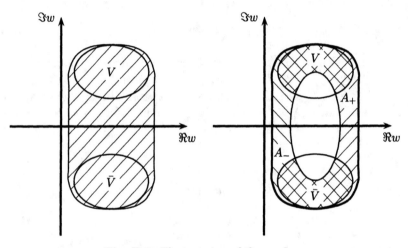

Fig. B.3. The covering of the w plane

Then A is the union of two simply connected sets A_+ and A_-, $A = A_+ \cup A_-$, and we have $A_+ \cap A_- = (A \cap V) \cup (A \cap \bar{V})$. The set $\Gamma^{-1}(U') \subset R_V$ contains two copies R_0 and R_1 of U', i.e., $\Gamma^{-1}(U') = R_0 \cup R_1$. The subset R_0 is a neighborhood of $w = \infty$ in the Riemann sphere S_0 and, similarly, R_1 is a neighborhood of $w = \infty$ in S_1. The intersection $R_0 \cup R_1$ on R_V can be identified with the intersection $A_+ \cap A_-$ in the complex w-plane.

Analogously, $\Gamma^{-1}(U)$ consists of two copies, which both contain the cut C and intersect in $\Gamma^{-1}(V \cup \bar{V})$. We denote these copies by R_2 and R_3 where the labelling is fixed by

$$R_0 \cap R_2 \subset \Gamma^{-1}(A_-) , \quad R_1 \cap R_3 \subset \Gamma^{-1}(A_+) . \tag{B.84}$$

In $\Gamma^{-1}(A_+ \cap A_-) \subset \Gamma^{-1}(V \cup \hat{V})$ the four covering sets are identified. But, over each point of the Riemann sphere they are in one sphere and, therefore, they are Hausdorff and may be used as coordinate patches for the basis R_V of the holomorphic vector bundles.

B.3.2 Patching Matrices for Real, Symmetric Framed Bundles

Let us now consider the holomorphic vector bundle $(E, R_V, \mathbb{C}^2)$ with typical fibre $\mathbb{C}^2$ and projection $\mathrm{pr} : E \to R_V$. This bundle is trivial if restricted to the contractible Hausdorff manifolds R_α. Since the R_α are Stein manifolds, we may choose globally a holomorphic frame e^α in the form of a 2×2-matrix

$$(e^\alpha) = (e_1^\alpha, e_2^\alpha) \ . \tag{B.85}$$

(Note that α enumerates the different covering sets and not the different vectors of one particular frame.) Then, the trivialization of E takes the form $\left(R_\alpha, y^i e_i^\alpha\right)$ for $\mathrm{pr}(y) \in R_\alpha$. In particular, we may have

$$e^0\big|_{\Lambda_0} = f^0 \ , \qquad e^1\big|_{\Lambda_1} = f^1 \ , \tag{B.86}$$

which is well defined because of $\Lambda_0 \in R_0$ and $\Lambda_1 \in R_1$. In an intersection $R_\alpha \cap R_\beta$ the patching matrix $P_{\alpha\beta}$ transforms holomorphic frames e^α into holomorphic frames e^β according to

$$e^\beta = e^\alpha P_{\alpha\beta} \ . \tag{B.87}$$

In general, there are six patching matrices which we denote as follows:

$$
\begin{aligned}
P_{01} &: A_+ \cap A_- &&\longrightarrow && \mathrm{Gl}(2;\mathbb{C}) \, , \\
P_{02} &: A_- &&\longrightarrow && \mathrm{Gl}(2;\mathbb{C}) \, , \\
P_{13} &: A_- &&\longrightarrow && \mathrm{Gl}(2;\mathbb{C}) \, , \\
P_{03} &: A_+ &&\longrightarrow && \mathrm{Gl}(2;\mathbb{C}) \, , \\
P_{12} &: A_+ &&\longrightarrow && \mathrm{Gl}(2;\mathbb{C}) \, , \\
P_{23} &: V \cup \hat{V} &&\longrightarrow && \mathrm{Gl}(2;\mathbb{C}) \ .
\end{aligned}
\tag{B.88}
$$

Since the covering of R_V by R_α suspends locally the twofold covering of $\mathbb{CP}^1$, the patching matrices depend only on w. The cocycle condition

$$P_{01} P_{12} P_{20} = I \tag{B.89}$$

fixes P_{01} on its whole domain $A_+ \cap A^-$ in terms of P_{12} and P_{02}, i.e., we are left with five transition matrices. This number may be reduced further by requiring particular properties of the holomorphic bundle E. Here the starting point is that the Yang matrix J which corresponds to a vacuum solution of the stationary and axisymmetric Einstein equations has to be real and symmetric, see (2.84). Recall that we have seen in Sect. B.2.2 what these requirements mean for the corresponding holomorphic vector bundles. In the sequel we shall assume that the conditions (B.82) and (B.83) for the holomorphic bundles are fulfilled. (How $\det J = -\varrho^2$ can be achieved will be discussed in Corollary B.37.) In order to transfer conditions like (B.82) and (B.83) to the patching matrices we first observe for J being symmetric the isomorphism between E^* and $i^*(E)$ can be understood as a family of invertible linear mappings depending holomorphically on $\Lambda \in R_V$

$$S_\Lambda : E_\Lambda \longrightarrow E^*_{i(\Lambda)} \ . \tag{B.90}$$

Since $i^2 = 1$ we have

$$S_{i(\Lambda)} : E_{i(\Lambda)} \longrightarrow E^*_\Lambda \ , \tag{B.91}$$

and we denote by $S^*_{i(\Lambda)}$ the map dual to $S_{i(\Lambda)}$

$$S^*_{i(\Lambda)} : E_\Lambda \longrightarrow E^*_{i(\Lambda)} \ . \tag{B.92}$$

Then we may define by

$$\sigma_\Lambda := \left(S^*_{i(\Lambda)} \right)^{-1} \circ S_\Lambda : E_\Lambda \longrightarrow E_\Lambda \tag{B.93}$$

a holomorphic section in the automorphism bundle $\mathrm{Aut}(E) = E \otimes E^*$ of E and have

Lemma B.31. $\sigma \equiv I$ on R_V.

Proof. We have $S_{\Lambda_0}(f^0) = \left(f^1 \right)^*$ and $S_{\Lambda_1}(f^1) = \left(f^0 \right)^*$ and, therefore, $\sigma = I$ over Λ_0 and Λ_1. Now we pull back the bundle E to the fibre $X \subset \mathbb{P}_V$

$$B|_X = \Pi^*(E) \tag{B.94}$$

and use the triviality of $B|_X$ and $\mathrm{Aut}(B|_X)$. Thus, $\Pi^*(\sigma)$ is a holomorphic section in a trivial bundle which is the identity map over $\lambda = 0$ and $\lambda = \infty$. By Liouville's theorem σ is then the identity on the whole of R_V.
□

Thus, we have for the standard cover of R_V with $R_0 = i(R_1)$ and $R_2 = i(R_3)$ and the corresponding frames

$$i^* \left(E_{i(\Lambda)} \right) = (i^*E)_\Lambda = E^*_\Lambda \ , \tag{B.95}$$

i. e.,

$$
\begin{aligned}
i^*(e^0) = \left(e^1 \right)^* \ i^*(e^1) = \left(e^0 \right)^* \ , \\
i^*(e^2) = \left(e^3 \right)^* \ i^*(e^3) = \left(e^2 \right)^* \ .
\end{aligned}
\tag{B.96}
$$

Analogously, for a vector bundle corresponding to a real solution of the Yang equation there exists an isomorphism

$$T_\Lambda : E_\Lambda \longrightarrow \bar{E}_{j(\Lambda)} \ , \tag{B.97}$$

and one defines

$$\tau_\Lambda := \bar{T}_{j(\Lambda)} \circ T_\Lambda \ . \tag{B.98}$$

Since τ is also a holomorphic section in the automorphism bundle $\mathrm{Aut}(E)$ of E, which is the identity mapping over Λ_0 and Λ_1, we find again, that τ is the identity map everywhere on R_V. Then we have

$$j^* \left(E_{j(\Lambda)} \right) = (j^* E)_\Lambda = \bar{E}_\Lambda \ , \tag{B.99}$$

i. e.,

$$j^* \left(e^\alpha \right) = \bar{e}^\alpha \ , \tag{B.100}$$

for all $\alpha = 0, \dots, 3$.

Using (B.96) and (B.100) we find for the patching matrices of a real, symmetric framed bundle the following.

Lemma B.32. *Let J be a symmetric real solution to the Yang equation (2.85) and $(E, R_V, Gl(2; \mathbb{C}))$ the corresponding holomorphic vector bundle over R_V with patching matrices as defined above. Then we have*

$$P_{\alpha\beta}(w) = \bar{P}_{\alpha\beta}(w) \ , \tag{B.101}$$

and

$$P_{02}(w) = (P_{31}(w))^T \ , \quad P_{03}(w) = (P_{21}(w))^T \ , \quad P_{23}(w) = (P_{23}(w))^T \ . \tag{B.102}$$

Proof. Let us show first (B.101). Since $e^\beta = e^\alpha P_{\alpha\beta}$, we also have $\bar{e}^\beta = \bar{e}^\alpha \bar{P}_{\alpha\beta}$ and, therefore,

$$j^* \left(e^\beta \right) = j^* \left(e^\alpha P_{\alpha\beta} \right) = \bar{e}^\alpha P_{\alpha\beta} \ , \tag{B.103}$$

and

$$j^* \left(e^\beta \right) = \bar{e}^\beta = \bar{e}^\alpha \bar{P}_{\alpha\beta} \ . \tag{B.104}$$

In order to prove (B.102), we first note that

$$i^* \left(e^2 \right) = \left(e^3 \right)^* = (P_{31})^T \left(e^1 \right)^* \ , \tag{B.105}$$

and

$$i^* \left(e^2 \right) = i^* \left(e^0 P_{02} \right) = (P_{02}) \left(e^1 \right)^* \tag{B.106}$$

is an identity, because the domains of P_{02} and P_{31} coincide. Furthermore, we have

$$i^* \left(e^3 \right) = \left(e^2 \right)^* = (P_{23})^T \left(e^3 \right)^* \ , \tag{B.107}$$

and

$$i^* \left(e^3 \right) = i^* \left(e^2 P_{23} \right) = (P_{23}) \left(e^3 \right)^* \ . \tag{B.108}$$

$\square$

According to the Lemma B.32 the framed bundle (E, f^0, f^1) corresponding to a real and symmetric solution of the Yang equation (2.85), is completely characterized by the following three patching matrices

$$F := P_{02} : \quad A_- \longrightarrow \mathrm{Gl}(2;\mathbb{C}) \,,$$
$$G := P_{12} : \quad A_+ \longrightarrow \mathrm{Gl}(2;\mathbb{C}) \,,$$
$$P := P_{23}|_V : V \longrightarrow \mathrm{Gl}(2;\mathbb{C}) \,. \tag{B.109}$$

Furthermore, we have

$$P_{23}|_{\bar{V}} = \bar{P}_{23}|_V = \bar{P} \,, \tag{B.110}$$

and we way obtain from the identities given above further requirements for the patching matrices. From the above Lemma we immediately find $P = P^T$. In addition, we have $G(w) = \bar{G}(w)$ and $F(w) = \bar{F}(w)$, i.e., on the real axis these matrices become real. Finally, on $A \cap V = A_- \cap V = A_+ \cap V$ we find

$$P_{23} P_{30} P_{02} = I = PG^T F \,. \tag{B.111}$$

We have

Proposition B.33. *The patching matrices are only determined up to gauge transformations of the form*

$$(P, F, G) \longrightarrow \left(LPL^T, KFL^{-1}, (K^T)^{-1} GL^{-1} \right) \,, \tag{B.112}$$

with the holomorphic and (in the above sense) real matrices $L : U \to \mathrm{Gl}(2;\mathbb{C})$ and $K : U' \to \mathrm{Gl}(2;\mathbb{C})$, fulfilling $L = \bar{L}$, $K = \bar{K}$ and $K(\infty) = I$. Here U and U' are the open subsets of the w-plane introduced at the beginning of Sect. B.3.1.

Proof. The proof follows by direct calculations. Since $P = P^T$ we have

$$\left(LPL^T \right)^T = LPL^T \,. \tag{B.113}$$

Furthermore, the cocycle condition (B.111) leads to

$$\left(LPL^T \right) \left((L^{-1})^T G^T \left((K^T)^{-1} \right)^T \right) (KFL^{-1}) = LPG^T FL^{-1} = LL^{-1} = I \,. \tag{B.114}$$

$$\square$$

Thus, we have a one-to-one correspondence between real symmetric solutions to the Yang equation (2.85) and equivalence classes of triples of patching matrices $[P, F, G]$ with properties discussed above.

B.3.3 The Axis-simple Case

As we have shown in Sect. 4.2, the solutions (4.19) to the Ernst equation are regular outside a contour which can correspond to the surface of a rotating body. It has been shown in [88] that under these circumstances there is a further reduction of the set of patching matrices $[P, G, F]$. To find this reduction one investigates the behavior of the holomorphic bundles by approaching the symmetry axis $\varrho = 0$.

Let $U \subset \mathbb{E}^2$ be defined as in the beginning of Sect. B.3.1. We denote by R_U the analogue of R_V, constructed for U. Let the projection

$$\eta : R_V \longrightarrow R_U \tag{B.115}$$

be defined by

$$\eta(w_\pm) := \begin{cases} w; & w \in U \setminus (V \cup \bar{V}) \\ w_\pm; & w \in V \cup \bar{V} \end{cases} , \tag{B.116}$$

for $w_\pm \in R_V$.

Definition B.34. *The triple (E, f^0, f^1) is called axis simple, iff there exists a triple $(E', f^{0'}, f^{1'})$ such that*

$$(E, f^0, f^1) = \eta^* \left(E', f^{0'}, f^{1'} \right) . \tag{B.117}$$

Proposition B.35. *The following statements are equivalent:*

(i) A bundle E' exists.

(ii) $E|_{\Gamma^{-1}(U)}$ is trivial.

(iii) If E is, furthermore, the holomorphic vector bundle of a real, symmetric solution to the Yang equation, then the patching matrix P can be analytically continued to a function on U, which is real on the symmetry axis.

Proof. (i) $\Longrightarrow$ (ii): Let E' exist, such that $\eta^*(E') = E$. Then we have for the fibre over $w_\pm \in \Gamma^{-1}(U \setminus (V \cup \bar{V}))$

$$E_{w_+} = E_{w_-} = \eta^*(E')_w . \tag{B.118}$$

Since U is contractible, $E'|_U$ is trivial and, therefore, also $E|_{\Gamma^{-1}(U)}$.

(ii) $\Longrightarrow$ (i): Let $E|_{\Gamma^{-1}(U)}$ be trivial. Then, there exists a biholomorphic mapping $\phi : \mathrm{pr}^{-1}(\Gamma^{-1}(U)) \to \Gamma^{-1}(U) \times \mathbb{C}^2$. We have to construct E' which is, being a bundle over a contractible manifold, trivial. Due to its triviality, E can globally be written as

$$E = \left\{ (p, e) | p \in \Gamma^{-1}(U), e \in \mathbb{C}^2 \right\} . \tag{B.119}$$

We construct E' by choosing the fibre $E|_p$ over $\eta(p)$. Then, the condition $\eta(p) = \mathrm{pr}(e)$ is fulfilled, and E' is trivial.

(iii) $\Longrightarrow$ (i): For vector bundles related to real symmetric solutions to the Yang equation, the patching matrix P is well defined and its analytic continuation $\left(\tilde{P} \right)$ yields a globally defined holomorphic section s in the vector bundle over $\Gamma^{-1}(U)$ putting:

$$s = \begin{cases} e_{(1)}^3(w_\pm) & ; \ w_\pm \in R_3 \\ e_1^2 \tilde{P}_{(i1)}(w_\pm) & ; \ w_\pm \in R_2 \end{cases} . \tag{B.120}$$

Here we sum over the index i of $\tilde{P}$ and make use of $e^3 = e^2 \tilde{P}_{23} \equiv e^2 \tilde{P}$ in $V \cup \bar{V}$. Then, $E|_{\Gamma^{-1}(U)}$ is trivial. Conversely, a global section as constructed above yields an analytic continuation of P. Using (B.96) respectively $P(w) = \bar{P}(w)$ we eventually find the reality on the real axis.

□

Since $R_U = S^0 \cup S^1$, E' is constructed out of two bundles $E'|_{S_0}$ and $E'|_{S_1}$ over the two spheres which form R_U. Being holomorphic bundles over the Riemann sphere these two bundles can, according to a theorem by Grothendieck (see e. g. [277] for a proof), be decomposed into direct sums of line bundles

$$E'|_{S_0} = L^p \oplus L^q \ ,$$
$$E'|_{S_1} = L^{-p} \oplus L^{-q} \ , \tag{B.121}$$

with

$$L^p = \underbrace{L \otimes \cdots \otimes L}_{p \text{ times}} \ , \tag{B.122}$$

and $L^{-1} = L^*$. Here L is the tautological bundle over $\mathbb{CP}^1$. The explicit form of $E'|_{S_1}$ for a given $E'|_{S_0}$ is a direct consequence of the symmetry of J (or $i^*(E') = E'^*$). Thus, the patching matrices of an axis-simple bundle corresponding to a real and symmetric solution to the Yang equation, have the following form

$$P : U \longrightarrow \text{Gl}(2; \mathbb{C}) \ , \tag{B.123}$$

with $P(w) = \bar{P}(w)$, $P(w) = P(w)^T$, can now be defined on U. Furthermore, we have

$$F : \left(\Gamma^{-1}(U) \cap \Gamma^{-1}(U') \big|_{S_0} \right) \longrightarrow \text{Gl}(2; \mathbb{C}) = \begin{pmatrix} (2w)^p & 0 \\ 0 & (2w)^q \end{pmatrix} \tag{B.124}$$

and, finally,

$$G : \left(\Gamma^{-1}(U) \cap \Gamma^{-1}(U') \big|_{S_1} \right) \longrightarrow \text{Gl}(2; \mathbb{C}) \ ,$$
$$G = P_{12} = \begin{pmatrix} (2w)^{-p} & 0 \\ 0 & (2w)^{-q} \end{pmatrix} \cdot P(w)^{-1} \ , \tag{B.125}$$

because of the cocycle condition (B.111).

Thus, in the axis-simple case, the holomorphic vector bundle E can be characterized by the triple (P, p, q) consisting of one patching matrix and two integers. It turns out that these parameters can be comparatively easily fixed and that we have a close relation between the value of $P(w)$ and the asymptotic behavior of $J(\varrho, \zeta)$ in the limit $\varrho \to 0$. The following proposition, for the proof of which we refer to [88] (for $p = q = 0$) and also [166], shows that the axis behavior of J can be read off from the patching data.

Proposition B.36. *Let $p > q$ and let $g, \hat{g}$ and Ω be functions of w, such that the first two do not vanish for $\varrho \to 0$, and let P denote the transition matrix in an axis simple vector bundle of the form*

$$P = \begin{pmatrix} g & -g\Omega \\ -g\Omega & g\Omega^2 + \hat{g}^{-1} \end{pmatrix} . \tag{B.126}$$

Then, the corresponding Yang matrix has for $\varrho \to 0$ the following behavior:

$$J(\varrho, \zeta) = \begin{pmatrix} \varrho^p & 0 \\ 0 & \varrho^q \end{pmatrix} \left(\begin{pmatrix} h(\zeta) & -h(\zeta)L \\ -h(\zeta)L & h(\zeta)L^2 + \left(\hat{h}(\zeta)\right)^{-1} \end{pmatrix} + \mathcal{O}\left(r^{p-q+1}\right) \right)$$
$$\times \begin{pmatrix} \varrho^p & 0 \\ 0 & \varrho^q \end{pmatrix} . \tag{B.127}$$

Here we have set $h(\zeta) = (-1)^p g(\zeta)$ and $\hat{h}(\zeta) = (-1)^q \hat{g}(\zeta)$ and

$$L = \frac{(-1)^p \varrho^{p-q}}{2^{p-q}(p-q)!} \left(\frac{d^{p-q}\Omega}{dw^{p-q}}\bigg|_{w=\zeta} \right) . \tag{B.128}$$

Then, for the determinants of the Yang matrix J and the patching matrix P it follows immediately

Corollary B.37. *Under the assumptions of the above proposition we have: If $\det P = 1$ then $\det J = \left(-\varrho^2\right)^{p+q}$.*

For our purposes more important is the converse, see [163, 164].

Proposition B.38. *Let $\mathcal{E} = e^{2U} + ib$ be a solution to the Ernst equation with regular symmetry axis. Then the patching matrix P is given by*

$$P(\zeta) = e^{-2U(0,\zeta)} \begin{pmatrix} 1 & -b(0,\zeta) \\ -b(0,\zeta) & e^{4U(0,\zeta)} + b(0,\zeta)^2 \end{pmatrix} . \tag{B.129}$$

Before we turn to the construction of the holomorphic vector bundles for the solutions of Chap. 4 we discuss the construction for several examples.
Examples:

(i) The Weyl class.

The solutions of this class have real Ernst potentials $\mathcal{E} = e^{2U}$, and we find for the patching matrix

$$P(\zeta) = \begin{pmatrix} e^{-2U(0,\zeta)} & 0 \\ 0 & e^{2U(0,\zeta)} \end{pmatrix} . \tag{B.130}$$

Let us consider the particular case of the Schwarzschild solution. From (1.8) with $\varphi = 0$ we get at the symmetry axis for $\zeta > m$

$$e^{2U(0,\zeta)} = \frac{\zeta - m}{\zeta + m} . \tag{B.131}$$

(ii) The Kerr metric.

Recall from (1.8) that the Ernst potential of this solution has the form

$$\mathcal{E} = \frac{e^{-i\varphi}r_+ + e^{i\varphi}r_- - 2m\cos\varphi}{e^{-i\varphi}r_+ + e^{i\varphi}r_- + 2m\cos\varphi} \;, \tag{B.132}$$

with $r_\pm = \sqrt{(\zeta \pm m\cos\varphi)^2 + \varrho^2}$. Then we find by a simple calculation for $\zeta > m\cos\varphi$

$$\mathcal{E}(0,\zeta) = \frac{\zeta^2 - m^2\cos^2\varphi}{(\zeta+m)^2 + m^2\sin^2\varphi} - i\frac{2m^2\sin^2\varphi}{(\zeta+m)^2 + m^2\sin^2\varphi} \;, \tag{B.133}$$

and we have for the patching matrix

$$P(\zeta) = \frac{1}{\zeta^2 - m^2\cos^2\varphi}\begin{pmatrix} (\zeta+m)^2 + m^2\sin^2\varphi & 2m^2\sin^2\varphi \\ 2m^2\sin^2\varphi & (\zeta-m)^2 + m^2\sin^2\varphi \end{pmatrix} \;. \tag{B.134}$$

B.4 Patching Matrices for the Class of Hyperelliptic Solutions

In Sect. 4.2 we have distinguished a class of solutions to the Ernst equation with physically interesting properties. The corresponding Ernst potentials are of the form

$$\mathcal{E}(\varrho,\zeta) = \frac{\Theta_{pq}(\omega(\infty^+) + u)}{\Theta_{pq}(\omega(\infty^+) - u)} \exp\left\{\frac{1}{2\pi i}\int_\Gamma \ln G(\tau)d\omega_{\infty^+\infty^-}(\tau)\right\} \;, \tag{B.135}$$

where the theta characteristic consists of blocks of the form $\frac{1}{2}\begin{bmatrix} 00 \\ 00 \end{bmatrix}$ and $\frac{1}{2}\begin{bmatrix} 00 \\ 11 \end{bmatrix}$. An essential property of these solutions is for $\Theta_{pq}(\omega(\infty^+) - u) \neq 0$ their regularity outside Γ_ξ, which can possibly be interpreted as the surface of the rotating body one is interested in. Thus, the solutions are regular on V being an open subset of $H \setminus \Gamma_\xi$, $V \subset H \setminus \Gamma_\xi$. In particular, the axis behavior is regular as long as

$$\Theta'_{p'q'}(u') \neq 0 \;, \tag{B.136}$$

which is the condition for an ergosphere not to hit the symmetry axis, see Sect. 7.3. Thus, we have

Proposition B.39. *Let $\mathcal{E}$ be a solution of the form (B.135). Furthermore, let R_V be the reduced twistor space for $V \subset H \setminus \Gamma_z$. Then the holomorphic vector bundle $E \to R_V$ corresponding to $\mathcal{E}$ is characterized by one patching matrix iff*

$$\Theta'_{p'q'}(u') \neq 0 \;.$$

Corollary B.40. *From Proposition B.38 it is obvious that the patching matrix is determined by the axis potential which is for the class of solutions (4.19) given by*

$$
\mathcal{E}(0,\zeta) = \frac{\Theta'_{\boldsymbol{p'q'}}\left(\omega'|^{\infty^+}_{\zeta^+} + \boldsymbol{u'}\right) + (-1)^\epsilon e^{-(\omega'_g(\infty^+)+u_g)}\Theta'_{\boldsymbol{p'q'}}\left(\omega'|^{\infty^+}_{\zeta^-} + \boldsymbol{u'}\right)}{\Theta'_{\boldsymbol{p'q'}}\left(\omega'|^{\infty^+}_{\zeta^+} - \boldsymbol{u'}\right) + (-1)^\epsilon e^{-(\omega'_g(\infty^+)-u_g)}\Theta'_{\boldsymbol{p'q'}}\left(\omega'|^{\infty^+}_{\zeta^-} - \boldsymbol{u'}\right)}
$$

$$
\exp\left\{\frac{1}{2\pi\mathrm{i}}\int_\Gamma \ln G(\tau)\mathrm{d}\omega'_{\infty^++\infty^-}(\tau) + u_g\right\} . \tag{B.137}
$$

Remark B.41.
(i) For the more general case of an algebro-geometric solution, obtained in Theorem 3.7, we have proven in Proposition 4.3, that the Ernst potential is regular on the axis except at the points where ξ coincides with the singularities of the Abelian integral of the second kind Ω, points of Γ and branch points, provided (B.136) holds (with $\boldsymbol{u'}$ replaced by $\boldsymbol{u'} + \boldsymbol{b'}$). If, e. g, a pole of Ω is on the axis then we have an essential singularity, see Proposition 4.1, and the solution is no longer axis-simple.
(ii) In proving Theorem 3.7 we have constructed explicitly a matrix Φ on $\mathcal{L}_0$, see (3.48), with column vectors given by

$$
\boldsymbol{X}(\hat{P}) = \begin{pmatrix} \psi(\hat{P}) \\ \pm\chi(\hat{P}) \end{pmatrix}, \quad \hat{P} \sim \xi^\pm , \tag{B.138}
$$

with ψ and χ being of the form (3.38) and (3.46). These vectors just form a basis in the fibre of E over any point of R_V.

References

1. E. D. Belokolos, A. I. Bobenko, V. Z. Enolskii, A. R. Its, V. B. Matveev: *Algebro-geometric approach to nonlinear integrable equations*, (Springer, Berlin 1994)
2. M. Audin: *Spinning tops*, (Cambridge University Press, Cambridge 1996)
3. S. V. Kovalevski: Acta Math. **12**, 177 (1889)
4. S. A. Chaplygin: *Collected works* **Vol. 1**, (Gostekhizdat, Moscow 1948)
5. V. I. Arnold: *Mathematical methods of classical mechanics*, (Springer, Berlin 1978)
6. W. Thirring: *Lehrbuch der mathematischen Physik* Bd. 1 (Springer, Wien 1977)
7. R. Abraham, J. E. Marsden: *Foundations of mechanics* (Addison–Wesley, Reading 1978)
8. A. M. Perelomov: *Integrable systems of classical mechanics and Lie algebras* Vol. 1, (Birkhäuser, Basel 1990)
9. C. S. Gardner, J. M. Greene, M. D. Kruskal, R. M. Miura: Phys. Rev. Lett **10**, 1095 (1967)
10. D. J. Korteweg, G. de Vries: Phil. Mag. **39**, 422 (1895)
11. M. Toda: *Nonlinear waves and solitons*, (Kluwer, Dordrecht 1989)
12. P. G. Drazin, R. S. Johnson: *Solitons: an introduction* (Cambridge University Press, Cambridge 1989)
13. P. D. Lax: Comm. Pure Appl. Math. **21**, 467 (1968)
14. L. D. Faddeev, L. A. Takhtajan: *Hamiltonian Methods in the Theory of Solitons*, (Springer, Berlin 1987)
15. A. C. Newell: *Solitons in Mathematics and Physics*, CBMS **48**, (SIAM, Philadelphia 1986)
16. F. Calogero, A. Degasperis: *Spectral Transforms and Solitons*, Studies in Mathematics and its Applications **13**, (North-Holland, Amsterdam-New York 1982)
17. M. Ablowitz, H. Segur: *Solitons and the inverse scattering transform*, SIAM Studies in Applied Mathematics **4**, (SIAM, Philadelphia 1981)
18. R. Penrose: Twistors as spin 3/2 charges. In *Gravitation and modern cosmology*, ed by A. Zichchi, N. Sánchez (Plenum Press, New York).
19. F. B. Estabrook, H. D. Wahlquist: J. Math. Phys. **16**, 1 (1975); J. Math. Phys. **17**, 1 (1976)
20. M. J. Ablowitz, A. Ramani, H. Segur: Lett. Nuovo Cim. **23**, 333 (1978)
21. M. J. Ablowitz, A. Ramani, H. Segur: J. Math. Phys. **21**, 715 (1980); J. Math. Phys. **21**, 1006 (1980)
22. J. B. McLeod and P. J. Olver, SIAM J. Math. Anal. **14**, 488 (1983)

238   References

23. J. Weiss, M. Tabor, G. Carnevale: J. Math. Phys. **24**, 522 (1983)
24. V. E. Zakaharov: *What is integrability?*, (Springer, Berlin 1991)
25. A. S. Fokas, U. Mugan, M. J. Ablowitz: Physica D **30**, 247 (1988)
26. F. Nijhoff, N. Joshi, A. Hone: Phys. Lett. A **267**, 147 (2000)
27. M. Gürses: Lett. Math. Phys. **11**, 59 (1986)
28. R. S. Ward: Phil. Trans. R. Soc. Lond. A **315**, 451 (1985)
29. K. K. Uhlenbeck: Commun. Math. Phys. **83**, 11 (1982)
30. R. S. Ward, R. O. Wells Jr.: *Twistor geometry and field theory*, (Cambridge University Press, Cambridge 1995)
31. M. F. Atiyah, N. J. Hitchin, V. G. Drinfeld, Yu. I. Manin: Phys. Lett. **A65**, 185 (1978)
32. L. J. Mason, N. M. J. Woodhouse: *Integrability, self-duality and twistor theory*, (Clarendon Press, Oxford 1996)
33. N. J. Hitchin: Introduction. In *Integrable systems–twistors, loop groups, and Riemann surfaces*, ed by N. M. J. Woodhouse, (Clarendon Press, Oxford 1999), pp 1
34. F. J. Ernst: Phys. Rev **167**, 1175 (1968)
35. H. Stephani, D. Kramer, M. MacCallum, C. Hoenselaers, E. Herlt: *Exact solutions of Einstein's field equations*, 2nd edn (Cambridge: Cambridge University Press 2003)
36. V. A. Belinski, E. Verdaguer: *Gravitational Solitons*, (Cambridge, Cambridge University Press 2001)
37. P. Forgács, Z. Horváth, L. Palla: Phys. Rev. Lett. **45**, 505 (1980)
38. P. Forgács, Z. Horváth, L. Palla: Phys. Rev. D **23** 1876 (1981)
39. P. Forgács, Z. Horváth, L. Palla: Phys. Lett. B **99** 232 (1981).
40. P. Forgács, Z. Horváth, L. Palla: Ann. Phys. **136** 371 (1981)
41. P. Forgács, Z. Horváth, L. Palla: Nucl. Phys. B **192** 141 (1981)
42. C. N. Yang: Phys. Rev. Lett. **38**, 1377 (1977)
43. S. Weinberg: Phys. Rev. **166**, 1568 (1968)
44. K. Meetz: J. Math. Phys. **10**, 65 (1969)
45. S. Coleman, J. Wess, B. Zumino: Phys. Rev. **177**, 2239 (1968)
46. S. Ketov: *Quantum Nonlinear Sigma–Models* (Springer, Berlin 2000)
47. W. J. Zakrzewski: *Low-dimensional Sigma Models* (Adam Hilger, Bristol 1989)
48. H. Eichenherr, M. Forger: Nucl. Phys. B, **155**, 381 (1979)
49. J. Eells, Jr., J. H. Sampson: Amer. J. Math., **86**, 109 (1964)
50. L. Bianchi: *Lezioni di Geometria Differenziale*, (Pisa 1909)
51. D. Korotkin: J. Math. Sciences, **94** 1177 (1999)
52. D. A. Korotkin: Theor. Math. Phys. **77**, 1018 (1989)
53. S. Novikov, S. V. Manakov, L. P. Pitaevskii, V. E. Zakharov: *Theory of solitons: the inverse scattering method*, (Consultants Bureau, New York 1984)
54. I. M. Gel'fand, B. M. Levitan: Izv. Akad. Nauk SSSR Ser. Mat. **15**, 309 (1951); A.M.S. Transl. (2) **1**, 253
55. V. A. Marchenko: Dokl. Akad. Nauk SSSR **104**, 695 (1955) (in Russian)
56. A. R. Its, V. Matveev: Theor. Math. Phys. **23**, 51 (1975)
57. A. R. Its, V. Matveev: Funct. Anal. Appl. **9**, 69 (1975)
58. V. Matveev: 30 years of finite-gap integration theory. In *30 Years of Finite Gap Integration*, ed by V. Kusnetsov, to appear in Phil. Trans. London Math. Soc., (2005)
59. G. B. Whitham: SIAM J. of App. Math. **14**, 956 (1966)

60. G. B. Whitham: *Linear and Nonlinear Waves*, (Wiley, New York/London/ Sidney 1999)
61. A. G. Gurevich, L. P. Pitaevskii: Sov. Phys. JETP **38** (2), 909 (1992)
62. P. Lax, D. Levermore: Commun. Pure Appl. Math. **36**, 253, 571, 809 (1983)
63. S. Venakides: Commun. Pure Appl. Math. **38**, 125 (1983)
64. R. Geroch: J. Math. Phys. **12**, 918 (1971)
65. A. Einstein, N. Rosen: J. Franklin Inst. **223**, 43 (1937)
66. H. Weyl: Ann. Physik **54**, 117 (1917)
67. R. P. Kerr: Phys. Rev. Lett. **11**, 237 (1963)
68. K. A. Khan, R. Penrose: Nature **229**, 185 (1971)
69. R. A. Matzner, C. W. Misner: Phys. Rev. **154**, 1229 (1967)
70. G. Weinstein: Comm. Pure Appl. Math. **43**, 903 (1990)
71. G. Weinstein: Comm. Pure Appl. Math. **45**, 1183 (1992)
72. G. Weinstein: Duke Math. J. **77**, 1183 (1995)
73. G. Weinstein: Amer. J. Math. **118**, 689 (1996)
74. M. Mars, J. M. Senovilla: Modern Phys. Lett. A **13**, 1509 (1998)
75. D. Maison: Phys. Rev. Lett. **41**, 521 (1978)
76. D. Maison, J. Math. Phys. **20**, 871 (1979)
77. V. A. Belinski, V. E. Zakharov: *Sov. Phys.-JETP* **48**, 985 (1978)
78. P. Breitenlohner, D. Maison: Ann. Inst. H. Poincaré **46** 215, (1987)
79. D. Korotkin: J. Phys. A **26**, 3823 (1993)
80. C. M. Cosgrove: J. Math. Phys. **21**, 2417 (1980)
81. R. Geroch: J. Math. Phys. **13**, 394 (1972)
82. I. Hauser, F. J. Ernst: J. Math. Phys. **22**, 1051 (1981)
83. I. Hauser, F. J. Ernst: `gr-qc/0002049` (2000)
84. B. Julia: Infinite dimensional algebras in physics. In: *Johns Hopkins workshop on current problems in particle physics: unified field theories and beyond,* (Johns Hopkins University, Baltimore 1981)
85. P. Breitenlohner, D. Maison: Explicit and Hidden Symmetries of Dimensionally Reduced (Super)Gravity Theories. In *Solutions to Einstein's Equations: Techniques and Results,* ed by C. Hoenselaers, W. Dietz, Lecture Notes in Physics **205**, (Springer, Berlin 1983) pp 276
86. B. Julia: In *Superspace and Supergravity*, ed. by S. Hawking, M. Roček, (Cambridge University Press, Cambridge 1981)
87. H. Nicolai: Two-dimensional gravities and supergravities as integrable systems. In *Recent aspects of quantum fields*, ed by H. Mitter, H. Gausterer, (Springer, Berlin 1991)
88. N. M. J. Woodhouse, L. J. Mason: Nonlinearity **1**, 73 (1988)
89. R. S. Ward: Comm. Math. Phys. **79**, 317 (1981)
90. D. Korotkin, V. Matveev: Leningrad Math. J. **1**, 379 (1990)
91. D. A. Korotkin: Commun. Math. Phys. **137** 383 (1991)
92. D. A. Korotkin: Class. Quant. Grav. **8** L219 (1991)
93. D. A. Korotkin: Class. Quant. Grav. **10** 2587 (1993)
94. D. Korotkin, V. Matveev: Lett. Math. Phys. **49**, 145 (1999)
95. D. A. Korotkin, V. B. Matveev: Funct. Anal. Appl. **34**, 18 (2000)
96. M. Babich, D. A. Korotkin: Lett. Math. Phys. **46**, 323 (1998)
97. J. B. Hartle, D. H. Sharp: Astrophys. J. **147**, 317 (1967)
98. L. Lindblom: Astrophys. J. **208**, 873 (1976)
99. R. Schödel et al.: Nature **419**, 694 (2002)

100. U. Heilig: Ann. Inst. Henri Poincaré, Physique Théorique **60** 457 (1994)
101. U. Heilig: Comm. Math. Phys. **166** 457 (1995)
102. H. D. Wahlquist: Phys. Rev. **172** 1291 (1968)
103. D. Kramer: Class. Quant. Grav. **1**, L3 (1984)
104. J. M. Senovilla: Phys. Lett. **123A**, 211 (1987)
105. M. Bradley, G. Fodor, M. Marklund, Z. Perjés: Class. Quant. Grav. **17**, 351 (2000)
106. J. Binney, S. Tremaine: *Galactic Dynamics*, (Princeton University Press, Princeton 1987)
107. U. M. Schaudt, H. Pfister: Phys. Rev. Lett. **77**, 3284 (1996)
108. U. M. Schaudt: Commun. Math. Phys. **190**, 509 (1998)
109. G. Neugebauer, R. Meinel: Astrophys. J. **414**, L97 (1993)
110. G. Neugebauer, R. Meinel: Phys. Rev. Lett. **73**, 2166 (1994)
111. G. Neugebauer, R. Meinel: Phys. Rev. Lett. **75**, 3046 (1995)
112. C. Klein, O. Richter: Phys. Rev. D **57**, 857 (1998)
113. C. Klein: Phys. Rev. **D63**, 064033 (2001)
114. C. Struck: Phys. Rep. **321**, 1 (1999)
115. V. C. Rubin, J. A. Graham, J. D. P. Kenney: Astrophys. J. **394**, L9 (1992)
116. H.-W. Rix, M. Franx, D. Fisher, G. Illingworth: Astrophys. J **400**, L5 (1992)
117. J. Bičák, T. Ledvinka: Phys. Rev. Lett. **71**, 1669 (1993)
118. C. Klein, O. Richter: Phys. Rev. Lett. **79**, 565 (1997)
119. J. Frauendiener, C. Klein: Phys. Rev. D **63**, 084023 (2001)
120. D. Maison: Duality and Hidden Symmetries in Gravitational Theories. In *Einstein's Field equations and Physical Implications*, Lecture Notes Phys. **540** ed by B. G. Schmidt (Springer, Berlin 2000) pp 1
121. J. Ehlers: Konstruktion und Charakterisierungen von Lösungen der Einstein'schen Gravitationsgleichungen, PHD thesis, Universität Hamburg, Hamburg (1957)
122. H. Röhrl: Math. Ann. **151**, 365 (1963)
123. C. Klein, O. Richter: J. Geom. Phys. **30**, 331 (1999)
124. I. M. Krichever: Itogi nauki i techniki, Current problems in mathematics, **23**, 79 (VINITI, Moscow 1983) (in Russian)
125. R. Meinel, G. Neugebauer: Phys. Lett. A **210**, 160 (1996)
126. D. A. Korotkin: Phys. Lett. A **229** 195 (1997)
127. C. Klein, D. Korotkin, V. Shramchenko: Math. Res. Lett. **9**, 27 (2002)
128. J. D. Fay: *Theta functions on Riemann surfaces*, Lecture Notes on Mathematics **352**, (Springer, New York 1973)
129. C. Klein, O. Richter: Phys. Rev. **D58**, CID 124018 (1998)
130. C. Klein, O. Richter: Phys. Rev. Lett. **83**, 2884 (1999)
131. J. Frauendiener, C. Klein: J. Comp. Appl. Math. **167**, 193 (2004)
132. www.lorene.obspm.fr
133. C. Klein: Theor. Math. Phys. **137**, 1520 (2003)
134. C. Klein: Phys. Rev. D **68**, 027501 (2003)
135. W. Kinnersley: J. Math. Phys. **18**, 1529 (1977)
136. B. Harrison: J. Math. Phys. **9**, 1744 (1968)
137. C. Klein: Phys. Rev. D **65**, 084029 (2002)
138. H. M. Farkas, I. Kra: *Riemann surfaces*, Graduate Texts in Mathematics, Vol. 71, (Berlin, Springer 1993).
139. D. Mumford: *Tata lectures on Theta* Vol. Vol. II, Progress in Mathematics, Vol. 43, (Birkhäuser, Boston 1993)

140. S. W. Hawking, G. F. R. Ellis: *The large scale structure of space-time*, (Cambridge University Press, Cambridge 1973)
141. N. S. Manton: Nucl. Phys. B **135**, 319 (1978)
142. V. A. Belinski, V. E. Zakharov: *Sov. Phys.-JETP* **50**, 1 (1979)
143. P. Houston, L. O. O'Raifeartaigh: Phys. Lett. B **93** 151 (1980)
144. S. Weinberg: *Gravitation and cosmology*, (John Wiley & Sons, New York 1972)
145. C. W. Misner, K. S. Thorne, J. A. Wheeler: *Gravitation*, (Freeman, New York 1973)
146. R. M. Wald: *General relativity*, (The University of Chicago Press, Chicago 1984)
147. H. Stephani: *Relativity*, 2nd edn (Cambridge University Press, Cambridge 2004)
148. J. L. Synge: *Relativity: the general theory*, (North-Holland Publishing Company, Amsterdam 1960)
149. W. Thirring: *Lehrbuch der mathematischen Physik* Vol. **2**, (Springer, Wien 1978)
150. H. Weyl: *Raum – Zeit – Materie*, (Springer, Berlin 1923)
151. W. Pauli: *Theory of relativity*, (Dover, New York 1981)
152. F. de Felice, C. J. S. Clarke: *Relativity on curved manifolds*, (Cambridge University Press, Cambridge 1990)
153. J. Stewart: *Advanced general relativity*, (Cambridge University Press, Cambridge 1990)
154. T. Lewis: Proc. Roy. Soc. London **A 136**, 176 (1932)
155. A. Papapetrou: Ann. Inst. H. Poincaré **A 4**, 83 (1966)
156. G. Neugebauer: J. Phys. A **12**, L67 (1979)
157. P. Fordy, J. C. Wood: Aspects Math. **23** (1993)
158. M. F. Atiyah: *The geometry of Yang-Mills fields*, (Lezione Fermiane, Scuola Normale Superiore Pisa 1979)
159. L. Bianchi: Ann. Math. **18**, 301 (1890)
160. S. Finikov: *Theorie der Kongruenzen*, (Akademie–Verlag 1959)
161. A. Bobenko: Aspects Mat. **23** (1993)
162. D. Korotkin, V. Reznik: Math. Notes **52**, 930 (1992)
163. J. Fletcher, N. M. J. Woodhouse: Twistor characterization of stationary axisymmetric solutions of Einstein's equations. In *Twistors in mathematics and physics*, ed by T. N. Bailey, R. J. Baston, (Cambridge University Press, Cambridge, 1990) pp 260
164. T. von Schroeter: *Twistor classification of type D vacuum space-times*, In Twistor Newsletter.
165. J. Fletcher: *More on the twistor description of the Kerr solution.* In: Twistor Newsletter.
166. S. Kolditz: *Zur Twistorformulierung einer speziellen Klasse von Lösungen der Ernst-Gleichung*, diploma thesis, University of Leipzig (2002)
167. L. Witten: Phys. Rev. **D19**, 718 (1979)
168. R. S. Ward: Gen. Rel. Grav. **15**, 105 (1983)
169. G. 't Hooft: Nucl. Phys. B **79**, 276 (1974)
170. A. M. Polyakov: JETP Lett. **20**, 194 (1974)
171. M. K. Prasad, C. M. Sommerfield: Phys. Rev. Lett. **35**, 760 (1975)
172. E. B. Bogomolny: Sov. J. Nucl. Phys. **24**, 861 (1976)
173. H. M. Farkas: Journ. d'Anal. Math. **44**, 205 (1984/85)

174. H. M. Farkas: Contem. Math. **136**, 161 (1992)
175. H. M. Farkas: Israel Math. Conference Proc. **9**, 231 (1996)
176. C. Poor: Proc. of the AMS **114**, 667 (1992)
177. A. Bobenko, L. Bordag: J. Phys. A **22**, 1259 (1989)
178. J. Frauendiener, C. Klein: to appear in Lett. Math. Phys. (2005).
179. H. E. Rauch: Comm. Pure Appl. Math. **12**, 543 (1959)
180. S. I. Zverovich: Russ. Math. Surv. (Uspekhi) **26**, 117 (1971)
181. Yu. L. Rodin: *The Riemann boundary problem on Riemann surfaces*, (Reidel, Dordrecht 1988)
182. N. I. Muskhelishvili: *Singular Integral Equations*, (Noordhoff, Groningen 1953)
183. G. Neugebauer: In *General Relativity*, ed by G. S. Hall, J. R. Pulham, (SUSSP Publication, Edinburgh, and IOP, London 1996), pp 61
184. C. Klein, O. Richter: J. Geom. Phys. **24**, 53 (1997)
185. H. Grauert: Math. Ann. **135**, 263 (1958)
186. B. A. Dubrovin: Russ. Math. Surv. (Uspekhi) **36**, 11 (1981)
187. H. Stahl: *Theorie der Abel'schen Funktionen*, (Teubner, Leipzig 1896)
188. P. A. Griffiths: Bull. Amer. Math. Soc. **76**, 228 (1970)
189. D. R. Morrison: Picard–Fuchs equations and mirror maps for hypersurfaces. In *Essays on mirror manifolds*, ed by S.-T. Yau, (Hong Kong: International Press, 1992)
190. F. Foucault: C. R. Acad. Sci. **314**, Série I, 617 (1992)
191. Yu. I. Manin: Izv. Akad. Nauk S.S.S.R. Ser. Mat. **22**, 737 (1958); *A.M.S. Transl. (2)* **37**, 59
192. B. B. Kadomtsev, V. Petviashvili: Dokl. Akad. Nauk **192**, 753 (1970)
193. I. M. Krichever: Funct. Anal. Appl. **9**, 84 (1975)
194. B. A. Dubrovin, R. Flickinger, H. Segur: Stud. Appli. Math., **99**(2), 137 (1997)
195. M. Jimbo, T. Miwa, K. Ueno: Physica D **2**, 306 (1981)
196. A. Yamada: Kodai Math. Journ. **3**, 114 (1980)
197. H. Müller zum Hagen: Proc. Camb. Phil. Soc. **68**, 199 (1970)
198. C. B. Morrey: Am. J. Math. **80**, 198 (1958)
199. H. Bateman, A. Erdélyi: *Higher transcendental functions*, (McGraw–Hill, New York 1955)
200. A. Tomimatsu, H. Sato: Phys. Rev. Lett. **29**, 1344 (1972)
201. A. Tomimatsu, H. Sato: Prog. Theor. Phys. **50**, 95 (1973)
202. C. Hoenselaers, W. Kinnersley, B. C. Xanthopoulos: J. Math. Phys. **20**, 2530 (1979)
203. D. Korotkin, H. Nicolai: Nucl. Phys. B **475**, 397 (1996)
204. T. Morgan, L. Morgan: Phys. Rev. **183**, 1097 (1969); Errata: **188**, 2544 (1969)
205. J. M. Bardeen, R. V. Wagoner: Astrophys. J. **167**, 359 (1971)
206. J. Bičák, D. Lynden-Bell, J. Katz: Phys. Rev. D, **47**, 4334 (1993)
207. J. Bičák, D. Lynden-Bell, C. Pichon: MNRAS **256**, 126 (1993)
208. C. Pichon, D. Lynden-Bell: MNRAS **280**, 1007 (1996)
209. W. Israel: Nuovo Cimento **44** B, 4349 (1966); Errata: **48** 2583 (1966); **44B** 1 (1966); **48B**, 463 (1967)
210. M. Ansorg, R. Meinel: Gen. Relativ. Gravit. **32**, 1365 (2000)
211. R. Geroch: J. Math. Phys. **11**, 2580 (1970)
212. R. Hansen: J. Math. Phys. **15**, 46 (1974)
213. G. Fodor, C. Hoenselaers, Z. Perjés: J. Math. Phys. **30**, 2252 (1989)
214. I. Hauser, F. Ernst: J. Math. Phys. **21**, 1126 (1980)

215. P. Kordas: Class. Quantum Grav. **12**, 2037, (1995)
216. R. Meinel, G. Neugebauer: Class. Quantum Grav. **12**, 2045, (1995)
217. C. Klein: Theor. Math. Phys. **134**, 72 (2003)
218. K. Thorne: Nonspherical Gravitational Collapse: A Short Review. In *Magic Without Magic: John Archibald Wheeler*, ed by J. Klauder (W. H. Freeman: San Francisco 1972) pp 231
219. C. L. Tretkoff, M. D. Tretkoff: Contemp. Math. **33**, 467 (1984)
220. M. Seppälä: Discrete Comput. Geom. **11**, 65 (1994)
221. M. Hoeij: J. Symb. Comput. **18**, 353 (1994)
222. P. Gianni, M. Seppälä, R. Silhol, B. Trager: J. Symb. Comp. **26**, 789 (1998)
223. B. Deconinck, M. van Hoeij, Physica D **152–153**, 28 (2001)
224. B. Deconinck, M. Heil, A. Bobenko, M. van Hoeij, M. Schmies: Math. Comp. **73**, 1417 (2004)
225. www-sfb288.math.tu-berlin.de/ jtem/
226. V. Z. Enolski, P. H. Richter: Periods of hyperelliptic integrals expressed in terms of theta-constants by means of Thomae formulae. In *30 Years of Finite Gap Integration*, ed by V. Kusnetsov, to appear in Phil. Trans. London Math. Soc., (2005)
227. A. Komar: Phys. Rev. **113**, 934 (1959)
228. E. Gourgoulhon, S. Bonazzola: Class. Quant. Grav. **11**, 443 (1994)
229. B. Fornberg: *A practical guide to pseudospectral methods*, (Cambridge University Press, Cambridge 1996)
230. W. L. Briggs, V. E. Henson: *The DFT, an owner's manual for the discrete Fourier transform*, (Siam, Philadelphia 1995)
231. M. Heil: Numerical Tools for the study of finite gap solutions of integrable systems, PhD thesis, TU Berlin, Berlin (1995)
232. E. Gourgoulhon, P. Haensel, R. Livine, E. Paluch, S. Bonazzola, J.-A. Marck: Astr. Astrophys. **349** 851 (1999)
233. C. Klein, Theor. Math. Phys. **127**, 767 (2001)
234. R. Meinel: The rigidly rotating disk of dust and its black hole limit. In *Recent Developments in Gravitation and Mathematical Physics*, ed by A. García, C. Lämmerzahl, A. Macías, T. Matos, D. Nuñes, (Science Network Publishing: Konstanz 1998), gr-qc/9703077
235. M. Rees: In *Black Holes and Relativistic Stars*, ed by R. Wald (Univ. Chicago Press, Chicago 1998) pp 79
236. A. Laor: Phys, Rep. **311**, 451 (1999)
237. S. Chakrabarti (ed.): *Observational Evidence for the Black holes in the Universe* (Kluwer, Dordrecht 1999)
238. C. DeWitt, B. DeWitt (ed.): *Black holes*, (Gordon and Breach, New York 1973)
239. O. Semerák: Towards gravitating disks around rotating black holes. In *Gravitation: Following the Prague Inspiration*, ed by O. Semerák, J. Podolský, M. Žofka, (World Scientific, Singapore 2002), pp 111
240. V. Karas, J.-M. Huré, O. Semerák: Class. Quant. Grav. **21**, R1 (2004)
241. B. Carter: *Black hole equilibrium states*. In *Black holes*, ed by C. DeWitt and B. S. DeWitt, (Gordon & Breach, New York 1973)
242. C. Klein: Ann. Phys. (Leipzig) **12**, 599 (2003)
243. C. Klein: Class. Quantum Grav. **14**, 2267 (1997)
244. J. Lemos, P. Letelier: Phys. Rev. D **49**, 5135 (1994)

245. F. J. Ernst: Phys. Rev **168**, 1415 (1968)
246. D. Maison: In *Developments in the Theory of Fundamental Interaction*, ed by L. Turko, A. Pekalski,
247. G. Weinstein, Comm. Part. Diff. Equ. **21**, 1389 (1996)
248. D. Maison: Stationary Solutions of the Einstein-Maxwell Equations. In *Nonlinear equations in classical and quantum field theory (Meudon/Paris, 1983/1984)*, Lecture Notes in Phys.
249. M. King, H. Pfister: Phys. Rev. D **65**, 084033 (2002)
250. M. King, H. Pfister: Class. Quant. Grav. **20**, 205 (2003)
251. J. Novak, E. Marcq: Class. Quant. Grav. **20**, 3051 (2003) **226**, (Springer, Berlin 1985) pp 125
252. G. Neugebauer, D. Kramer: J. Phys. A **16**, 1937 (1983)
253. D. Korotkin: Math. Ann. **329** 335 (2004)
254. A. Kitaev, D. Korotkin: Int. Math. Res. Not. **17**, 877 (1998)
255. A. Kokotov, D. Korotkin: math-ph/0310008 (2003)
256. G. Springer: *Introduction to Riemann surfaces*, (Addison–Wesley, Reading 1957)
257. L. Alfors, L. Sario: *Riemann surfaces*, (Princeton University Press, N.J. 1960)
258. M. Atiyah: *Riemann Surfaces and spin structures*, (Annales Scientifiques de l'Ecole Normale Supérieure, 1971).
259. A. F. Beardon: *A Primer on Riemann Surfaces*, London Mathematical Society Lecture Notes **78**, (Cambridge University Press, Cambridge 1984)
260. J.-B. Bost: Introduction to Compact Riemann Surfaces, Jacobians and Abelian Varieties. In *From Number theory to Physics*, ed by M. Waldschmidt, P. Moussa, J.-M. Kuck, C. Itzykson, (Springer, Berlin 1992)
261. J. Jost: *Compact Riemann Surfaces*, (Springer, Berlin 19997)
262. J. Lewittes: Acta Math. **111**, 35 (1964)
263. A. Hurwitz, R. Courant: *Funktionentheorie*, (Springer, Berlin 1964)
264. P. Griffiths, J. Harris: *Principles of algebraic geometry*, (John Wiley & Sons, New York 1978)
265. S. S. Chern: *Complex manifolds without potential theory*, (Springer, New York 1995)
266. R. C. Gunning: *Lectures on Riemann surfaces*, (Princeton University Press, Princeton 1967)
267. A. Krazer, W. Wirtinger: Abelsche Funktionen und allgemeine Thetafunktionen. In *Encyklopädie der mathematischen Wissenschaften* **II (2)**, Heft 7, (Teubner, Leipzig 1915)
268. A. I. Markushevich: *Introduction to the classical theory of Abelian functions*, (American Mathematical Society, Providence 1992).
269. M. Koecher, A. Krieg: *Elliptische Funktionen und Modulformen*, (Springer, Berlin 1998)
270. A. Krazer: *Lehrbuch der Thetafunktionen*, (Teubner, Leipzig 1903)
271. O. Richter, C. Klein: Algebro-geometric approach to the Ernst equation I. In *Mathematics of Gravitation*, Banach Center Publications **41**, ed by P.T. Chruściel, (PAS, Inst. Math., Warsaw 1997) pp 195
272. I. A. Taimanov: Russian Math. Surveys **52**, 147 (1997)
273. S. Kobayashi, K. Nomizu: *Foundations of differential geometry* Vol. **I**, (Plenum, New York 1963)
274. M. F. Atiyah: *The geometry of Yang–Mills fields*, Lezione Fermiane, Scuola Normale Superiore Pisa (1979)

275. M. F. Atiyah, N. J. Hitchin, I. M. Singer, Proc. R. Soc. Lond. A **362**, 425 (1978)
276. H. Baum: Self-dual connections and holomorphic bundles. In *Self-dual Riemannian Geometry and Instantons*, ed by T. Friedrich, (Teubner, Leipzig 1981) pp 105
277. N. J. Hitchin: Riemann surfaces and integrable systems. In *Integrable systems–twistors, loop groups, and Riemann surfaces* 1, ed by N. M. J. Woodhouse, (Clarendon Press, Oxford 1999)
278. G. Dell'Antonio, S. Doplicher, Jona-Lasinio: *Mathematical Problems in Theoretical Physics* Lecture Notes in Physics, **80**, (Springer, Berlin et al. 1978)

Index

Lecture Notes in Physics

For information about earlier volumes
please contact your bookseller or Springer
LNP Online archive: springerlink.com